Introduction to Physical Hydrology

Cover design: hydraulic jump

A **hydraulic jump** is the result of fast-moving water in a stream reaching slower-moving water downstream. At the hydraulic jump, water is swept upwards and falls back on itself like a crashing wave, falling in the upstream direction. In other words: at the hydraulic jump, kinetic energy is explosively converted into turbulence and potential energy.

At the hydraulic jump, water flow changes from **supercritical** to **subcritical**. We will learn more about these flow types and the hydraulic jump in Chapter 5 of this book.

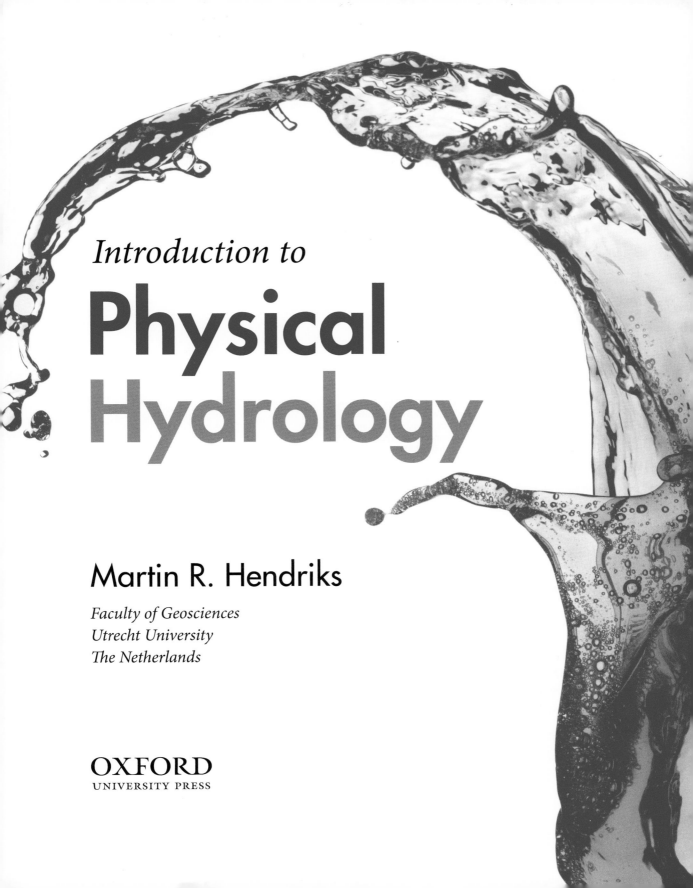

Introduction to

Physical Hydrology

Martin R. Hendriks

Faculty of Geosciences
Utrecht University
The Netherlands

OXFORD
UNIVERSITY PRESS

OXFORD
UNIVERSITY PRESS

Great Clarendon Street, Oxford OX2 6DP

Oxford University Press is a department of the University of Oxford.
It furthers the University's objective of excellence in research, scholarship,
and education by publishing worldwide in

Oxford New York

Auckland Cape Town Dar es Salaam Hong Kong Karachi
Kuala Lumpur Madrid Melbourne Mexico City Nairobi
New Delhi Shanghai Taipei Toronto

With offices in

Argentina Austria Brazil Chile Czech Republic France Greece
Guatemala Hungary Italy Japan Poland Portugal Singapore
South Korea Switzerland Thailand Turkey Ukraine Vietnam

Oxford is a registered trade mark of Oxford University Press
in the UK and in certain other countries

Published in the United States
by Oxford University Press Inc., New York

British Library Cataloguing in Publication Data

Data available

Library of Congress Cataloging-in-Publication Data

Hendriks, Martin R.
  Introduction to physical hydrology / Martin R. Hendriks.
      p.   cm.
  Includes bibliographical references and index.
  ISBN 978-0-19-929684-2 (alk. paper)
  1. Hydrology.   I. Title.
  GB661.2.H46 2010
  551.48—dc22
                                        2009037758

Typeset by MPS Limited, A Macmillan Company
Printed in Great Britain by Ashford Colour Press Ltd

ISBN 978–0–19–929684–2

5 7 9 10 8 6

To Anneke

Acknowledgements

I wish to thank Jonathan Crowe, Editor in Chief in the Higher Education Department of Oxford University Press, for his continued support of the project; his comments on the first drafts (especially for putting me on the right track by using boxes and equation annotations); the discussions of a number of matters; for managing the logistics such as approaching external reviewers; plus for all matters regarding the production of the book – all very much appreciated! Writing a book is a personally rewarding, yet lone enterprise: it was always a pleasure to receive your supportive e-mails!

Professor Cees van den Akker is thanked for his stimulating lectures on groundwater hydraulics in the mid-1990s: his lectures on the blackboard, using only a piece of chalk, showed me the art and beauty of selecting well-designed strategies for solving groundwater hydraulics problems – these strategies have found their way into section 3.15. Good use has also been made of the excellent syllabi on groundwater by Professor Co de Vries. Dr Thom Bogaard and Dr Theo van Asch are thanked for the many stimulating discussions over the years on a wide variety of hydrology subjects – much of this has found its way into this book. Dr Anne Marie van Dam is thanked for her useful comments and extensive feedback on the November 2008 draft text used during lectures; Laura Nieuwenhoven is thanked for her feedback on the November 2007 lecture text on groundwater; and Henk Mark for his feedback on the Penman–Monteith equation. Professor Marc Bierkens is thanked for his support and for allowing me to invest time for writing this textbook. Dr Hans Renssen, Dr Albert Klein Tank, Dr Hans van der Kwast, Professor Steven de Jong, Dr Hanneke Schuurmans, Aline Duine, Professor Ruud Schotting, Dr Rens van Beek, Professor Majid Hassanizadeh, Loes van Schaik, Arien Lam, Dr Derek Karssenberg, Dr Maarten Kleinhans, and Marcel van Maarseveen are all thanked for providing useful information for the book.

I wish to thank two anonymous external reviewers for their useful comments and feedback on the November 2008 draft; following their comments I have written a number of useful additions to the book. All at Oxford University Press are thanked for their cooperation and advice, including Emma Lonie for production-editing the book, and Holly Edmundson for her help in clearing permissions for the use of a number of figures; the design team of Oxford University Press is thanked for their excellent work on the layout of the book. Geoff Palmer is thanked for copy-editing, Graham Bliss for proofreading, and Jonathan Burd for providing the index of the book.

I wish to thank my father for helping me out on the analyses of earlier work, which has also found its way into this book, as well as for his continued interest. Eric and Ingmar: thank you for being there and cheering things up. Last, but not least, I would like to thank my wife Anneke for letting me wander off to the study, excusing me from household duties, all for the greater cause of writing a hydrology textbook, and for taking care. It has all provided the solidity for me to keep on … and keep on tapping away at my computer keyboard. I could not have written this book without you!

MRH
Abcoude, October 2009

Contents

Welcome to the book

Water plays an important role in almost all natural processes on our Planet Earth. Indeed, we find evidence of the past workings of water on Solar Planet Mars: in 2008, the robotic spacecraft *Phoenix* – on a NASA mission to Mars – confirmed the existence of water, our elixir of life, by uncovering ice in the shallow Martian subsurface.

Hydrology concerns itself with water on and under the earth's surface. We live on a planet with much water and we know that our lives depend on water in many ways. So there are many reasons for learning more about water. Water is an astonishing liquid – for instance, becoming lighter when it freezes. The first reason for studying hydrology, therefore, quite simply is … because it is an amazing branch of science!

Hydrology as an applied science is flourishing and a great many people are involved in water management and related innovations. Traditionally, students of the earth sciences, physical geography, environmental sciences, and civil engineering learn about the principles of hydrology during their studies. However, hydrology also plays an important role in social studies – for instance, with regard to water management, the availability of fresh water, and sustainable access to safe drinking water, or with regard to adequate sanitation services. Any student may thus want to know about hydrology!

Physical hydrology emphasizes the physical aspects of hydrology (see 'What moves water?' below). This book aims to make physical hydrology accessible to undergraduate students who have not studied hydrology previously and have limited knowledge of physics and mathematics. The book may also be used by students who wish to enter a master's programme in hydrology, and who have a background in natural sciences but limited knowledge of hydrology. Importantly, the book is also set up to suit self-directed study.

Water and water-related problems are all-important and will continue to be so. Therefore hydrology is a **no-regrets study**, whatever the background of a student or reader of this book may be.

What moves water?

Physical hydrology encapsulates the art of understanding what, physically, makes water flow.

Water does not flow from a high to a low position by law of nature, although we often observe it to do just that. Neither does water flow from moist to dry spots in soil or from high to low pressure by law of nature. What, then, makes water flow?

The answer lies in a range of fascinating concepts, which we explore throughout this book. For example: you may have seen Darcy's law before – but why is there a minus sign in the equation? What is persistence and how does it come about? How can a water pressure be negative? What is hysteresis? What is preferential flow? What can ripples in the water teach us?

These are just some of the questions that are answered in this book. Read this book and you are sure to find out what moves water!

Didactic concept: a novel approach to physical hydrology

This book is written from a clear didactic concept, developed over almost two decades of teaching hydrology to both undergraduate and graduate students in physical geography, environmental sciences, and the earth sciences.

Chapter 1 of this book starts with the central concept of hydrology – the **hydrological cycle** – the central unit in hydrology – the **drainage basin** – and the central equation in hydrology – the **water balance**.

The weather – and, on another scale, **climate change** – affect our water system on earth. Chapter 2 considers the hydrological aspects of atmospheric water, and some reference is also made to **climate hydrology** or **global hydrology**, contemporary issues related to the hydrology of large areas or to the earth as a whole.

This book then takes a different approach from many other textbooks on hydrology. Because the general physical laws used in hydrology are best explained for steady groundwater flow, our journey into hydrology then continues with **groundwater** (and not with soil water or surface water).

The western part of The Netherlands – Holland – has a reversed landscape: rivers flow above the land surface and groundwater can be demonstrated to flow vertically upwards. This man-made landscape, reclaimed from the sea in earlier days, is the ideal landscape in which to explore and discover the basic laws of hydrology (or 'what moves water'), as one is not easily tricked into wrongfully assuming what seems all too obvious in an ordinary landscape.

Basic laws discussed in Chapter 3 are **Bernoulli's law** (the energy equation) and **Darcy's law** (the flow equation): steady groundwater flow is shown to be described or modelled simply from combining Darcy's law and the water balance (equation).

Chapter 4, on **soil water**, teaches us that applying the energy and flow equation for soil water is complicated by pores in the soil containing not only water, but also air: the latter causes water pressures in the soil water zone (unsaturated zone) to be negative with respect to atmospheric pressure.

In the final chapter, Chapter 5, we deal with **surface water**, water of rivers and brooks that is visible to the naked eye – but also water that may flow haphazardly or chaotically, further complicating matters.

Learning from this book

This book has been written from my conviction that students and people interested in the role of water in nature and the environment, but who may lack a firm understanding of basic physics and mathematics (or lack some confidence in this field!), are best served with a treatment of hydrology that (re-)introduces these basic physical and mathematical concepts.

Also, to come to a full understanding of the physics and mathematics involved, it is important to work through exercises. This may seem like a struggle at first, but answers and in-between steps explaining how to get to the right answer are provided in the Answers section at the end of this book. So, see the exercises as a challenge to be tackled, with the knowledge that a supportive, helping hand (in the form of worked solutions) is there to guide you.

Learning features

This book includes a number of learning features to help make your learning as effective as possible.

Important terms and key conclusions

Important terms and key conclusions are highlighted in the text; to make for easy and quick discovery of the information being sought, the important terms also feature in the index at the back of the book.

The zone below this water table is the **saturated zone** and the water stored there is **groundwater**.

Annotated equations

Many equations in this book are annotated, providing quick and easy reference to the meaning and units of measurement of the variables used.

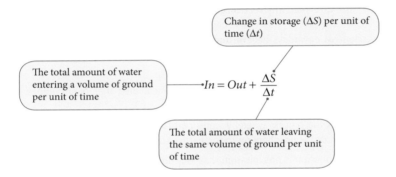

Boxes

Each chapter features a number of boxes: these supplement the main text, providing background information, further detailed information, or an interesting aside.

BOX 3.3 Drinking water for the city of Amsterdam

Seepage intensities in the Horstermeerpolder and the Bethunepolder near the Utrechtse Heuvelrug amount to more than 20 mm day^{-1}. Seepage water in the Horstermeer has a high chloride content. Seepage water in the Bethunepolder is fresh and is used, together with water from the Loosdrechtse Plassen (Loosdrecht Lakes) and Amsterdam-Rijnkanaal (Amsterdam–Rhine Canal), to provide drinking water for the city of Amsterdam (Kosman 1988). Besides this, pre-treated water from the River Rhine is infiltrated into the coastal dunes south of Zandvoort, also as a source of drinking water for Amsterdam (Van Til and Mourik 1999).

Exercises

The book contains many exercises. Answers and in-between steps, which show how to get to the right answer, are given in the Answers section at the end of the book. However, do not look at the answers too hurriedly, but try the exercises yourself first! In course evaluations, students consistently claim that struggling with the exercises was very worthwhile in coming to grips with the study material.

> **Exercise 3.10.3** A sandy formation consists of four layers of equal thickness. Values for the hydraulic conductivity of these layers are 1, 5, 10, and 50 m day^{-1}.
> Determine the substitute horizontal and vertical hydraulic conductivity of the formation.

Summaries

Each chapter ends with a summary, where the chapter's key learning points are summarized. Use these summaries as quick and easy reference to what has been learned and, if necessary, re-read any sections about which you are unsure.

- Many scientists nowadays believe that current global warming induces a speeding up of the hydrological cycle, and that this in turn may lead to more extreme weather conditions on parts of the earth.
- A drainage basin or catchment is the geographical area that drains into a river or reservoir.
- Precipitation is the process whereby water particles, both liquid (rain) and solid (snow, hail), fall from the atmosphere on to the earth's surface, and thus on land or water.
- Channel precipitation is precipitation that falls directly on a stream or river channel.

Section 3.15

Section 3.15, on groundwater hydraulics, offers in-depth information for solving a number of groundwater flow problems. It is good practice to solve these exercises by using only Table 3.3 as a memory aid! The Answers section at the end of this book provides answers to in-between steps, to highlight the best strategy for solving the exercises. (However, should one decide to skip section 3.15 and move on to Chapter 4 on soil water, then no harm will be done to the build-up of the story of this book as a whole.)

Conceptual toolkit

For students feeling uneasy with mathematics, the Conceptual Toolkit at the back of the book provides basic information: this information will come in particularly handy for solving the exercises of section 3.15.

Mathematics toolboxes

For students interested in the mathematical background to the equations given in Table 3.3, further information is provided in the Mathematics Toolboxes at the end of the book. Also, further information on the Richards equation and the open channel flow equations is presented there.

Website material

This book is accompanied by an **Online Resource Centre** at www.oxfordtextbooks.co.uk/orc/hendriks/, which features:

- For everyone:
  - Spreadsheets, available to download.
  - A test bank of multiple-choice questions, plus an example of a groundwater hydraulics exam for section 3.15 on groundwater hydraulics.
- For the lecturer:
  - Figures from the book, available in electronic format to support your lecture preparation.

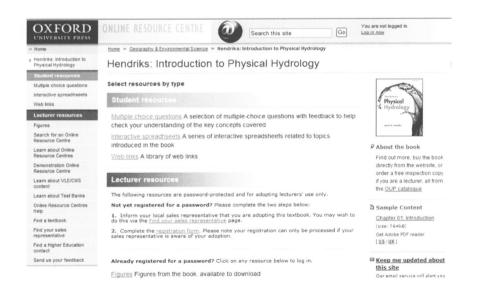

Table of SI units

SI base units

amount of substance	mole	mol
electrical current	ampere	A
length	metre	m
luminous intensity	candela	cd
mass	kilogram	kg
temperature	kelvin	K
time	second	s

Examples of SI derived units

area	square metres	m^2
volume	cubic metres	m^3
velocity	metres per second	$m\,s^{-1}$
acceleration	metres per second squared	$m\,s^{-2}$
density	kilograms per cubic metre	$kg\,m^{-3}$

SI derived units with special names and symbols

force	newton	N	$kg\,m\,s^{-2}$
pressure	pascal	Pa	$N\,m^{-2} = kg\,m^{-1}\,s^{-2}$
energy, work	joule	J	$N\,m = kg\,m^2\,s^{-2}$
power	watt	W	$J\,s^{-1} = kg\,m^2\,s^{-3}$
potential difference (electrical)	volt	V	$W\,A^{-1} = kg\,m^2\,s^{-3}\,A^{-1}$
resistance (electrical)	ohm	ω	$V\,A^{-1} = kg\,m^2\,s^{-3}\,A^{-2}$
conductivity (electrical)	siemens	S	$A\,V^{-1} = A^2\,s^3\,kg^{-1}\,m^{-2}$
Celsius temperature	degree Celsius	$^\circ C$	$K - 273.15$
dynamic viscosity	pascal second	Pa s	$N\,m^{-2}\,s = kg\,m^{-1}\,s^{-1}$
surface tension	newton per metre	$N\,m^{-1}$	$kg\,s^{-2}$

Multiple	Prefix	Symbol
10^{12}	tera	T
10^{9}	giga	G
10^{6}	mega	M
10^{3}	kilo	k
10^{2}	hecto	h
10^{1}	deca	da
10^{-1}	deci	d
10^{-2}	centi	c
10^{-3}	milli	m
10^{-6}	micro	μ
10^{-9}	nano	n
10^{-12}	pico	p
10^{-15}	femto	f

About the author

Dr Martin R. Hendriks is Associate Professor of Physical Hydrology at the Faculty of Geosciences, Utrecht University, where he has lectured for more than two decades. He studied Physical Geography at the University of Amsterdam and obtained a Doctorate in Hydrology from the VU University Amsterdam. He has written in a number of international, scientific journals.

Figure acknowledgements

1.2 Reproduced from Ward, R.C. and Robinson, M. (2000). Principles of Hydrology. Fourth Edition. McGraw-Hill. **B1.2** Based on data by Professor W. Broecker. Modified by Dr E. Maier-Reimer. **2.5** Reproduced from Schmidt, F.H. (1976). Inleiding tot de meteorologie. Aula-boeken 112, Het Spectrum. **2.7** Reproduced from Ward, R.C. and Robinson, M. (2000). Principles of Hydrology. Fourth Edition. McGraw-Hill. **2.11** Reproduced from Schuurmans, J.M., Bierkens, M.F.P., Pebesma, E.J., and Uijlenhoet, R. (2007). Automatic prediction of high-resolution daily rainfall fields for multiple extents: the potential of operational radar. Journal of Hydrometeorology 8, 1204–1224. © 2009 American Meteorological Society (AMS). **2.12** Reproduced from Shuttleworth (1993). Evaporation. In: Maidment, D.R. (ed.). Handbook of Hydrology. McGraw-Hill. **2.14** From Van der Kwast, J. and De Jong, S.M. (2004). Modelling evapotranspiration using the Surface Energy Balance System (SEBS) and Landsat TM data (Rabat region, Morocco). EARSeL Workshop on Remote Sensing for Developing Countries, Cairo, 1–11. Reproduced with kind permission from Professor Steven de Jong. **B2.5** Based on data from Hils, M. (1988). Einfluss des langfristiger Klimaschwankungen auf die Abflüsse des Rheins unter besonderer Berücksichtigung der Lufttemperatur. Diplomarbeit, Bundesamt für Gewässerkunde, Koblenz. **B2.9** Reproduced from Schmidt, F.H. (1976). Inleiding tot de meteorologie. Aula-boeken 112, Het Spectrum. **B2.12.2** Reproduced with kind permission from Douglas Parker, University of Leeds. **3.12** From De Vries, J.J. and Cortel, E.A. (1990). Introduction to Hydrogeology. Lecture notes. Institute of Earth Sciences, VU University Amsterdam, The Netherlands. Reproduced with kind permission from Professor Co de Vries. **3.22** Reproduced with kind permission from IF Technology B.V., Arnhem, The Netherlands. **3.36** Reproduced from De Vries, J.J. (1980). Inleiding tot de Hydrologie van Nederland. Rodopi, Amsterdam. **3.39** Reproduced from Haitjema, H.M. (1995). Analytical Element Modeling of Groundwater Flow. San Diego, California. Academic Press. By kind permission of Professor Henk Haitjema. **3.40** Tóth, J., A theoretical analysis of groundwater flow in small drainage basins, Journal of Geophysical Research, 68(16), 4795–4812. © American Geophysical Union. (1963). Reproduced by permission of American Geophysical Union. **3.41** Reproduced from Hubbert (1940) The theory of groundwater motion. Journal of Geology 48, 785–944. © University of Chicago. **3.42** Reproduced from Hendriks, M.R. (1990), Regionalisation of hydrological data: effects of lithology and land use on storm runoff in east Luxembourg. PhD thesis, VU University Amsterdam, The Netherlands, ISBN 90-6266-079-7 (thesis). Also available as Netherlands Geographical Studies 114, Utrecht: Royal Dutch Geographical Society (KNAG), ISBN 90-6809-124-7 (NGS). **B3.7** Reproduced from De Vries, J.J. (1980). Inleiding tot de Hydrologie van Nederland. Rodopi, Amsterdam. **4.12** Reproduced from Held, R.J and Celia, M.A. (2001). Modelling support of functional relationships between capillary pressure, saturation, interfacial area and common lines. Advances in Water Resources 24, 325–343. © Elsevier. **4.13** Reproduced from Bouma, J. (1977). Soil survey and the study of water in the unsaturated zone. Soil Survey Paper 13. Netherlands Soil Survey Institute, Wageningen. With kind permission from Professor Johan Bouma. **4.21** Reproduced from Philip, J (1964). The gain, transfer and loss of soil water. Water Resources Use and Management, Melbourne University Press,

257–275. **4.22** Reproduced from Horton, R.E. (1939). Analysis of runoff-plat experiments with varying infiltration capacity. Transactions of American Geophysical Union 20, 693–711. **4.31** Reproduced from Rubin, J. (1966), Theory of rainfall uptake by soils initially drier than their field capacity and its applications, 2, 739–749. © Water Resources Research. **4.33** Reproduced from Wellings, S.R. and Bell, J.P. (1982). Physical controls of water movement in the unsaturated zone. Quarterly Journal of Engineering Geology 15(3), 235–241. **4.34** Reproduced from Vachaud, G., Vauclin, M., Khanji, D., and Wakil, M. (1973). Effects of air pressure on water flow in an unsaturated stratified vertical column of sand, 9, 160–173. © Water Resources Research. **4.36** © Cornell University (2002). **4.37** © Cornell University (2002). **5.3** Reproduced from Van Rijn, L.C. (1994). Principles of fluid flow and surface waves in rivers, estuaries, seas and oceans. Second Edition. Oldemarkt: Aqua Publications. **5.18** Reproduced from Gregory, K.J. and Walling, D.E. (1973). Drainage Basin Form and Process: a geomorphological approach. Edward Arnold Ltd, London. **5.39** Reproduced from Kirkby, M.J., Naden, P.S., Burt, T.P. and Butcher, D.P. (1987). Computer Simulation in Physical Geography, Wiley. **5.42** Reproduced from Ward, R.C. and Robinson, M. (2000). Principles of Hydrology. Fourth Edition. McGraw-Hill. **5.43** Troch, P.A. (2008). Land Surface Hydrology. Chapter 5 in Bierkens, M.F.P., Dolman, A.J. and Troch, P. (eds.), Climate and the Hydrological Cycle. Special Publication 8, 99–115. Reproduced from © IAHS. **B5.10.2** Reproduced from Ward, R.C. and Robinson, M. (2000). Principles of Hydrology. Fourth Edition. McGraw-Hill.

Introduction

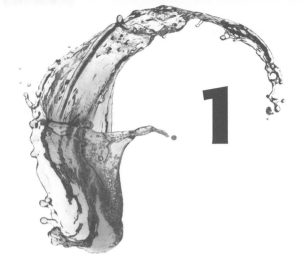

Physical hydrology is the study of the occurrence, movement, and physical properties of water on and below the earth's surface, with the exception of oceanic water. Oceans are studied in the science field of oceanography. This book deals with the principal rules governing the flow of water on the land part of the earth. The major water types are atmospheric water (Chapter 2), groundwater (Chapter 3), soil water (Chapter 4), and surface water (Chapter 5). This first chapter introduces these major water types and deals with essential concepts and definitions in hydrology. Attention is paid to the central concept in hydrology – the hydrological cycle, the central unit in hydrology – the drainage basin, and to the central equation in hydrology: the water balance. The major hydrological processes are introduced, and some reference is made to issues of **climate hydrology** or **global hydrology**, contemporary issues related to the hydrology of large areas or the earth as a whole.

1.1 Major water types

A blanket of air, which we call the atmosphere, surrounds the earth. The water contained in the air above the land surface is thus **atmospheric water**. This water is typically water vapour, but it may also be liquid water as in rain, or solid water as in snow (ice crystals). Atmospheric water is especially studied in the science field of meteorology. Water at the surface, stored in ponds, lakes, streams, or rivers, is called **surface water**.

At some depth below the land surface, pores (minute holes) within the soil, sediment, or rock are saturated with water. If we (are able to) dig a large pit, the level to which water from the ground rises in the pit is called the **water table** (Figure 1.1). The zone below this water table is the **saturated zone** and the water stored there is **groundwater**.

Below the land surface, but above the water table, pores in the soil may contain both air and water. This zone is known as the **unsaturated zone**. The water stored there is called **soil water**.

The estimated percentage storages of water on earth are presented in Table 1.1. The total amount of water on earth is about 1.4×10^{18} m³ (Maidment 1993). Fresh water is most important to mankind as a source of drinking water. Only about 2.5% of all available water on earth is fresh water. Of all available fresh water, 69% is contained in

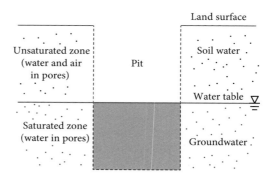

Figure 1.1. A vertical cross-section showing definitions of hydrological terms

polar ice (Box 1.1), 30% in fresh groundwater, and 1% in surface water, soil water, and atmospheric water taken together.

Because much groundwater resides in former sea deposits or is in contact with sea-water through underground water-bearing layers, more than half of the total available groundwater on earth is brackish (a mix of fresh and saline water) or saline.

Table 1.1 Estimated volumes and percentages for different kinds of water storage on earth (simplified after Maidment 1993)

	Volume in 10^9 m^3	% of all water	% of fresh water
Oceans	1 338 000 000	96.5	
Polar ice	24 023 500	1.7	68.6
Fresh groundwater	10 530 000	0.8	30.1
Brackish and saline groundwater	12 870 000	0.9	
Surface water, soil water, and atmospheric water	475 710	0.03	1.4
Saline surface water	85 400	0.006	
Total water	1 385 984 610	100	
Fresh water	35 029 210	2.5	100

BOX 1.1 What if all the polar ice were to melt?

Important happenings for **sea-level change** through geological time are the start and finish of ice ages. During an ice age, much water is contained in ice on land. Because of this, the sea level during the peak of the last ice age was 120 m lower than today. During the warm period preceding the last ice age, when the Northern Hemisphere was about four degrees warmer than at present, for a short period of time the sea level was 6 m higher than today, as evidenced by wave-cut notches at this height along cliffs in the Bahamas and the Orkney Islands.

We can make a rough estimate of the sea-level rise that would take place if all of the polar ice were to melt. On and near the poles we have two major ice sheets that could contribute significantly to sea-level rise: Antarctica and Greenland.

Ice contains gases and particles that are trapped inside. Thus the resulting water volume when ice melts is slightly less than the ice volume. The conversion rate from ice to water volume is about 0.9. Using ice volume data (km^3) from the IPCC (2001), we can thus estimate the equivalent water volume (km^3) by multiplying the ice volume by 0.9. When we divide this water volume (km^3) by the estimated area of the earth's oceans and seas of 362×10^6 km^2, we obtain a rough estimate of a sea-level rise of 71 m if all the polar ice were to melt (see Table B1.1).

Table B1.1 The estimated sea-level rise if all polar ice were to melt, using ice volume data from the IPCC (2001, Table 11.3)

	Antarctica	Greenland	Both
Ice volume (10^6 km^3)	25.7	2.9	28.6
Water volume (10^6 km^3)	23.1	2.6	25.7
Sea-level rise (m)*	64	7	71

* Assuming an area for the earth's oceans and seas of 362×10^6 km^2

When we allow for isostatic rebound, a slow lifting of the ground when the ice sheet (mass) disappears, and for sea water replacing grounded ice, the sea-level rise is slightly less. The IPCC (2001) then calculates a 61 (Antarctica) + 7 (Greenland) = 68 m sea-level rise; similarly, the contribution to sea-level rise due to the melting of glaciers and small ice caps has been calculated as 0.5 m, which falls well within the error of uncertainty of the sea-level rise due to the melting of the ice sheets of Antarctica and Greenland.

How quickly could all this ice melt?

The Antarctic and Greenland ice sheets lie above the snow-line, defined for this purpose as the altitude above which there (currently!) is a permanent covering of snow throughout the year. It would thus probably take several millennia to melt all of this ice, especially for the Antarctic ice sheet. The sea-level rise of 0.5 m due to a further melting of glaciers and small ice caps is thought to be proceeding more quickly, within a millennium or perhaps even within the current century.

Sea-level rise in the last century amounted to 0.2 m and a further global rise of the order of 0.2–0.6 m by 2100 is predicted by the IPCC (2007), with the largest contribution (75%) to sea-level rise this century predicted to arise from the **thermal expansion** of the oceans due to global warming.

One should note that the processes that could cause a rapid breakdown of ice sheets are not fully understood at present, and that the above answer is therefore speculative.

1.2 The hydrological cycle

The **hydrological cycle** (Figure 1.2) is a central concept within hydrology. The hydrological cycle describes the way in which the water of the ocean is heated by the sun, evaporates, and is carried over the earth in atmospheric circulation as water vapour.

Water vapour is the most abundant **greenhouse gas** in the atmosphere and most important in establishing the earth's **climate**, which is the weather averaged over a long period of time, usually 30 years. Greenhouse gases such as water vapour (H_2O), carbon dioxide (CO_2), and others allow much of the sun's warming energy to pass through as **shortwave radiation**, but absorb the **longwave**, infrared **radiation** emitted by the earth's surface. This effectively sets the air temperatures at the earth's surface to its current values, on average 15°C, a phenomenon known as the **greenhouse effect**.

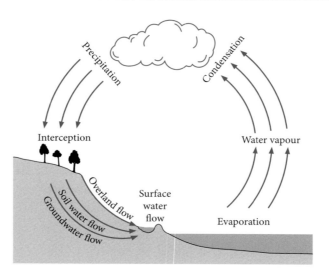

Figure 1.2. The hydrological cycle (after Ward and Robinson 2000)

The greenhouse effect as such is not to be confused with the enhanced greenhouse effect, which describes an out-of-control warming of an atmosphere by an overabundance of greenhouse gases. Such a runaway effect has most likely caused the current surface temperatures of the solar planet Venus to lie well in excess of 400°C.

$1\ \mu m = 1$ micrometre $= 10^{-6}$ m.

The average residence time is the average amount of time for which a water molecule resides in a particular place or with a particular system.

Without water vapour, which accounts for 60–70% of the greenhouse effect, and other greenhouse gases, the surface air temperatures of the solar planet earth would be well below freezing, on average at –18°C.

On the other hand, there is increasing concern nowadays that increased amounts of CO_2 and other greenhouse gases in the earth's atmosphere, related to human activities such as the burning of fossil fuels, land clearing, and agriculture, may lead to an increase of the current greenhouse effect, with **global warming** as a consequence.

In the earth's atmosphere, the water vapour condenses to form very small (1–100 μm) water droplets, recognizable as clouds. The **average residence time** of water in the ocean is of the order of thousands of years, whilst the average residence time of water in the atmosphere is only about ten days. Water precipitates from clouds as rain, snow, or hail, infiltrates soils, recharges groundwater, discharges into streams, and is eventually returned to the ocean.

Evaporation from the ocean is an important process: by leaving salts in the ocean (Box 1.2), it purifies the water, causing water vapour, cloud droplets, and consequently precipitation on the continents to consist of fresh water.

Of course, there are many short cuts and feedback mechanisms to be added to the general scheme of Figure 1.2. For instance, water vapour may return to the ocean as rain before being carried over land (short cut), or rainfall may be intercepted by vegetation and evaporated back into the atmosphere, from where it may return as rain later on (feedback); the feedback mechanisms are manifold and are therefore not included in Figure 1.2. Also, part of the water may manage to escape from the hydrological cycle for long periods of time; for instance, when stored in polar ice or as deep-seated groundwater. The average residence time for groundwater, including very deep groundwater, is approximately 20 000 years.

The overall cycle, however, proceeds endlessly, with climate change inducing changes in the speed of the processes involved. In line with this, many scientists nowadays believe that current **global warming** induces a **speeding up of the hydrological cycle**, and that this in turn may lead to more extreme weather conditions on parts of the earth.

BOX 1.2 Can global warming cause an ice age?

There is a hypothesis concerning the onset of an ice age involving global warming and the hydrology of Arctic areas that is too intriguing to be left unspoken here. The hypothesis, after work by Broecker (1997), is that due to global warming an increased fresh meltwater inflow from Arctic areas into the North Atlantic Ocean will impede the downflow of water of the North Atlantic Drift, an extension of the Gulf Stream: this will effectively halt the transport of warm ocean water from the equator to higher latitudes.

The largely wind-driven **Gulf Stream** and its deeper-seated **thermohaline*** extension, the **North Atlantic Drift,** can best be compared with a pump. Warm ocean water travels from Florida to north-west Europe. On its journey, the warm ocean water becomes more saline due to evaporation. Direct cooling

of this saline water by air south of Spitsbergen further increases the density of the water, causing it to become heavier than the water below it and to sink to greater depths, from where it is carried as a deep cold return flow southwards to the tropics again (Figure B1.2). Because of the Gulf Stream and North Atlantic Drift, winters in western Europe are mild, especially when compared to winters at the same latitude in western Canada and Alaska.

Global warming is hypothesized to cause an increased meltwater inflow from the in part permanently frozen (**permafrost**) Arctic: fresh water flows into the North Atlantic Ocean, decreases the ocean water's salinity, causes the North Atlantic Drift water to sink less deeply, and thus impedes the pumping mechanism. With a weaker pump at work, less

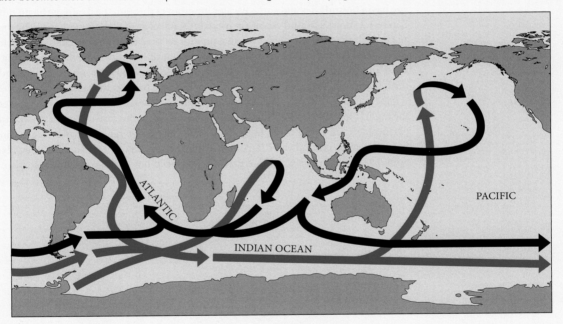

Figure B1.2. The ocean conveyor belt circulation: cold, deep currents are shown in blue; warm, shallow currents are shown in black (based on data by Professor W. Broecker; modified by Dr. E. Maier-Reimer)

warm water – which, importantly, becomes less saline – is transported to the North Atlantic Ocean, causing the pumping mechanisms to become weaker still. When this impeding mechanism starts to reinforce itself, a **positive feedback**, the transport of Gulf Stream and North Atlantic Drift water to higher latitudes effectively comes to a halt. Because of this shutdown of the so-called **Atlantic Meridional Overturning**, temperatures in the Northern Hemisphere will drop dramatically, causing the onset of a new ice age.

Researchers believe that at the end of the last ice age, when the global climate was warming, a shift to glacial conditions between 12 900 and 11 500 years BP (before present) may have evolved from a similar mechanism, which started with the emptying of a large meltwater lake into the North Atlantic Ocean (Broecker 2006): this shift in climate is known as the **Younger Dryas event**.

During our present interglacial, around 8200 years BP, meltwater outbursts from large proglacial lakes may again have led to a disturbance of down-welling water in the North Atlantic Ocean, now with, as a consequence, a fall in temperature of a few degrees for some hundreds of years, an event known as the **8.2 ky event** (Ellison *et al.* 2006).

Drijfhout (2007), however, concludes, from modelling the global ocean–atmosphere interaction, that the effects of a shutdown of the thermohaline circulation are much larger in a cold than a warm climate, thus weakening the hypothesis. Contrasting with the hypothesis, Weaver and Hillaire-Marcel

(2004) even argue that in light of the palaeoclimate record, a widespread collapse of the Atlantic Meridional Overturning is highly unlikely, and that 'it is safe to say that global warming will not lead to the onset of a new ice age'.

* The density of ocean water depends on both temperature (thermo = heat) and salinity (haline = sea salt). Differences in density cause a global thermohaline circulation of ocean water called the ocean conveyor belt circulation (Figure B1.2).

1.3 Drainage basin hydrological processes

A central unit within hydrology is the **drainage basin** or **catchment**. This is the geographical area that drains into a river or reservoir. Water flow is contained within **drainage basin boundaries** and surface water flow moves through the **drainage basin outlet**. Figure 1.3 shows a **digital elevation model (DEM)** of a small drainage basin. The drainage basin is the most convenient spatial unit for relating the water output of an area to its water input (precipitation). It is an important unit of investigation when managing the water quantity and/or quality of rural areas, cities, or even larger areas. Understanding the hydrological processes that act within a drainage basin is thus important.

Since surface water flows obey gravity, drainage basin boundaries are often taken as the highest locations within the landscape. However, in the case of, for instance, a tilted permeable rock layer, these highest locations need not represent the true drainage boundaries for groundwater flow, as is evident from Figure 1.4. Thus, drainage basin boundaries for surface water and groundwater need not necessarily coincide. Also, the

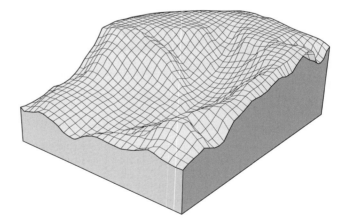

Area shown: 2.66 km^2
Altitude range: 268–421 m
View is from the NW
29 (WE) × (NS) point altitude matrix
Vertical exaggeration: 4×

Figure 1.3. A block diagram of the digital elevation model (DEM) of the Kribsbaach drainage basin, Gutland, Luxembourg, computed from a 50×50 m point altitude matrix

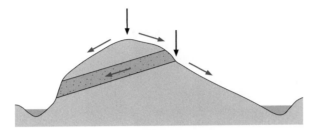

Figure 1.4. Drainage basin boundaries in cross-section
The tilted rock layer (indicated by the small blue dots) has
a higher permeability than the surrounding rocks. The blue
arrows indicate the direction of water flow. The black arrows
point to the boundaries of the left drainage basin for surface
water flow (left arrow) and groundwater flow (right arrow)

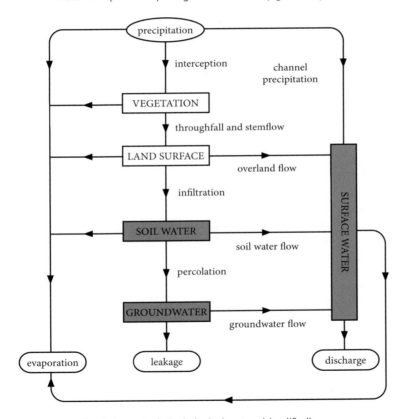

Figure 1.5. The drainage basin hydrological system (simplified)
Ovals represent input or output processes; lower-case characters represent
hydrological processes; rectangles and upper-case characters show various kinds of
water storage; the blue background shows the major types of water storage within
a drainage basin (for average conditions)

amount of **leakage**, the loss of quantities of groundwater through underground
water-bearing layers, is often unknown and may cause a problem when using the drainage basin as a unit. Despite these practical difficulties, the drainage basin is a most useful
unit when studying the hydro-environment – as, for instance, in pollution studies.

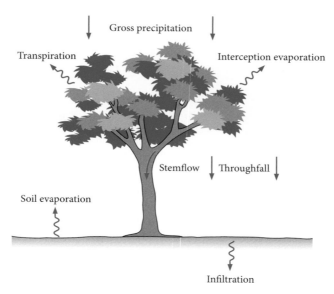

Figure 1.6. Interception
Interception = gross precipitation − net precipitation
Interception = interception storage change per unit of time +
interception evaporation
Net precipitation = stemflow + throughfall

Precipitation is the process whereby water particles, both liquid (rain) and solid (snow, hail), fall from the atmosphere on to the earth's surface, and thus on land or water (Figure 1.5). Precipitation that falls directly on to a stream or river channel is called **channel precipitation**.

On its way down, the precipitation may be intercepted by vegetation or buildings. This process is known as **interception** (Figure 1.6). The precipitation falling on to the top of a vegetation canopy is the **gross precipitation**. The precipitation that reaches the ground surface is the **net precipitation**. Net precipitation is made up of through-fall and stemflow. **Throughfall** consists of raindrops that fall through spaces in the vegetation canopy and that drip from wet leaves and branches. **Stemflow** is the water that runs down the main stem of a tree.

On contact with the soil, water may trickle into pores or cracks in the soil, sediment, or rock, a process called **infiltration**. Within the unsaturated zone, water may move further down to the groundwater table, a process known as **percolation**.

Evaporation of water is the change of water in a liquid or solid state to water vapour. Water may evaporate from an ocean, any wet surface - for instance, a soil surface (**soil evaporation**) - or from the stomata (numerous small pores) of living leaves of plants. In the latter case, it is also called **transpiration**. Evaporation of water involves a transfer, sometimes referred to by hydrologists as a loss, of water into the atmosphere. Precipitation that is intercepted by vegetation or buildings may evaporate back into the atmosphere. This is called **interception evaporation** (or interception loss). It is important to note that transpiration involves evaporation principally through leaf stomata, and thus from water that resides within the plant, whilst interception evaporation is evaporation from water that is stored on plant surfaces, such as branches and leaves. The amount of water stored on plant surfaces, as intercepted rain, snow, or ice, is referred to as **interception storage**.

The term interception loss is misleading in the sense that there are no losses within the hydrological cycle, as all water is recycled.

Before infiltrating the soil, precipitation may also be stored at the land surface; for instance, within litter layers built up of fallen tree leaves, in land surface depressions, or as snow or ice. The amount of water thus stored is called **surface storage**. Water that is released from this storage and runs down over a sloping land surface is called **overland flow** (Figure 1.2).

The water that is stored in the unsaturated zone as soil water, or deeper down, in the saturated zone, as groundwater, may also flow – usually very slowly – to lower areas in the landscape. These water flows are **soil water flow** and **groundwater flow** (Figure 1.2) and will be dealt with in more detail in the coming chapters. Taken together, these flows are often referred to as **subsurface flow**.

All of the flows mentioned above may contribute to the **surface water flow** of brooks or rivers (Figure 1.2), where the water is further discharged to eventually reach a sea or ocean, or a **closed basin lake**, which is a lake with no outlets, where the water can only evaporate, such as the Great Salt Lake in Utah, in the United States.

1.4 The water balance

The behaviour of water can be studied at many scales; for example, from the infiltration of water in a certain type of soil to the movement of water for large continents. For both research and water management purposes, it is important to know first of all the amounts of water entering and leaving a certain area or volume of ground, and to know the amount of water that is taken into or released from storage. A central equation when studying hydrological processes in the landscape is the **water balance (equation)**, which in its most simple form is written as follows:

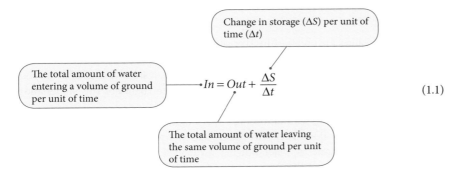

$$In = Out + \frac{\Delta S}{\Delta t} \qquad (1.1)$$

Change in storage (ΔS) per unit of time (Δt)

The total amount of water entering a volume of ground per unit of time

The total amount of water leaving the same volume of ground per unit of time

Total amount of water stored in the ground = water stored as surface water + water stored as soil water + water stored as groundwater.

The terms *In* and *Out* in the above equation have absolute values (greater than or equal to zero) irrespective of the direction of the processes involved. ΔS (in $\Delta S/\Delta t$) can have a zero, positive, or negative value as explained below (Δt is positive).

$\Delta S = 0$ means no change in water storage within the studied area for the period of time for which the water balance is set up. The total amount of water stored in the ground at the end and at the start of the selected time period is the same, irrespective of the manner in which the water is stored. For instance, part of the stored water may have changed from soil water to groundwater, but the total amount of water stored within the volume of ground has remained the same.

A positive change in storage – that is, $\Delta S > 0$ – signifies that water is taken into storage. The total amount of water stored in the ground is larger at the end of the selected time period than at the beginning, again irrespective of the way in which the water is stored. A positive change in storage in a drainage basin under natural conditions is manifest from wetter vegetation, a higher surface water level in streams or a lake, a wetter soil, and/or an elevated water table at the end of the selected time period.

A negative change in storage – that is, $\Delta S < 0$ – signifies a decrease of the total water storage in the area. In a drainage basin under natural conditions, this is manifest from drier vegetation, a lower surface water level in streams or lakes, a drier soil, and/or a lowered water table at the end of the selected time period.

ΔS may be set at zero when long-term averages for the ingoing (*In*) and outgoing (*Out*) terms are used in the water balance.

The water balance components are often estimated on a yearly basis, but in principle any time period may be taken; for example, half-yearly, monthly, daily, for the duration of a flooding event, and so on. It is of importance that the water balance is set for a strictly defined area, drainage basin, or volume of ground and for a strictly defined period of time. The same units of measurement must be used to the left and right of the equals sign (=). These are usually either a volume per time – for example, m³ per year (m³ year⁻¹) – or a length per time – for example, mm per day (mm day⁻¹).

As an example, the water balance of The Netherlands – a country bordering the sea, and one that has large rivers flowing through it – may be estimated as follows:

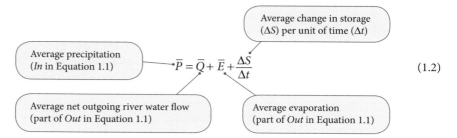

$$\overline{P} = \overline{Q} + \overline{E} + \frac{\Delta S}{\Delta t} \tag{1.2}$$

Data on $\overline{P}$ and $\overline{E}$ are rounded 1971–2000 values from KNMI (2002); the division of $\overline{Q}$ over the summer and winter half-years is taken as in De Vries (1980).

The average values are obtained by averaging over a long period of time.

A **hydrological year** has a duration of one year and starts and ends at the end of the dry season, when storage is at its lowest; or, alternatively, it starts and ends at the end of the wet season, when storage is at its highest. The water balance of The Netherlands for an average hydrological year is presented in Table 1.2.

Because we are dealing with an average hydrological year, the average change in storage $\overline{\Delta S} = 0$ and $\overline{Q} = \overline{P} - \overline{E}$ is approximately 240 mm year⁻¹.

Discharge is a volume of water flowing through a cross-section in a particular time interval. Note that the average net outgoing river water flow $\overline{Q}$ of 240 mm year⁻¹ (= 0.24 m year⁻¹) is equal to the average net outgoing discharge $\overline{Q}$ in m³ year⁻¹ divided by the area of The Netherlands, which is approximately 40×10^9 m²; thus the average net outgoing discharge $\overline{Q}$ of The Netherlands should be of the order of 9.6×10^9 m³ year⁻¹.

The example of Table 1.2 shows us that in summer, evaporation exceeds precipitation (on average, by 460 − 380 = 80 mm). Furthermore, $\overline{\Delta S} = -150$ mm, or, in words, the average change in storage in the summer half-year amounts to −150 mm; 150 mm of water is released from storage.

Since $\overline{\Delta S} = 0$ mm on a yearly basis, this must mean that the average change in the winter half-year amounts to +150 mm, and that in winter 150 mm of water is taken into storage.

Table 1.2 The water balance of The Netherlands for an average hydrological year, calculated using Equation 1.2: data in millimetres (per year or half-year); summer half-year, April–September; winter half-year, October–March

Equation 1.2	$\bar{P}$	$\bar{Q}$	$\bar{E}$	$\overline{\Delta S}$
Year	800	240	560	0
Summer half-year	380	70	460	−150
Winter half-year	420	170	100	+150

The example of Table 1.2 further shows us that in winter, precipitation exceeds evaporation (on average, by 420 − 100 = 320 mm), that there is slightly more precipitation in winter than in summer (on average, 420 − 380 = 40 mm more), that the evaporation in winter is lower than in summer with its higher temperatures (on average, by 460 − 100 = 360 mm), and that the net outflow from rivers in The Netherlands is higher in winter than in summer (on average, by 170 − 70 = 100 mm).

The average net outgoing discharge $\bar{Q}$ is the average discharge of rivers leaving The Netherlands minus the average discharge of rivers entering The Netherlands.

The water balance is an important tool for estimating the effects of **land-use change** and **climate change** scenarios, and for water management purposes in general.

For instance, a land-use change can influence the water balance due to a change in evaporation.

Another example is that **global warming** may lead to higher winter temperatures in mountainous areas, and thus to a rise of both the snowline and the 0°C isotherm, and to less storage of snow in these areas in winter. This, in turn, leads to lower meltwater discharges in spring and summer of rivers flowing from these mountains. Such an effect is predicted by Kwadijk (1991) for the summer discharge of the River Rhine, which flows from the Swiss Alps to the North Sea. Such a decrease in discharge could have serious, unwanted economic consequences if it was to hinder cargo transport by boats along the Rhine.

Water management for which the water balance is an important investigative tool may, for instance, involve counteracting the effects of intensive irrigation and/or drainage schemes or of the pumping up of groundwater, all of which may have a general drying out or desiccating effect. A well-known example is the Aral Lake in Central Asia, which was transformed from a large freshwater lake to a small saltwater lake due to a dramatic fall in the water level during the twentieth century. This in turn was caused by the large irrigation schemes from the two main rivers feeding the Aral Lake, the Amu Darya and Syr Darya rivers.

Exercise 1.4.1 A drainage basin bordering the sea has an area of 7500 km^2. The average precipitation for this drainage basin is 900 mm year^{-1}. The average surface water flow at the outlet of the drainage basin equals 22.5×10^8 m^3 year^{-1}. The average groundwater flow to the sea is 100 mm year^{-1}. The averages are determined for 30 hydrological years.

a. Draw up the water balance.
b. Determine the average actual evaporation in mm year^{-1} and m^3 year^{-1}.

Exercise 1.4.2 A gentle hill slope with an area of 10^4 m^2 receives rain for a period of 40 minutes. The average rainfall intensity equals 30 mm hour^{-1}. The total surface water flow during the 40 minutes is 15×10^4 litres. The quantities of water involved in evaporation and percolation to the groundwater are negligible.
Determine the change in soil water storage in mm and m^3.

Summary

- Physical hydrology is the study of the occurrence, movement, and physical properties of water on and below the earth's surface, with the exception of oceanic water.

- If we dig a large pit, the level to which water from the ground rises in the pit is called the water table. The zone below this water table is the saturated zone and the water stored there is groundwater. Below the land surface, but above the water table, pores in the soil may contain both air and water. This zone is known as the unsaturated zone. The water stored there is called soil water.

- Only about 2.5% of all available water on earth is fresh water. Of all available fresh water, 69% is contained in polar ice, 30% in fresh groundwater, and 1% in surface water, soil water, and atmospheric water taken together. More than half of the total available groundwater on earth is brackish (a mix of fresh and saline water) or saline.

- Water vapour is the most abundant greenhouse gas in the atmosphere and most important in establishing the earth's climate.

- Greenhouse gases such as water vapour (H_2O), carbon dioxide (CO_2), and others allow much of the sun's warming energy to pass through as shortwave radiation, but absorb the longwave, infrared radiation emitted by the earth's surface. This effectively sets the air temperatures at the earth's surface to their current values, on average 15°C, a phenomenon known as the greenhouse effect.

- The enhanced greenhouse effect is an out-of-control warming of an atmosphere by an overabundance of greenhouse gases.

- Evaporation of water is the change of water in a liquid or solid state to water vapour.

- Evaporation from the ocean is an important process: by leaving salts in the ocean it purifies the water, causing water vapour, cloud droplets, and consequently precipitation on the continents to consist of fresh water.

- Many scientists nowadays believe that current global warming induces a speeding up of the hydrological cycle, and that this in turn may lead to more extreme weather conditions on parts of the earth.

- A drainage basin or catchment is the geographical area that drains into a river or reservoir.

- Precipitation is the process whereby water particles, both liquid (rain) and solid (snow, hail), fall from the atmosphere on to the earth's surface, and thus on land or water.

- Channel precipitation is precipitation that falls directly on a stream or river channel.

- Interception is the process of vegetation or buildings intercepting precipitation. The precipitation falling on to the top of a vegetation canopy is the gross precipitation. The precipitation that reaches the ground surface is the net precipitation. Net precipitation is made up of throughfall and stemflow. Throughfall consists of raindrops that fall through spaces in the vegetation canopy and that drip from wet leaves and branches. Stemflow is the water that runs down the main stem of a tree.

- On contact with the soil, precipitation may trickle into pores or cracks of the soil, sediment, or rock, a process called infiltration. Within the unsaturated zone, water may move further down to the groundwater table, a process known as percolation.

- Water that evaporates from the stomata (numerous small pores) of living leaves of plants is called transpiration.

- Precipitation that is intercepted by vegetation or buildings may evaporate back into the atmosphere. This is called interception evaporation (or interception loss).

- The amount of water stored on plant surfaces, as intercepted rain, snow, or ice, is referred to as interception storage.

- The amount of water stored at the land surface - for instance, within litter layers built up of fallen tree leaves, in land surface depressions, or as snow or ice - is called surface storage.

Water that is released from this storage and runs down over a sloping land surface is called overland flow.

- Water balance (equation): $In = Out + \Delta S/\Delta t$.

- The terms *In* and *Out* in the above equation have absolute values (greater than or equal to zero) irrespective of the direction of the processes involved. (ΔS can have a zero, positive, or negative value.)

- A positive change in storage – that is, $\Delta S > 0$ – signifies that water is taken into storage. The total amount of water stored in the ground is larger at the end of the selected time period than at the beginning, irrespective of the way in which the water is stored. A positive change in storage in a drainage basin under natural conditions is manifest from wetter vegetation, a higher surface water level in streams or lakes, a wetter soil, and/or an elevated water table at the end of the selected time period.

- A negative change in storage – that is, $\Delta S < 0$ – signifies a decrease of the total water storage in the area. In a drainage basin under natural conditions, this is manifest from drier vegetation, a lower surface water level in streams or lakes, a drier soil, and/or a lowered water table at the end of the selected time period.

- ΔS may be set at zero when long-term averages for the ingoing (*In*) and outgoing (*Out*) terms are used in the water balance.

- It is of importance that the water balance is set for a strictly defined area, drainage basin, or volume of ground, and for a strictly defined period of time. The same units of measurement must be used to the left and right of the equals sign (=).

- A hydrological year has a duration of one year and starts and ends at the end of the dry season, when storage is at its lowest; or, alternatively, it starts and ends at the end of the wet season, when storage is at its highest.

- The water balance is an important tool for estimating the effects of land-use change and climate change scenarios, and for water management purposes in general.

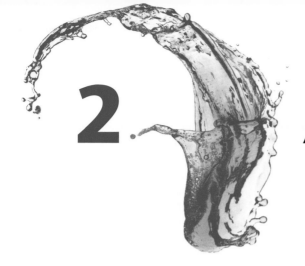

2. Atmospheric water

Introduction

Atmospheric water is linked with soil water (Chapter 4) and surface water (Chapter 5) by the processes of precipitation and evaporation. As shown in Table 1.1, surface water, soil water, and atmospheric water only account for 0.03% of all water on earth. Despite this low percentage, this resource is very important to us as a source of drinking, agricultural, and industrial water.

The amount of water vapour in the earth's atmosphere is estimated to be equivalent to a liquid water layer of only 25 mm. Since the average precipitation over the globe is estimated as 1000 mm per year, a simple calculation teaches us that the atmosphere loses its water vapour on average $\frac{1000}{25} = 40$ times per year and that the **average residence time** of water vapour in the atmosphere therefore equals $\frac{365}{40} = 9$ days. This rapid average turnover rate means that atmospheric water is important both as input and output to soil water (Chapter 4) and surface water (Chapter 5).

The current understanding of atmospheric processes above land masses such as Africa and the central United States further suggests that a significant fraction of precipitation in these areas originates from local evaporation. This causes a **soil moisture – precipitation feedback** (Bierkens and Van den Hurk 2008): precipitation recharges local soil water, which in part returns to the atmosphere by evaporation, from which it returns in part to the soil by precipitation. This reinforcing mechanism (**positive feedback**) may lead to the **persistence** of either wet or dry conditions on a seasonal or multi-year timescale, such as the multi-year droughts in the Sahel region of Africa.

The following topics with regard to precipitation are introduced in this chapter: the generation of precipitation from clouds, different types of precipitation, the measurement of precipitation at a point in the landscape, and the estimation of areal precipitation. After this, evaporation from both vegetated land surfaces and open water is discussed. At the end of the chapter, the reader is shown the strength of using the Penman–Monteith evaporation model as a tool to gain understanding of the effects that a **land-use change** from forest to grassland may have for two different settings: a dry climate and a wet climate.

2.1 Cloud formation

Heating of the land and of the air directly above it causes the air to ascend, a direct result of a heated pocket of air being less dense than the air that surrounds it.

Physics teaches us that **dry air** that ascends in the atmosphere expands and cools down at a fixed rate of 1°C for every 100 m it rises, which is called the **dry adiabatic lapse rate**. (An **adiabatic process** is a process during which a body of air cools or warms without there being an exchange of energy with the surrounding air.) This cooling down is simply a consequence of the heat being distributed over a larger column of air. The upward movement of the heated pocket of air continues until, at some greater height, the pocket of air cools down to the temperature of the surrounding air, which then effectively halts a further upward movement of the air.

Heating of the sea or a wet land surface causes evaporation and now **moist air** to ascend. Cool air cannot hold as much water vapour as warm air, as indicated by the lower saturation vapour pressure for cooler air shown in Figure 2.1. **Saturation vapour pressure** is the pressure exerted by water vapour molecules in the air, when the air is saturated with water vapour; the physics background is provided in Boxes 2.1 and 2.2. At some greater height, the continued cooling of an ascending pocket of moist air may cause it to become saturated with water vapour before it reaches the same temperature as the surrounding air. The relative humidity of the air is then 100%, and the temperature has reached its **dew point** (T_d) if this temperature is greater than 0°C, or its **frost point** (T_f) if the temperature is 0°C or less. At the dew point, **condensation** starts. This means that part of the water vapour in the rising air turns into liquid water, which is visible as a cloud. Likewise, at frost point **sublimation** starts, which is the change from water vapour to ice without passing through the liquid state, and which is also visible as a cloud. A **cloud** can thus be defined as a visible mass of very small water droplets (1–100 µm) and/or ice crystals (when temperatures are low) floating in the atmosphere (Box 2.3).

Condensation is the opposite of evaporation. Water vapour has a higher energy than liquid water, which in turn has a higher energy than ice. Thus, with condensation, freezing, and sublimation, energy in the form of heat is released

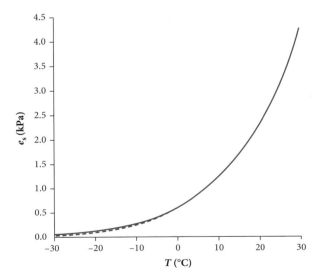

Figure 2.1. Saturation vapour pressure e_s (kPa) and temperature T (°C). The full curve is the curve for water – Equation B2.2.3 is an approximate equation for this curve; the broken curve is the curve for ice

BOX 2.1 Some underlying physics: force, mass, and weight

Force can be defined as an external cause responsible for any change of a physical system. The unit of both force and weight is the newton, abbreviated as N. A **newton** is defined as the amount of force required to accelerate a mass of one kilogram (kg) at a rate of one metre per second each second (m s^{-2}): 1 N = 1 kg m s^{-2}.

Note that in the physical sciences an important distinction is made between mass and weight. The base unit of **mass**, the **kilogram** (kilo = 1000), is a constant, and has been defined as being equal to the mass of the international prototype of the kilogram, kept in Sèvres, France. The **weight** of an object equals its mass multiplied by the acceleration due to gravity. The **acceleration due to gravity**, though often approximated to lie around 9.8 m s^{-2}, is not constant in reality, but varies with latitude, altitude, and location on earth. If we drop an object, and assume air resistance to be negligible, then a value of 9.8 m s^{-2} for the acceleration due to gravity tells us that the object's speed after one second is 9.8 m s^{-1}, after two seconds 9.8 + 9.8 = 19.6 m s^{-1}, and after three seconds 19.6 + 9.8 = 29.4 m s^{-1}, and so on. Also, objects with different masses fall at the same rate when air resistance is negligible. Thus, a well-known experiment shows us that a feather and a piece of lead, dropped from the same height and at the same time in a vertical, vacuum-pumped cylinder, hit the bottom of the cylinder at exactly the same time.

BOX 2.2 Some further physics: air pressure, vapour pressure, and relative humidity

Pressure p is defined as the magnitude of a normal force F (N) divided by the surface area A (m^2) over which this normal force acts:

$$p = \frac{F}{A} \qquad (B2.2.1)$$

The word 'normal' in this context means that the force acts in a direction perpendicular to the surface. Thus, from the above definition, the unit of pressure is N divided by m^2, also known as a **pascal** (Pa): 1 Pa = 1 N m^{-2}.

With **air pressure**, also called **atmospheric pressure**, the force is exerted by the weight of the molecules (tiny particles) in the air. Since weight is a force caused by gravity, we can substitute weight W (N) for force F (N) in the above equation (Figure B2.2):

$$p = \frac{W}{A} \qquad (B2.2.2)$$

It is common practice to present air pressure data in **hectopascals** (hPa), which is simply 100 times the value of a pascal (hecto = 100). A hectopascal is the same as a **millibar** (mb or mbar).

Standard atmospheric pressure is defined as 1013.25 hPa (= 1 atmosphere), and this value is roughly equal to the average air pressure at sea level. In reality, air pressures, also at sea level, vary from place to place and moment to moment on the globe.

Water vapour in the air also exerts a pressure, albeit only a very small part of the total air pressure. With vapour pressure, the force is exerted by the weight of the water vapour

molecules in the air. A distinction can be made between the **actual vapour pressure** (e_a), the actual partial pressure of the water vapour molecules in the air at a certain temperature, and the **saturation vapour pressure** (e_s), the partial pressure of the water vapour molecules when the air is saturated with water vapour, and thus when the air holds the maximum amount of water vapour possible at that temperature. Vapour pressure is usually presented in **kilopascals** (kPa = 1000 Pa) or hPa. The change of the saturation vapour pressure with temperature is presented in Figure 2.1.

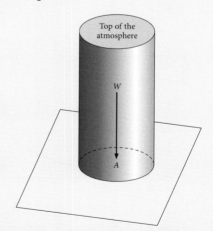

Figure B2.2. Air pressure is the weight W (N) of a column of air divided by the earth's surface area A (m^2)

An approximate equation for this curve is as follows:

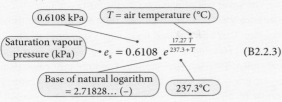

$$e_s = 0.6108 \; e^{\frac{17.27\,T}{237.3+T}} \qquad \text{(B2.2.3)}$$

The **relative humidity** *RH* is defined as the ratio of the actual vapour pressure e_a (kPa) to the saturation vapour pressure e_s (kPa) at the same temperature:

$$RH = \frac{e_a}{e_s} \qquad \text{(B2.2.4)}$$

BOX 2.3 San Francisco's summer fog

San Francisco, California is not only famous for its earthquake(s), but also for its summer fog, 'which can bring the temperature down in a hurry and make the tourists' knees visibly shake in their short pants. To be prepared for any kind of weather, dress in layers and never venture out without a sweatshirt – or you'll probably end up buying one' (Lonely Planet 2006).

In summer, heating by the sun causes the land surface temperature of San Francisco to rise. The land surface heats the air above it. Because warm air is less dense, the air rises, causing low air pressure at San Francisco's land surface. The air above the ocean is drawn into this lower-pressure zone, perceptible as an ocean breeze.

The air above the ocean is initially warm, but is cooled by the California current, which transports cool ocean water southwards along the coast of California. Often, this cooling is sufficient to lower the temperature of the air to its dew point. As there are abundant salt particles floating in the air above the ocean for water vapour to condense on, clouds are generated above the ocean's surface (Figure B2.3). **Fog** consists of clouds in contact with the earth's surface. The fog caused by the cooling mechanism just described is transported by the ocean breeze into San Francisco, causing parts or all of hilly San Francisco to be shielded from the sun and the temperature in these foggy parts to drop.

Figure B2.3. Fog at the Golden Gate Bridge, San Francisco
© mr_focus/istockphoto

In winter, the contrast between land and sea temperatures is less extreme; as a consequence, there is less cooling of the air above the ocean and the air temperature does not fall to its dew point as often, making the occurrence of fog less frequent.

---

to the atmosphere; with evaporation and melting, heat energy is absorbed (Box 2.4). Heat related to a change in phase of water is called **latent heat**. If, after initial condensation, the air continues to ascend and expand, the cooling down takes place at a rate of 0.6°C for every 100 m it rises. This lesser cooling rate compared to dry adiabatic cooling is due to latent heat being released during condensation. Consequently, this lapse rate is known as the **saturated adiabatic lapse rate** (Box 2.5).

The water vapour condenses on small particles, such as dust, sea salt, or chemical substances that float in the air and act as condensation nuclei, causing clouds to form. **In fact, condensation of water vapour can only take place at a relative humidity of 100% if these particles are present.** The condensation nuclei are **hygroscopic**; that is, they attract water in much the same way as kitchen salt gets wet under unfavourable

---

BOX 2.4 Sprinkling water as frost protection in orchards

There are many fruits and other crops that will be damaged during frost conditions.

Installations that sprinkle water are widely used to provide **frost protection** for these plants when night frost is predicted. Due to the sprinkling of water, a layer of ice forms around the blossom of young plants. **Freezing** of the supplied water releases heat that keeps the ice–water mixture at about 0°C, and thus prevents young fruit or crops from getting damaged. The applied water must be plentiful in order to supply enough heat by freezing to more than compensate for all the heat losses due to radiation, air movement, and evaporation: water should drip slowly but continuously from the ice attached to the plants. Also, the sprinkling installation should remain working for all of the frost period, since failure of the system would lead to irreversible frost damage to the fruit or crops. For an informative account on the numerous aspects of frost protection, the reader is referred to Evans (1999).

When only light frost at night is predicted – a rather common situation in The Netherlands, for instance, with its climate influenced by the sea – continuous sprinkling of water can have the desired effect that the air that drops in temperature during night time becomes saturated with water vapour before freezing sets in. With only a slight further reduction of the temperature, **condensation** at dew point then delivers the heat energy needed to effectively prevent the light frost conditions from setting in.

BOX 2.5 The sensitivity of the River Rhine discharge to global warming

A **hypsographical curve** is a curve of the percentage of area below or above certain altitudes for a selected area. Figure B2.5 is the hypsographical curve for the drainage basin of the River Rhine upstream of Rheinfelden (near Basel), Switzerland.

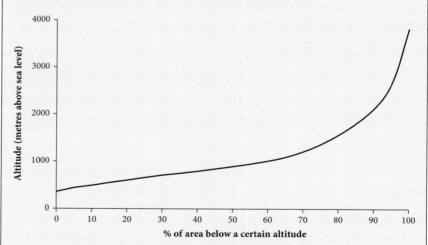

Figure B2.5. Hypsographical curve for the drainage basin of the Rhine upstream of Rheinfelden, Switzerland (34,550 km²) (after Kwadijk 1991, based on data by Hils 1988)

A presumed rise in mean winter temperature of 4°C in the Alps upstream of Rheinfelden will cause both the mean altitude of the **snowline**, defined for this purpose as the altitude above which snow accumulates on a seasonal basis, and the 0°C isotherm in winter to rise by 700 m as calculated below:

$$\frac{4°C}{\left(\dfrac{0.6°C}{100\ m}\right)} \approx 700\ m$$

Rise in winter temperature (→ 4°C)

Saturated adiabatic lapse rate (→ denominator)

The mean altitude of the snowline under the presumed scenario of a 4°C rise in mean winter temperature can simply be found by adding 700 m to the present-day mean altitude of the snowline: the reduction in snow-covered area under the presumed scenario can be calculated from the hypsographical curve.

Kwadijk (1991), in a water balance study of the Rhine drainage basin and using the method described above, investigated the sensitivity of the Rhine discharge to global warming. He investigated a scenario with a rise in annual temperature of 2°C, assuming this to be made up of a 4°C rise in winter, a 2°C rise in spring and autumn, and an unchanged mean summer temperature. Using these values, which lie within the range of global warming scenarios due to a doubled CO_2 concentration in the atmosphere, he predicted a reduction in the summer discharge of the Rhine of 10–15%.

circumstances. Of particular importance as **cloud condensation nuclei** are **atmospheric aerosols** (Box 2.6) in the range from 0.001 to 1 μm, especially the so-called **Aitken nuclei**, particles in the range from 0.01 to 0.1 μm, of which many are electrically charged to bond with water.

BOX 2.6 Atmospheric aerosols, global dimming, and 9/11

Atmospheric aerosols are solid and liquid particles, 0.001–10 μm in size, floating in the air that originate from the land, sea, or chemical reactions in the atmosphere. They are mineral dust, volcanic ash, sea salt, sulphate, carbon-containing compounds, and so on. Aerosols occur naturally, but about 10% of the total amount of aerosols in the atmosphere is currently estimated as being due to human activities. Atmospheric aerosols both absorb and scatter solar radiation and thus play an important role in regulating the amount of solar radiation to reach the surface of the earth, a consequence described in the literature as radiative forcing. Radiative forcing is the change in the balance between radiation coming into the atmosphere and radiation going out. A positive radiative forcing tends on average to warm the earth's surface, and a negative radiative forcing tends on average to cool the earth's surface. Atmospheric aerosols are thought to provide a negative forcing and thus a masking effect on global warming. This masking effect is also known as global dimming, the observed worldwide reduction of the order of 5% over the three decades from 1960 to 1990 of sunlight reaching the surface of the earth due to atmospheric aerosols and cloud changes. Some climate researchers claim that without aerosols, the current average temperature of the earth due to global warming would already have risen by more than 5°C.

In line with this, the three-day absence of condensation trails from jet aircraft due to the grounding of all commercial aircraft in the United States after the 9/11 terrorist attacks resulted in a measurable increase in the daily temperature range, the difference between the daytime maximum and night-time minimum temperatures, of about 1.1°C across the United States, not including Alaska and Hawaii (Travis et al. 2002).

Aviation accounts for some 2% of the global emissions of human-related CO_2. However, with regard to the impacts of aviation – more specifically, **contrails** (condensation trails behind airplanes) – on radiative forcing and climate, large uncertainties remain (Wuebbles 2007).

2.2 Generation of precipitation

Precipitation starts when large water droplets or ice crystals fall to the earth due to gravity. In many clouds, the water droplets or ice crystals are not heavy enough to overcome **updraft** of the air, which is the vertical upward movement of the air as described in section 2.1. Also, precipitation may evaporate as it falls into the warmer air below the clouds. Thus, in order for precipitation to reach the ground, water droplets or ice crystals must become large and heavy enough to counteract both the effect of updraft and evaporation. There are two main processes that generate precipitation (Figure 2.2).

In **warm clouds**, clouds with temperatures above freezing point, turbulent atmospheric mixing can cause water droplets of varying size to grow through a **collision–coalescence process**, as schematized in the left-hand part of Figure 2.2. By collision of rising and falling droplets in a cloud, the droplets coalesce to produce larger and larger droplets. Eventually, the droplets are large and heavy enough to overcome the air currents in the cloud and the updraft beneath it. This method of raindrop production is the primary mechanism in tropical regions. It typically produces **drizzle**, a fairly steady, light rain.

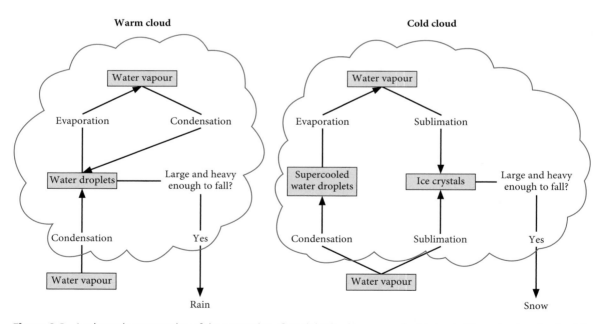

Figure 2.2. A schematic presentation of the generation of precipitation in warm and cold clouds. Warm cloud: water droplets go through many condensation–evaporation cycles whilst circulating in the cloud before they coalesce into large drops of water that are sufficiently heavy to fall through the cloud base as rain. Cold cloud: ice crystals are favoured over supercooled water droplets in attracting water vapour; they grow larger until they are large and heavy enough to fall through the cloud base as snow

Cold clouds, clouds with temperatures below freezing point, can contain both ice crystals and supercooled water droplets, as shown on the right-hand side of Figure 2.2. Supercooled water is liquid water that is chilled below its freezing point, without it becoming solid. As with condensation, the processes of freezing and sublimation also need nuclei in order to get started. As freezing nuclei (the term includes sublimation), ice crystals themselves are very efficient in attracting water (vapour). The saturation vapour pressure over ice, which is the pressure of water vapour in close contact and equilibrium with ice, is only slightly lower than that over liquid water (Figure 2.1), but it is sufficiently lower to favour the ice crystals (over the supercooled water droplets) in capturing water vapour from the air. The ice crystals sublimate the water vapour and grow larger. This reduces the amount of water vapour in the cloud: the air in the cloud becomes drier. The lower vapour pressure causes other supercooled water droplets to also evaporate, and then, as a follow-up, the newly formed water vapour to also sublimate on the ice crystals. By this positive feedback mechanism, known as the Bergeron–Findeisen process, all supercooled water is thus effectively transformed into ice crystals that grow larger and heavier; this is schematized on the right-hand side of Figure 2.2. The feedback is called positive because the process of sublimation itself, via another process (evaporation), strengthens sublimation. Also, the ice crystals may collide with other ice crystals and grow larger still through collision–coalescence. Eventually, the ice crystals will have grown large and heavy enough to fall from the cloud as snow (ice crystals) or, when they melt on their way down, as cold rain. The Bergeron–Findeisen process (Box 2.7) is the dominant process for the generation of precipitation from cold clouds in the middle and upper latitudes of the earth.

BOX 2.7 VIPs in understanding cloud physics

In his 1911 book *Thermodynamik der Atmosphäre*, Alfred Wegener, famous for his continental drift theory, proposed a theory of ice crystal growth based on the difference in saturated vapour pressure between ice crystals and supercooled water droplets, and suggested that such growth was possible in a cloud. In 1928, the Swedish meteorologist Tor Bergeron was the first to recognize that this could lead to precipitation (Schultz and Friedman 2007). After experimental confirmation by the German meteorologist Walter Findeisen in 1938, the generation of precipitation in a cloud containing both ice crystals and supercooled water droplets has subsequently become known as the Bergeron–Findeisen process, the Bergeron process, or the Wegener-Bergeron-Findeisen process.

2.3 Precipitation types

Rising, expansion, and cooling of air and the formation of clouds by condensation are preliminary steps for the generation of precipitation. The processes leading to the uplift of air and the scale at which these processes act may, however, differ. Following this, a number of different types of precipitation can be distinguished.

Precipitation as a result of local heating of the air at the earth's surface, as described in some detail in section 2.1, is called convective precipitation (Figure 2.3).

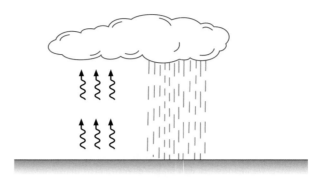

Figure 2.3. Convective precipitation

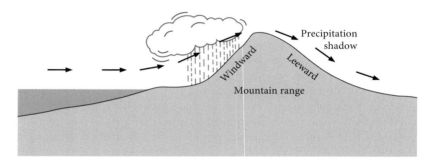

Figure 2.4. Orographic precipitation

This process is active in tropical areas and in the interior of continents. Convective precipitation is local and often intense (for example, during thunderstorms).

When horizontal air currents are forced to rise over natural barriers such as mountain ranges, **orographic precipitation** may occur, as shown in Figure 2.4. Uplift of the air causes precipitation to fall on the windward side of a mountain range. This results in drier air on the descending (often warming) **leeward** or downwind side of the mountain range, where a **precipitation shadow**, an area without or with only slight precipitation, is formed.

Unequal heating of the earth's surface creates differences in air pressure. **Wind**, movement of air parallel to the ground, initially moves in the direction of low air pressure. However, in large weather systems the direction of the movement is affected by the earth's rotation (the **Coriolis effect**; Box 2.8) and at or near the earth's surface by friction. As a consequence, **Buys Ballot's law** (1857) states that when you stand

An *isobar* connects points of equal air pressure.

BOX 2.8 The Coriolis effect

Imagine that you are sitting on a horizontal, merry-go-round-like disc that rotates in an anticlockwise fashion when viewed from above. You are sitting near the outer edge and a friend of yours is sitting near the centre of the disc. You are both firmly attached and facing each other. You are both asked to roll a tennis ball over the disc's smooth surface in each other's direction. Observed from where you sit, your ball curves to the right. Observed from where your friend sits, his/her ball also curves to the right (Figure B2.8). Why is this?

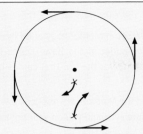

Figure B2.8. Curved deflection to the right of rolling balls on an anticlockwise rotating merry-go-round as analogue for moving air parcels on the earth's Northern Hemisphere (see text for explanation)

You are sitting near the outer edge, so both you and the ball experience a high velocity (distance per unit time). Your friend near the centre experiences a lower velocity. Your ball rolls inwards over parts of the disc with slower velocities. However, the ball carries its original, faster velocity with it, and from where you sit you see your ball curve to the right.

Now think from your friend's perspective. His/her ball rolls over parts of the disc with increasingly faster velocities. His/her ball also carries its original, but now slow, velocity with it. Your friend sees his/her ball also curve to the right.

Actually, you may sit anywhere on the rotating disc and roll the ball in any direction: viewed from wherever you are sitting on the disc, your ball will always curve to the right. For instance, when you sit midway along the disc's radius, facing the centre, and roll your ball to your right, you add the ball's velocity to the disc's velocity, causing the ball to deflect to an area of higher velocity, which is to the right of the ball's path of motion. If you had rolled the ball to your left, you would subtract from the disc's velocity, causing the ball to deflect to

an area of lower velocity, which, again, is to the right of the ball's path of motion.

Now replace the rotating disc with earth's Northern Hemisphere and the rolling ball with a moving air parcel, and you can easily imagine that an earth observer who studies the path of an air parcel in the Northern Hemisphere observes a similar effect, a curved deflection to the right. (Watching from a stationary position in space, an observer would not see the air parcel deflect but, rather, would see the air parcel moving in a straight line, with the earth rotating out from under it.) To an earth observer, the curved deflection on the Southern Hemisphere, which has a clockwise rotation when viewed from 'under' the South Pole, is to the left. At the earth's equator, where both hemispheres meet, the deflection is zero, with the deflection increasing towards the poles. At the poles, the deflection is at a maximum and a perfect spin against the rotation of earth itself.

The mechanism leading to curved deflections of moving air parcels over our planet Earth is known as the **Coriolis effect** and is named after the French scientist Gaspard Gustave de Coriolis, a fanatical billiard player, who physically described the movement of billiard balls on a rotating billiard table in 1835: the mechanism itself was first identified in 1778 by Pierre-Simon Laplace (1749–1827).

The Coriolis effect is related to the earth's rotation, one rotation per day, and acts on relatively long-lived phenomena such as mid-latitude cyclones and large vortices in the ocean. In contrast, the water that drains from your kitchen sink or bathtub rotates every few seconds and flows fast. Contrary to what some people may want to make you believe, the Coriolis effect is much too weak to determine the direction in which water drains from your kitchen sink or bathtub.

with your back to the wind in the Northern Hemisphere, low air pressure is in front of you on your left and high air pressure is behind you on your right (Box 2.9): in the Southern Hemisphere, left and right must be reversed. In other words, wind travels anticlockwise near the earth's surface around areas of low pressure (100–2000 km across) in the Northern Hemisphere, as shown in Figure 2.5, and clockwise in the Southern Hemisphere. The air does so in an attempt to counteract the low pressure at the centre of the **low-pressure area (cyclone)**. Importantly, this converging air movement near the earth's surface can only take place when there is a possibility for the air to escape upwards from the centre of the low-pressure area. In the centre of a low-pressure area the air rises, expands, and cools, causing the formation of clouds by condensation at a higher level. Precipitation associated with this **convergence** and uplift of the air is called **cyclonic precipitation**.

The movement of air around and in high-pressure areas is the opposite of what happens around and in low-pressure areas. The air movement near the earth's surface in a **high-pressure area (anticyclone)** diverges away from the centre (Figure 2.5).

BOX 2.9   Buys Ballot's law

Christophorus Buys Ballot (1817–1890), professor in Meteorology and Physical Geography at Utrecht University, The Netherlands and founder of the Royal Dutch Meteorological Institute (KNMI) in 1854, is probably best known for the law that carries his name, Buys Ballot's law (1857):

When you stand with your back to the wind in the Northern Hemisphere, low air pressure is in front of you on your left and high air pressure is behind you on your right.

Buys Ballot's law is highlighted in Figure B2.9. This figure shows the direction of the wind *v* at or near the earth's surface, together with the balancing forces that determine this direction. The wind *v* can be seen to move at an angle with the isobars in the direction of low pressure in front and to the left.

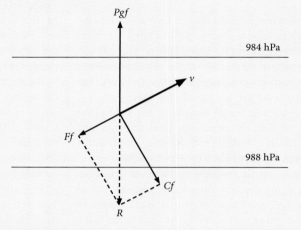

Figure B2.9.   Direction of the wind *v* at or near the earth's surface on the Northern Hemisphere, together with the balancing forces: pressure gradient force *Pgf*, Coriolis 'force' *Cf* and friction force *Ff* (the isobar values are given as examples) (after Schmidt 1976). The pressure gradient force *Pgf* acts perpendicular to the isobars in the direction of lower air pressure; the Coriolis 'force' *Cf* acts at right angles to *v* and is directed to the right of *v* in the Northern Hemisphere; the friction force *Ff* acts to slow down wind velocity and thus acts to oppose *v*. *Pgf*, *Cf* and *Ff* are in balance, with *Pgf* equal and opposite to *R*, which is the resultant of *Cf* and *Ff*.

Shortly after formulating his law, Buys Ballot discovered that he had empirically verified a theoretical relation deduced one year earlier by the American meteorologists J.H. Coffin and William Ferrel (1817–1891). Ferrel gracefully declined Buys Ballot's later offer to have the law renamed.

This **divergence** near the earth's surface is due to descending air at the centre of the high-pressure area. Air that descends warms up. Since warmer air can contain more water vapour than cooler air, a high-pressure area usually brings us fair weather.

The earth receives much heat at the equator and very little at the poles, which causes transport of heat in the direction of the poles and of cold air in the direction of the equator. Because of the earth's rotation (the Coriolis effect) and also the uneven distribution of land masses and oceans, a number of **global circulation cells** are created (Figure 2.6).

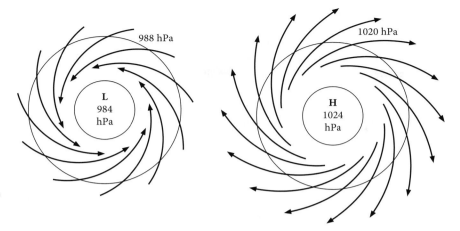

Figure 2.5. Air movement near the earth's surface around a low- (L) and high- (H) pressure area in the Northern Hemisphere (the isobar values are given as examples) (after Schmidt 1976)

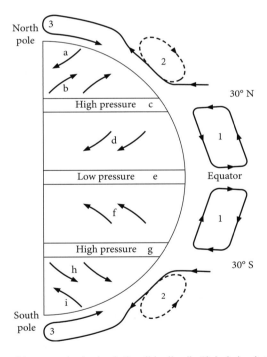

Figure 2.6. The earth's atmospheric circulation (idealized). Global circulation cells: 1, Hadley cell; 2, Ferrel cell; 3, polar cell. Lower atmosphere: a, polar easterlies; b, southwesterly winds; c, subtropical high pressure (Tropic of Cancer; horse latitudes); d, northeasterly trade winds; e, equatorial low pressure (area with no steady surface winds, called the doldrums); f, southeasterly trade winds; g, subtropical high pressure (Tropic of Capricorn; horse latitudes); h, northwesterly winds; i, polar easterlies. Note the deflection of the air movement in the lower atmosphere to the right in the Northern Hemisphere and to the left in the Southern Hemisphere due to the Coriolis effect

At the equator, warm air rises and starts to flow in the direction of the poles at an altitude of around 15 km. At latitudes of about 30°, the air descends. In the Northern Hemisphere, this descending air is responsible for deserts such as the Sahara and the build-up of the semi-permanent Azores anticyclone. Figure 2.6 shows the descending air splitting at the earth's surface. Some of it flows back to the equator, whilst another part continues to flow towards the poles. Meanwhile, at the poles cold air sinks and is transported over the earth's surface in the direction of the equator. In the mid-latitudes (around 60°), cold air from the poles and warmer air that has continued its route towards the poles meet, as shown in Figure 2.6.

When cold and warm air masses meet, **frontal precipitation** is formed (Figure 2.7a). At a cold front, the advancing cold air forces the warm air to rise along a steep slope, causing rapid lifting and intense rain of short duration. At a warm front, the advancing warm air, being lighter, easily rides up over the cold air. This leads to a gentle slope between the air masses, which causes a gradual lifting and cooling, and moderate rainfall of long duration.

Of course, it is possible to get combinations of the above-mentioned types of precipitation. Both convection and orographic uplift can cause cloud development and precipitation in the mountains on a summer afternoon, and large-scale weather

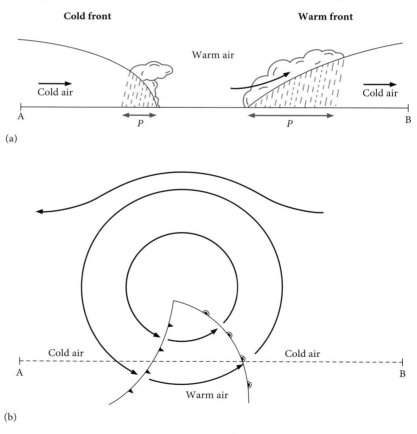

Figure 2.7. A cold front (left) and a warm front (right): (a) a side view, showing the resulting frontal precipitation; (b) a plan view, showing the cold and warm fronts as part of a cyclone in the Northern Hemisphere (after Ward and Robinson 2000)

systems (typically > 500 km across) with both frontal and cyclonic precipitation are common at the mid-latitudes (Figure 2.7b).

A last precipitation type that may be mentioned is **artificial precipitation**. With **cloud seeding**, chemicals such as **silver iodide** are cast into the atmosphere from an aircraft to allow water droplets or ice crystals to form more easily. The effectiveness of cloud seeding in increasing the amount of precipitation from a cloud remains uncertain, in part because it is difficult to assess how much precipitation would have occurred had cloud seeding not been applied.

2.4 Measuring precipitation

One may simply think of **rainfall depth** as the real depth of a layer of water on a flat, impermeable surface after rainfall. In Box 2.10, it is shown how one can easily obtain an estimate of the rainfall depth at a point in the landscape. However, to precisely measure rainfall over short time intervals, say 5 minutes, one needs a **recording raingauge**. In landscape erosion studies, it may be important to know the maximum rainfall depth in 5 minutes. Likewise, in an urban area it may be important to have a record of the rainfall depths in, say, 30 minutes, to aid in the design of the sewer system. When a rainfall depth

---

BOX 2.10 A low-budget raingauge

A low-budget way to make a simple **collector-type, non-recording raingauge** for measuring the rainfall depth is to buy a plastic bottle in a shop, drink or dispose of the bottle's contents, cut the top from the bottle at about two-thirds of the height of the bottle, turn the top around, and attach the turned-over top to the lower two-thirds of the bottle (Figure B2.10). The top of the bottle acts as an orifice (opening) that catches rainfall and the lower part as a cylindrical container where rainfall is stored. Finally, dig a small hole in the ground, and tightly fit the bottom of this raingauge in the hole.

Figure B2.10. A low-budget raingauge

To determine the rainfall depth, proceed as follows. Determine the diameter (equal to twice the radius) of the orifice in centimetres (cm). Determine the surface of catch of the orifice (cm^2), which equals πr^2, where $\pi = 3.14159...$ and r is the radius of the orifice (cm). Empty the bottle after a rainfall event or at fixed time intervals, and measure the volume of rainfall in millilitres (cm^3). Simply divide the volume of rainfall (cm^3) by the surface of catch of the orifice (cm^2), and multiply this result by 10 to obtain the rainfall depth in millimetres (mm).

is measured in a short period of time, it is customary to speak in terms of the **rainfall intensity**, which is a rainfall depth per unit of time, often expressed in mm hour^{-1}. For instance, a rainfall intensity of 24 mm hour^{-1} in 5 minutes ($= \frac{1}{12}$ hour) means that ($\frac{1}{12}$ hour $\times$ 24 mm hour^{-1} =) 2 mm of rain has actually fallen in this 5 minute period.

There are three main types of recording raingauges: the weighing, the float and siphon, and the tipping bucket raingauge (Figure 2.8).

In a **weighing raingauge**, the mass of the water in the gauge is continuously measured and registered. Since 1 litre of water has a mass of (approximately) 1 kg, any difference in mass can easily be converted to a difference in rainfall volume and thus rainfall depth.

In a **float and siphon raingauge**, the depth of the water in the cylindrical water collector is continuously measured and registered using a float resting on the water.

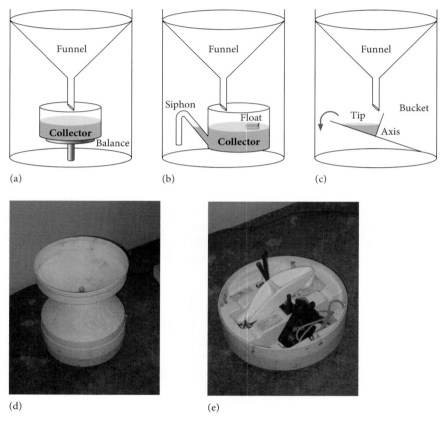

Figure 2.8. Weighing (a), float and siphon (b), and tipping bucket (c) raingauges. (d) and (e) show a tipping bucket raingauge with and without the funnel fitted, while (e) shows the tipping bucket tipped to the left. Weighing raingauge: the rainfall depths are deduced from differences in the mass of water as measured with a balance. Float and siphon raingauge: the position of the float over time is registered; after a certain volume of rain, the siphon empties most of the water from the collector, causing the registration to continue from a lower level – rainfall depths and intensities can be calculated from the water-level increments over time. Tipping bucket raingauge: one half of the bucket fills and tips after a certain mass (volume) of rain has been collected; the tipping causes the other half of the bucket to be positioned under the funnel outlet – the timing of the tips is registered, and rainfall depths and average intensities can be calculated accordingly

A siphon empties the water collector once the water has reached its maximum height, which corresponds to the position of the bending end of the siphon. In this way, rainfall depths and intensities can be determined.

In a **tipping bucket raingauge**, the bucket tips over each time a certain volume of rain has fallen, usually corresponding to a 0.2 mm rainfall depth. By registering the timing of the tips, the average rainfall intensity each time the bucket tips, as well as rainfall depths over longer periods of time, can be determined.

It is probably useful to mention a number of the problems that one can encounter when measuring precipitation. When placing a raingauge in the field, one should take care to ensure that the raingauge is well exposed; for instance, that the distance from a high object is sufficiently far as not to be measuring in the high object's rain shadow. The raingauge is an obstacle to wind, causing turbulence, and as a consequence, raindrops may be blown out of the orifice, leading to an underestimation of the rainfall depth. The higher the orifice, the larger the wind speed, plus the smaller the water droplets, the more pronounced the undercatch of the raingauge will be. Snowflakes are easily carried off by wind. Thus, the **water equivalent** of snow, being the equivalent water depth of snow when melted, is particularly prone to underestimation when measured in a raingauge. **Hail** consists of supercooled water droplets transformed into irregular lumps of ice. Because the hard lumps of hail splash out of the gauge's orifice, the water equivalent of hail may be seriously underestimated when measured in a raingauge. In collector-type raingauges, evaporation of the collected rainwater may be a problem. High humidity of the air or frost action can also cause serious problems for equipment in the field. Also, one should be aware that the rainfall intensity during extreme events may sometimes be too high for the tipping bucket or siphoning system to keep track. In Box 2.11, a number of measures of good practice are presented that can in part prevent a number of the problems just mentioned.

BOX 2.11 Good practice when measuring precipitation

- As a rule of thumb, the distance of any object from the raingauge should be at least twice and preferably four times the height of the object.

- To prevent underestimation of the rainfall depth, it is best to measure rainfall depth in a raingauge where the top of the raingauge is placed at ground level at the centre of a grid and brush network that prevents the splashing of raindrops from the ground into the gauge's orifice (Figure B2.11). Another reason may be that a hydrologist usually wants to know the rainfall depth entering the soil, rather than the rainfall depth at some standard height above the ground.

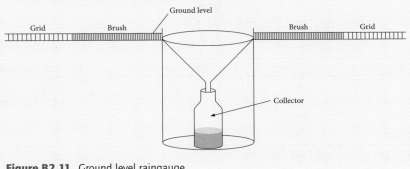

Figure B2.11. Ground level raingauge

- The best way to measure the water equivalent of snow is to use a snow pillow. This is an envelope of stainless steel or synthetic rubber (with an area of about 1.5 m²) filled with an antifreeze solution. Snow that accumulates on the pillow exerts a pressure on the solution that is converted into an electrical reading of the snow's water equivalent.

- To measure water equivalent totals of a hailstorm, an ordinary bucket placed outside will perform better than a fancy raingauge. This is because the rectangular cross-section of a bucket causes fewer outward splashes than the tilting side of a gauge's orifice.

- Evaporation of rainwater from a collector-type raingauge may be countered by the prior application of light oil or even kerosene, aviation fuel, to the collector. These substances provide a film on top of the water that prevents evaporation once the collector fills up with water.

- Rice is hygroscopic; that is, it attracts moisture from the air. Placing an opened sack of rice within a raingauge can help to prevent the malfunctioning of mechanical parts of the raingauge due to the high humidity of the air.

- Heating of a raingauge is an option to prevent negative effects from frost action.

- It is always good practice to let no water flow away from the raingauge, and thus to collect all the rainwater. One can then easily check how well the tipping bucket or siphoning system functioned during high rainfall intensities.

2.5 Areal precipitation

In many hydrological or environmental studies, the interest lies with areal precipitation rather than precipitation at a single point in the landscape. This involves the design of a network of raingauges. If the area under study is flat, the raingauges can be distributed evenly over the area. Usually, not many raingauges are needed. For estimating the areal precipitation, the **arithmetic mean precipitation depth** or the use of Thiessen polygons may then suffice.

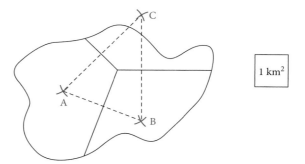

Raingauge	Area (km²)	Precipitation depth (mm)	Product (km² mm)
A	8.7	26	8.7 × 26 = 226.2
B	6.7	18	6.7 × 18 = 120.6
C	3.1	15	3.1 × 15 = 46.5
Totals	18.5		393.3

Areal precipitation = 393.3/18.5 km² mm km⁻² = 21.3 mm

Figure 2.9. The computation of areal precipitation using Thiessen polygons

Thiessen polygons (Figure 2.9) are built up by drawing midlines, perpendicular bisectors, between the raingauges on a map. Perpendicular bisectors of any three raingauges meet in one point. To each area that surrounds a raingauge, the value of the precipitation depth measured in the raingauge is assigned. Multiplying the precipitation depth by its representative area, summing the products for all raingauges, and then dividing by the total area under study gives a weighted average of the precipitation depth over the area. Figure 2.9 shows a worked-out example. This method is more accurate than calculating the arithmetic mean, but it is rather inflexible, since missing data from any one of the gauges calls for a new network to be constructed. Also, the Thiessen polygons do not cope well with orographic effects in precipitation.

When precipitation is spatially highly variable, such as in mountainous areas, or if a high accuracy of estimate is required, more raingauges are needed and their positioning in the landscape becomes more important. When using the **isohyetal method** (Figure 2.10), lines with the same precipitation depth (**isohyets**) are constructed. Although the drawing of isohyets is slightly arbitrary and may involve some experience, the **topography** (the land described in terms of elevation, slope, and orientation) and also knowledge of the storm pattern can be well accounted for when using this method. Figure 2.10 shows a worked-out example.

Statistical **interpolation** involves the construction of new data points from observed data. Statistical interpolation methods, with names such as trend surface analysis, the reciprocal-distance-squared method, and kriging, are also sometimes used to estimate areal precipitation depths from point data.

Nowadays, good use is made of radar and satellites in estimating areal precipitation depths. Both radar and satellites do not directly measure precipitation. Radar is an acronym for **radio detection and ranging**. With **radar**, an empirical relation between the radar reflectivity and rainfall is used. More reflectivity indicates a larger likelihood of precipitation. When using **satellite images**, information can be gathered concerning cloud-top brightness and cloud-top temperature. Both a brighter cloud top and a lower temperature indicate a greater likelihood

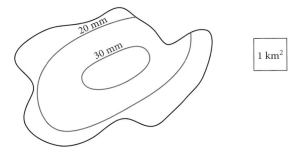

Location	Area (km²)	Average precipitation depth (mm)	Product (km² mm)
Outer area	6.8	20⁻: (10 + 20)/2 = 15	6.8 × 15 = 102
Middle area	9.7	(20 + 30)/2 = 25	9.7 × 25 = 242.5
Inner area	2.0	30⁺: (30 + 36)/2 = 33	2.0 × 33 = 66
Totals	18.5		410.5

Areal precipitation = 410.5/18.5 km² mm km⁻² = 22.2 mm

Figure 2.10. The computation of areal precipitation using isohyets

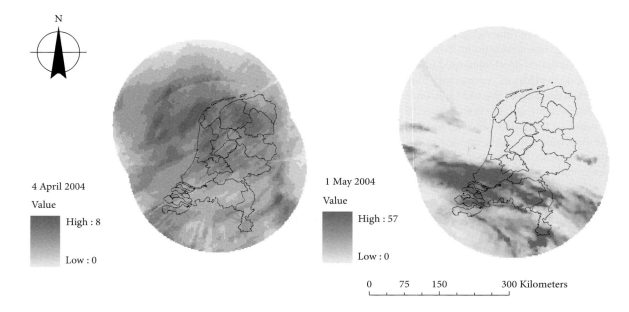

Figure 2.11. Daily precipitation values for The Netherlands on 4 April 2004 (left) and 1 May 2004 (right), estimated from observations by two radar stations (taken from Schuurmans *et al.* 2007). The 4 April event shows low precipitation values over a large area, whilst the 1 May event shows high precipitation values over a small area: the latter is most likely to be convective precipitation

of precipitation. Radar can produce estimates of precipitation for time intervals as small as 5 minutes and for areas ranging from 4 to 138 000 km². As an example, Figure 2.11 shows daily precipitation values estimated from radar observations for an area of 82 875 km² including The Netherlands. Satellite images are generally better suited for producing estimates for larger areas, such as (parts of) continents. With both methods, the magnitude of the output needs to be linked (**calibrated**) and checked (**validated**) with precipitation data at the ground, the so-called **ground truth**.

2.6 Evaporation types and measurement

There are three key types of evaporation. Precipitation that is intercepted by vegetation (in part) evaporates back into the atmosphere (**interception evaporation**); water evaporates from a wet soil surface (**soil evaporation**), or through plant stomata (**transpiration**). On a global scale, about 57% of all precipitation on land evaporates: 112% (!) of the precipitation that falls on the oceans evaporates. This percentage is larger than 100 because rivers deliver additional water to the oceans. Also, in warm, dry climates up to 96% of the yearly precipitation may evaporate. Evaporation thus constitutes an important part of the hydrological cycle.

A twofold distinction of evaporation at the land surface can be made. **Potential evaporation** is the maximum evaporation rate (mm day^{-1}) when the moisture content of the soil and vegetation conditions do not limit evaporation – in other words, under unstressed conditions. **Actual evaporation** is the evaporation rate (mm day^{-1}) under existing atmospheric, soil, and vegetation conditions. The atmosphere may be humid, a soil may be dry, and/or plant stomata may be closed – all circumstances that limit evaporation. The actual evaporation is therefore always less than or equal to the potential evaporation.

The **reference crop evaporation** is the potential evaporation (mm day^{-1}) of an idealized grass crop that serves as a reference value for determining the potential evaporation of other crops.

A low-budget and direct way to obtain some measure of the evaporation rate (on days with no precipitation) is to use a pan filled with water and to measure the height of the water in the pan for two consecutive days at exactly the same time (Figure 2.12). The difference in height (mm) divided by the time interval (day) yields the **pan evaporation** (mm day^{-1}). The pan evaporation in mm day^{-1} will generally be higher than the **open-water evaporation** (the rate of liquid water transformation to vapour from open water) in mm day^{-1} from a lake nearby. One of the reasons for this is that the small size of the pan ensures a warming effect of the sides of the pan by solar radiation. This will significantly enlarge the evaporation output from the pan because of the warming up of the water inside the pan by the pan itself. Thus, to obtain open-water evaporation rates from pan evaporation data, the pan evaporation values need to be multiplied by a **pan coefficient**, a fractional value (larger than zero and smaller than one) that depends on the type of pan used and the time of use (season).

When the weather is warm and the relative humidity of the air is low, the difference between pan evaporation and open-water evaporation will be larger than when the weather is cold and the relative humidity of the air is high. Thus pan coefficients for the same type of pan are usually smaller (that is, a larger correction is needed) in summer than in winter. For instance, the pan coefficients that relate reference crop evaporation to the 'U.S. Weather Bureau Class A pan' evaporation range from 0.35 to 0.85 depending on relative humidity, wind speed, and the length of the upwind distance of green crop or dry fallow (Doorenbos and Pruitt 1977). Sometimes, sunken pans are preferred in crop evaporation studies, as these have a water level at or slightly below ground level and higher pan coefficients under comparable circumstances than the ordinary type of pan shown in Figure 2.12.

A good way to obtain values of actual or potential evaporation is by installing a lysimeter. A **lysimeter** (Figure 2.13) is a device made of steel, concrete, or synthetic material dug into the terrain, in which a volume of soil, usually with vegetation, is isolated hydrologically, preventing leakage from the device. The position of the water table in the lysimeter is monitored and can be regulated by pumping measured amounts of water in or out. In a weighing lysimeter, the change in water storage is

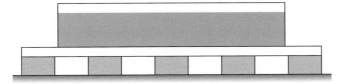

Figure 2.12. An evaporation pan. As an example, the US Weather Bureau Class A pan is circular, 1.21 m in diameter, and 25.5 cm deep: the pan must be level, with its bottom 15 cm above ground level; it is filled with water to 5 cm below the rim and the water level should not drop to more than 7.5 cm below the rim (Shuttleworth 1993)

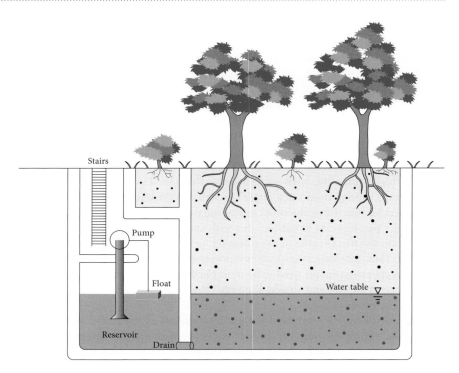

Figure 2.13. A lysimeter

determined by the difference in mass of the lysimeter; the evaporation is estimated from the **water balance**. The diameter of a lysimeter typically varies from 0.5 to 2.0 m, with larger devices having surface areas of the order of 35×25 m². Lysimeters are difficult and expensive to install, but are most useful in experimental research for obtaining empirical equations to estimate evaporation from meteorological variables.

2.7 Estimating evaporation: Penman–Monteith

Due to the difficulties involved in measuring evaporation directly, it is generally estimated from meteorological data. The evaporation rate is controlled by the energy needed to diffuse water vapour into the atmosphere, the **earth's surface energy balance** (Box 2.12), and the **atmospheric demand** for water vapour (Box 2.13).

The **Penman–Monteith equation**, named after the work of the British physicists Howard L. Penman (1909–1984) in 1948 and John L. Monteith in 1973, is a physically based method for estimating the actual evaporation E_a (or, under unstressed conditions, the potential evaporation E_p) based on both the energy balance and atmospheric demand. To estimate evaporation using the Penman–Monteith equation, one needs to know the net radiation at the earth's surface R_n (MJ m⁻² day⁻¹; Box 2.12), and both the air temperature T (°C) and the relative humidity of the air RH (−), as these determine the atmospheric demand or saturation deficit $e_s - e_a$ (kPa; Box 2.13). Also, one needs to know the resistances (s m^{-1}) encountered by the evaporation process (these will be introduced later in the text).

In the following text, the Penman–Monteith equation is first presented in its full physical form (Equation 2.1) where, even without any knowledge of the stated variables (but see also Boxes 2.12 and 2.13), the units of measurement to the left and right of the equals sign (=) can be checked to be balanced, and then as a simplified version (Equation 2.2), where one should be aware that the constant values in the equation have hidden units. After this, a short introduction to the variables used in the Penman–Monteith equation is given.

BOX 2.12 The earth's surface energy balance

Energy is a measure of a physical system's ability to perform mechanical work or to produce or absorb heat. An **energy balance** is a systematic presentation of energy flows and transformations in a system. The theoretical basis for an energy balance is the first law of thermodynamics, according to which energy cannot be created or destroyed, only modified in form. In Figure B2.12.1, a simplified energy balance at the earth's surface is presented. Energy terms that are usually small, such as the loss of energy associated with horizontal air movement, are not included in the figure. All variables used in this exposé have absolute values (greater than or equal to zero), irrespective of the direction of the processes involved, unless there is a specific statement to the contrary.

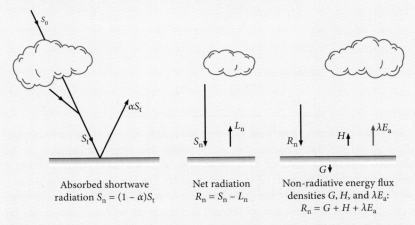

| Absorbed shortwave radiation $S_n = (1 - \alpha)S_t$ | Net radiation $R_n = S_n - L_n$ | Non-radiative energy flux densities G, H, and λE_a: $R_n = G + H + \lambda E_a$ |

Figure B2.12.1. Simplified energy balance at the earth's surface (explanation: see text)

A body with a high temperature, such as the sun, emits shortwave radiation and a body with a low temperature, such as the earth, radiates longwave radiation. The **net radiation** at the earth's surface R_n can be measured with a net radiometer or determined as the net incoming shortwave radiation at the earth's surface S_n minus the net outgoing longwave radiation at the earth's surface L_n:

$$R_n = S_n - L_n \qquad (B2.12.1)$$

R_n, S_n, and L_n are energy flux densities in megajoules per square metre and per day (MJ m^{-2} day^{-1}). An **energy flux density** is the amount of energy that traverses a small area perpendicular to the direction of the energy flow in a time interval, divided by

1 MJ = 1 megajoule = 10^6 joules.

that area and by that time interval. A **joule** (J) is the energy required, or the work done, to make an object on which a force of one newton acts move a distance of one metre in the direction of the force. A joule thus is the same as a **newton metre** (J = N m = kg m^2 s^{-2}). A joule is also a **watt second** (W s), the equivalent energy of one watt of power (an amount of work per unit of time) radiated or dissipated for one second.

The net incoming shortwave radiation at the earth's surface S_n equals that part of the incoming shortwave radiation at the earth's surface S_t (MJ m^{-2} day^{-1}) that is absorbed, not reflected. Since the **albedo** α gives the fraction of the incoming radiation that is reflected, with typical values of around 0.08 for open water, 0.15 for forest, 0.23 for grass and crops, and 0.90 for freshly fallen snow (see Table B2.12.1), $(1 - \alpha)$ gives the fraction of the incoming shortwave radiation S_t that is absorbed. Thus:

$$S_n = (1 - \alpha) S_t \qquad (B2.12.2)$$

Table B2.12.1 Possible mean values of albedo α (–) for a number of natural surfaces

Open water	0.08
Bare clay soil	0.11
Forest	0.15
Grass and crops	0.23
Bare sandy soil	0.31
Fresh snow	0.90

Only part of the sun's shortwave radiation incident at the top of the earth's atmosphere S_0 (MJ m^{-2} day^{-1}) reaches the earth's surface. The shortwave radiation can be absorbed, reflected, or scattered by the atmosphere and its floating particles; part of it reaches the ground. The incoming shortwave radiation that reaches the ground and that we already know as S_t (MJ m^{-2} day^{-1}) can be measured using a pyranometer or determined from S_0 (Figure B2.12.2), the number of bright sunshine hours per day n – determined, for instance, by a Campbell–Stokes sunshine recorder – and the day length N (Table B2.12.2) by means of the following empirical (experimental) equation:

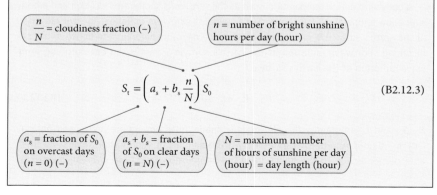

$$S_t = \left(a_s + b_s \frac{n}{N} \right) S_0 \qquad (B2.12.3)$$

$\frac{n}{N}$ = cloudiness fraction (–)

n = number of bright sunshine hours per day (hour)

a_s = fraction of S_0 on overcast days ($n = 0$) (–)

$a_s + b_s$ = fraction of S_0 on clear days ($n = N$) (–)

N = maximum number of hours of sunshine per day (hour) = day length (hour)

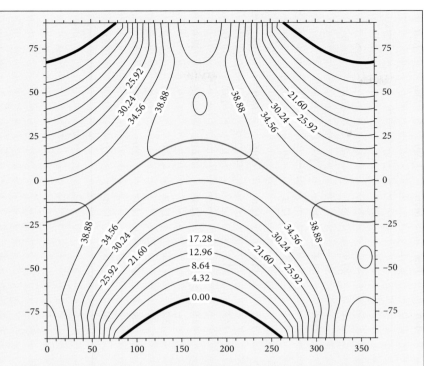

Figure B2.12.2. The sun's shortwave radiation incident at the top of the earth's atmosphere S_0 (MJ m^{-2} day^{-1}) as a function of latitude (vertical axis) and day of the year (horizontal axis); the blue line indicates the latitude at which the sun is directly overhead at noon; the thick contours mark the edges of the polar nights, where $S_0 = 0$ (reproduced with kind permission from Douglas Parker, University of Leeds)

Combining Equations B2.12.2 and B2.12.3 gives

$$S_n = (1 - \alpha)\left(a_s + b_s \frac{n}{N}\right)S_0 \tag{B2.12.4}$$

The net outgoing longwave radiation at the earth's surface L_n can be determined from the air temperature, the actual vapour pressure, and, again, the cloudiness fraction by means of the following empirical equation:

σ = the Stefan–Boltzmann constant = 4.903×10^{-9} MJ m^{-2} °C^{-4} day^{-1}

$$L_n = \sigma(T + 273.2)^4 \left(a_e + b_e \sqrt{e_a}\right)\left(a_c + b_c \frac{n}{N}\right) \tag{B2.12.5}$$

T = air temperature (°C) e_a = actual vapour pressure (kPa)

The actual vapour pressure e_a (kPa) can be determined from the relative humidity and the air temperature using Equations B2.2.4 and B2.2.3. The relative humidity RH (−) can be measured with a relative humidity sensor and the air temperature T (°C) can simply be measured using a thermometer.

$a_s = 0.25$
$b_s = 0.50$
$a_e = 0.34$
$b_e = -0.14$ kPa$^{-0.5}$
$a_c = 0.25$
$b_c = 1 - a_c = 0.75$

Table B2.12.2 Mean day length N (hour)

Northern latitudes	Jan	Feb	Mar	Apr	May	June	July	Aug	Sept	Oct	Nov	Dec
Southern latitudes	July	Aug	Sep	Oct	Nov	Dec	Jan	Feb	Mar	Apr	May	June
90° (pole)	0.0	0.0	0.0	24.0	24.0	24.0	24.0	24.0	24.0	0.0	0.0	0.0
80°	0.0	0.0	10.9	24.0	24.0	24.0	24.0	24.0	15.3	5.2	0.0	0.0
70°	0.0	7.3	11.5	16.1	24.0	24.0	24.0	18.4	13.6	9.1	3.1	0.0
60°	6.7	9.0	11.7	14.5	17.1	18.6	17.9	15.5	12.9	10.1	7.5	5.9
50°	8.5	10.0	11.8	13.7	15.3	16.3	15.9	14.4	12.6	10.7	9.0	8.1
40°	9.6	10.7	11.9	13.3	14.4	15.0	14.7	13.7	12.4	11.2	10.0	9.3
30°	10.4	11.1	12.0	12.9	13.6	14.0	13.9	13.2	12.4	12.0	10.6	10.8
20°	11.0	11.5	12.0	12.6	13.1	13.3	13.2	12.8	12.3	12.0	11.2	10.9
10°	11.6	11.8	12.0	12.3	12.6	12.7	12.6	12.4	12.1	12.0	11.6	11.5
0° (equator)	12.0	12.0	12.0	12.0	12.0	12.0	12.0	12.0	12.0	12.0	12.0	12.0

Parameter/variable values for average conditions are given in the margin, and the net radiation at the earth's surface R_n can be determined for these average conditions as follows:

$$R_n = S_n - L_n$$

with

$$S_n = (1-\alpha)\left(0.25 + 0.50\frac{n}{N}\right)S_0 \qquad \text{(B2.12.6)}$$

$$L_n = 4.903 \times 10^{-9}(T+273.2)^4(0.34-0.14\sqrt{e_a})\left(0.25+0.75\frac{n}{N}\right)$$

In conclusion, R_n for different natural surfaces (Table B2.12.1) can be estimated by determining n (using a Campbell–Stokes sunshine recorder), N (from Table B2.12.2), S_0 (from Figure B2.12.2), T (using a thermometer), and e_a (using a relative humidity sensor).

This net radiation energy at the surface R_n is used for warming the soil below the earth's surface by soil heat transfer G, warming the air by sensible heat transfer from the earth's surface H, and/or latent heat transfer, which is evaporation from the earth's surface λE_a (Figure B2.12.1). G, H, and λE_a (all in MJ m^{-2} day^{-1}) are non-radiative energy flux densities and are assigned a positive value when directed away from the surface:

$$R_n = G + H + \lambda E_a \qquad \text{(B2.12.7)}$$

Non-radiative energy flux densities directed away from the surface are of positive sign; non-radiative energy flux densities directed towards the surface are of negative sign.

λ (in λE_a) actually is $\rho\lambda/1000$, a multiplication factor (MJ dm^{-3}) to transfer E_a from mm day^{-1} to MJ m^{-2} day^{-1} ($\rho \approx 1000$ kg m^{-3}, and the unit of 1000 in the denominator is mm m^{-1}).

Soil heat transfer G is the transmission of heat across matter. Heat is transferred downwards when the earth's surface is warmer than the subsurface (soil, rock, or water body) and G has a positive value. If the subsurface is warmer than the surface, then heat is transferred upwards and G has a negative value. Since the magnitude of G over 10- to 30-day periods is relatively small, it can often be neglected in hydrological applications (Shuttleworth 1993).

Sensible heat H is heat energy that is transported by a body that has a temperature higher than that of its surroundings. Sensible heat energy is transferred between the earth's surface and the atmosphere when there is a difference in temperature between them. Heat is transferred upwards when the surface is warmer than the overlying air and H has a positive value. If the overlying air is warmer than the surface, then heat is transferred downwards and H has a negative value.

A special case of H having a negative value is the oasis effect. This effect can occur in hot areas such as deserts. When warm, dry air in equilibrium with dry soil reaches a wet surface such as a lake (oasis), the evaporation rate increases and sensible heat is used to maintain this high rate. This effect can already be manifest in irrigated crop fields and sprinkled golf courts under moderately dry conditions, as observed by Van der Kwast and De Jong (2004) in a combined satellite remote sensing and ground measurement study of the earth's surface energy balance near Rabat, Morocco.

The latent heat λE_a is energy that is used for evaporation. It is called 'latent' because the energy is stored in the water molecules to be released later, during the condensation process. The energy cannot be sensed or felt, as it does not raise the temperature of the water molecules. In the case of evaporation at the earth's surface, λE_a is directed away from the surface and has a positive value. In the case of condensation at the earth's surface, λE_a has a negative value.

BOX 2.13 Atmospheric demand

Besides the net radiation R_n as a source of energy, the **atmospheric demand** for water vapour also plays an import role in evaporation. This is shown by $e_s - e_a$ in Equation 2.1: e_s is the saturation vapour pressure at the surface (kPa) and e_a is the actual vapour pressure of the air above the evaporating surface (kPa), as shown in Figure B2.13; e_s and e_a are estimated for the same air temperature T (°C). The difference between e_s and e_a ($e_s - e_a$) is called the **saturation deficit** (kPa). The larger the saturation deficit, the higher the evaporation rate will be. In the absence of wind, evaporation may cause overlying air to become saturated with water vapour. In that case, the saturation deficit would become zero and evaporation would be effectively halted by the evaporation process itself, which is a **negative feedback** mechanism. Both the effect of **wind speed** and **surface roughness** in turbulently mixing moist air with drier air from elsewhere is important in keeping the actual vapour pressure of the overlying air low, the atmospheric demand high, and thus in maintaining evaporation.

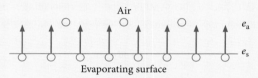

Figure B2.13. The evaporation rate is determined by the difference between the saturation vapour pressure e_s at the surface and the actual vapour pressure e_a of the air above the evaporating surface

Atmospheric demand also provides an additional reason for pan evaporation at the land surface in mm day^{-1} to be generally higher than the open-water evaporation from a nearby lake in mm day^{-1} presented earlier. The additional reason is that air being delivered by wind over a pan at the land surface will be generally drier than the air being delivered by wind over a lake. The larger the lake, the more prominent the difference in evaporation in mm day^{-1} will be. Also, when wind blows over a series of small lakes with land in between, given that their total surface area is equal to the surface area of a large lake, then – all other conditions being equal – the series of small lakes will show a larger evaporation rate in mm day^{-1} than the large lake. This is due to the recovery (lowering) of the actual vapour pressure over the land parts in between the lakes.

1000 = constant (mm m^{-1})

ρ = water density ≈ 1000 kg m^{-3}

λ = latent heat of vaporization ≈ 2.45 MJ kg^{-1}

G = heat transfer into a soil, rock, or water body (MJ m^{-2} day^{-1})

$86\,400$ = constant (s day^{-1})

ρ_a = air density ≈ 1.2 kg m^{-3} at sea level

c_p = specific heat of air at constant pressure = 1.013×10^{-3} MJ kg^{-1} °C^{-1}

γ = psychometric constant ≈ 0.067 kPa °C^{-1}

$1000/\rho\lambda$ = factor (dm^3 MJ^{-1}) to convert from MJ m^{-2} day^{-1} to mm day^{-1}

The Penman–Monteith equation is stated as follows:

$$E_a = \frac{1000}{\rho\lambda} \times \frac{\Delta(R_n - G) + \dfrac{86\,400 \times \rho_a c_p (e_s - e_a)}{r_a}}{\Delta + \gamma\left(1 + \dfrac{r_s}{r_a}\right)} \qquad (2.1)$$

The heat transfer into a soil, rock, or water body G (MJ m^{-2} day^{-1}) is relatively small and can often be neglected. If we set G to zero and insert values for the constants and near-constants into the Penman–Monteith equation, we obtain:

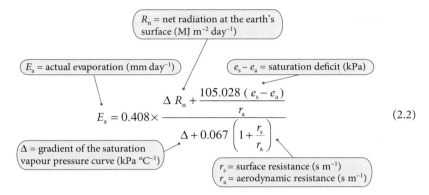

$$E_a = 0.408 \times \frac{\Delta R_n + \dfrac{105.028\,(e_s - e_a)}{r_a}}{\Delta + 0.067\left(1 + \dfrac{r_s}{r_a}\right)} \tag{2.2}$$

where:
- R_n = net radiation at the earth's surface (MJ m^{-2} day^{-1})
- E_a = actual evaporation (mm day^{-1})
- $e_s - e_a$ = saturation deficit (kPa)
- Δ = gradient of the saturation vapour pressure curve (kPa °C^{-1})
- r_s = surface resistance (s m^{-1})
- r_a = aerodynamic resistance (s m^{-1})

The **net radiation at the earth's surface** R_n (MJ m^{-2} day^{-1}) can be estimated as stated in Box 2.12; the saturation vapour pressure e_s (kPa) can be estimated from the air temperature using Equation B2.2.3, and the actual vapour pressure e_a (kPa) from the relative humidity using Equation B2.2.4.

Δ is the **gradient of the saturation vapour pressure curve** (kPa °C^{-1}), which is the slope of the tangent line to the curve of the saturation pressure for liquid water in Figure 2.1. Δ can be seen to increase with air temperature T (°C). Mathematically, it can be derived from Equation B2.2.3 as follows (section C2.5):

$$\Delta = \frac{4098\,e_s}{(237.3 + T)^2} \tag{2.3}$$

where 237.3°C and 4098°C are constants.

The **aerodynamic resistance** r_a is the resistance encountered by water vapour in diffusing into the air from a vegetation surface (interception evaporation) or water body, and is reciprocal to the **roughness of the earth's surface** and the **wind speed** near the surface. Both a rougher surface (forest has a rougher surface than grass, which has a rougher surface than open water) and more wind cause more turbulent mixing of the air and thus a smaller resistance to evaporation. The aerodynamic resistance under unstressed conditions commonly varies from 5–10 s m^{-1} for forest and 50–70 s m^{-1} for grass to 110–125 s m^{-1} for open water (Table 2.1).

Table 2.1 Possible ranges of values of aerodynamic resistance r_a (s m^{-1}) and surface resistance r_s (s m^{-1}) under unstressed conditions (actual evaporation = potential evaporation) for a number of land uses

Land use	r_a (s m^{-1})	r_s (s m^{-1})
Forest	5–10	80–150
Grass	50–70	40–70
Open water	110–125	0

The **surface resistance** r_s is a physiological resistance imposed by vegetation stomata on the movement of water vapour by transpiration: r_s varies with water availability (soil moisture content). If r_s is the minimum surface resistance for a vegetated surface (circumstances are unstressed), then Equations 2.1 and 2.2 estimate the potential evaporation. The surface resistance under unstressed conditions commonly varies from 40–70 s m^{-1} for grass to 80–150 s m^{-1} for forest (see Table 2.1). For an open-water surface, $r_s = 0$ s m^{-1}.

When estimating the evaporation at the earth's surface using Equation (2.2), one needs to know the air temperature T (°C), the relative humidity RH (–), the sun's shortwave radiation incident at the top of the earth's atmosphere S_0 (MJ m^{-2} day^{-1}), the **cloudiness fraction** n/N (Box 2.12), the albedo α (Table B2.12.1), the aerodynamic resistance r_a (s m^{-1}) and the surface resistance r_s (s m^{-1}).

Table 2.2 shows an example of calculation results using the Penman–Monteith equation (Equation 2.2) for estimating evaporation from forest, grassland, and open-water surfaces. Such calculations can easily be made in a spreadsheet following the information provided here in the text and in Box 2.12.

Open water has the highest evaporation rate (mm day^{-1}) in the example in Table 2.2 due to the low albedo value (low reflection of solar radiation) and, of course, the absence of a surface resistance r_s.

Table 2.2 The estimation of evaporation from forest, grassland, and open-water surfaces using the Penman–Monteith equation

Equation parameters/variables		
a_s (–)	0.25	
b_s (–)	0.50	
a_e (–)	0.34	
b_e (kPa$^{-0.5}$)	−0.14	
a_c (–)	0.25	
b_c (–)	0.75	$b_c = 1 - a_c$
Data		
Air temperature, T (°C)	20.0	Thermometer
Relative humidity, RH (–)	0.75	Relative humidity sensor
Sun's shortwave radiation, S_0 (MJ m^{-2} day^{-1})	37.50	Figure B2.12.2
Number of bright sunshine hours per day, n (hour)	8.15	Campbell–Stokes sunshine recorder
Total day length, N (hour)	16.3	Table B2.12.2

Table 2.2 *Continued*

Land-use data	Forest	Grassland	Open water	
Albedo, α (−)	0.15	0.23	0.08	Table B2.12.1
Aerodynamic resistance, r_a (s m^{-1})	5	69	110	Table 2.1
Surface resistance, r_s (s m^{-1})	150	69	0	Table 2.1
Calculations				
Cloudiness fraction, n/N (−)		0.5		n and N from above
Saturation vapour pressure, e_s (kPa)		2.338		Equation B2.2.3
Actual vapour pressure, e_a (kPa)		1.754		Equation B2.2.4
Net radiation at the earth's surface, R_n (MJ m^{-2} day^{-1})	12.44	10.94	13.75	Equation B2.12.6
Gradient of the saturation pressure curve, Δ (kPa °C^{-1})		0.145		Equation 2.3
Evaporation (mm day^{-1})	**2.6**	**3.6**	**4.9**	Equation 2.2
Latent heat transfer from the earth's surface (MJ m^{-2} day^{-1})	6.33	8.87	12.03	Multiply the evaporation by λ
Sensible heat transfer from the earth's surface (MJ m^{-2} day^{-1})	6.10	2.07	1.72	Equation B2.12.7

A **land-use change** from forest to grassland could cause a significant decrease in convective precipitation. When we compare forest and grassland in our example, the albedo α for forest is lower. This causes the net radiation at the earth's surface R_n to be larger for forest. Despite this higher value of R_n, the evaporation rate (mm day^{-1}) is lower for forest. Since the aerodynamic resistance is low for forest (due to turbulent mixing), the low forest evaporation rate can only be due to the high value of the surface resistance r_s of the forest. As a consequence of the higher value of R_n and the lower value of latent heat transfer (which is the evaporation in units of MJ m^{-2} day^{-1}), the forest has a significantly higher value for sensible heat transfer than grassland in our example. In reality, this surface heating can give rise to upward movement of air and convective precipitation (as described earlier).

In wet climates, grassland evaporation is not necessarily higher than forest evaporation and the effects of a **land-use change** from forest to grassland may be much less manifest. In wet forests the evaporation as interception evaporation, which is related to the aerodynamic resistance, can be more important than the evaporation as transpiration, which is related to both the surface and aerodynamic resistance. If we reduce the surface resistance r_s for forest in our spreadsheet example to 80 s m^{-1}, the evaporation rate will be 4.5 mm day^{-1}, which is higher than the evaporation rate of grassland. Also, the sensible heat transfer for forest (calculated as 6.10 MJ m^{-2} day^{-1} in Table 2.2), would reduce to 1.47 MJ m^{-2} day^{-1} in the spreadsheet calculation, and upward movement of air and the effects of a land-use change from forest to grassland would be much less manifest.

In Box 2.14, the **reference crop evaporation** and the use of **crop coefficients** in estimating potential evaporation for different crops is explained: also, the **FAO Penman–Monteith equation**, a standard method to compute the reference crop evaporation E_{rc} (mm day^{-1}), is shown. Box 2.15 introduces two radiation-based empirical equations for estimating the reference crop evaporation that are in fact simplifications of

---

BOX 2.14 Reference crop evaporation, crop coefficients, and the FAO Penman-Monteith equation

An important standard evaporation rate is the reference crop evaporation E_{rc}, which is defined as the rate of evaporation from an idealized grass crop with a fixed crop height of 0.12 m, an albedo of 0.23, and a surface resistance of 69 s m^{-1}: this closely resembles previous definitions of a reference crop; namely, an extensive surface of short green grass cover of uniform height, actively growing, completely shading the ground, and not short of water (Shuttleworth 1993).

If there are values available of the surface resistance r_s and the aerodynamic resistance r_a that are appropriate for a land surface and a particular crop under unstressed conditions, then the potential evaporation may be estimated using the Penman-Monteith equation (Equation 2.1). In practice, however, trustworthy values for these variables are only available for research situations. In order to estimate the potential evaporation, it is therefore common practice to calculate the above-defined reference crop evaporation (for which r_s and r_a are prescribed) and to multiply this by the **crop coefficient** K_c of the particular crop (as determined from research – for instance, by using a lysimeter); the value of K_c changes with the growth stage of the crop:

K_c = crop coefficient of a particular crop at a certain growth stage (–)

E_p = potential evaporation (mm day^{-1})

$$E_p = K_c \times E_{rc} \qquad (B2.14.1)$$

E_{rc} = reference crop evaporation (mm day^{-1})

Allen *et al.* (1998) recommend the **FAO Penman-Monteith equation** as the sole standard method for computation of the reference crop evaporation E_{rc} (mm day^{-1}). This equation is the Penman-Monteith equation (Equation 2.1) for a grass crop height of 0.12 m, a standardized height for wind speed, temperature, and humidity measurements at 2 m above ground level, an albedo of 0.23, and an aerodynamic resistance r_a (s m^{-1}) of

$208/u_2$, where u_2 is the wind speed (m s^{-1}) at 2 m, and a surface resistance of 69 s m^{-1} (all variable notations have been presented earlier and their units are also the same as earlier):

$$E_{rc} = \frac{0.408\,\Delta(R_n - G) + \gamma\,\dfrac{900}{T+273}\,u_2\,(e_s - e_a)}{\Delta + \gamma\,(1 + 0.34\,u_2)} \qquad \text{(B2.14.2)}$$

Allen *et al.* (1998) also provide abundant information on crop coefficients.

BOX 2.15 Empirical equations for estimating the reference crop evaporation

Over very large areas, the energy balance largely governs evaporation, and Equation 2.1 may be replaced by a simpler radiation-based equation with fewer variables, the Priestley–Taylor equation:

E_{PT} = Priestley–Taylor reference crop evaporation (mm day^{-1})

$$E_{PT} = C_{PT} \times \frac{1000}{\rho\lambda} \times \frac{\Delta}{\Delta + \gamma}\,(R_n - G) \qquad \text{(B2.15.1)}$$

$C_{PT} = 1.26$ (humid climate) or 1.74 (arid climate) (–)

C_{PT} in the Priestley–Taylor equation introduces an empirical (experimental) factor into the equation.

Another example of a radiation-based equation involving the measurement of only two meteorological variables, the air temperature T (°C) and the incoming shortwave radiation at the earth's surface S_t (MJ m^{-2} day^{-1}), is the Makkink equation, which is used by the Royal Dutch Meteorological Society (KNMI) and which performs well for the Netherlands:

E_{MK} = Makkink reference crop evaporation (mm day^{-1})

$$E_{MK} = C_{MK} \times \frac{1000}{\rho\lambda} \times \frac{\Delta}{\Delta + \gamma}\,S_t \qquad \text{(B2.15.2)}$$

$C_{MK} = 0.65$ (humid climate) (–)

There are many other types of empirical equations for estimating the reference crop evaporation or potential evaporation; for instance, the Blaney-Criddle method, the Hargreaves equation, or the Thornthwaite method. These empirical methods are conceptually simple, need only a few variables as input, and may perform well in a given setting (location and climate). Since these methods are site specific, one should take care not to use them outside the setting for which they were developed.

---

the Penman–Monteith equation: the **Priestley–Taylor equation** and the **Makkink equation**.

For areas of some 30 × 30 km², progress is nowadays made in combining radiance and reflectance data from **satellite remote sensing** with ground station data

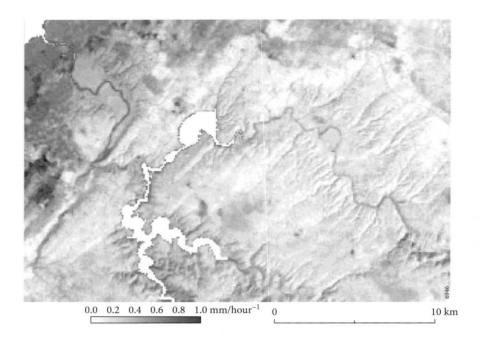

0.0   0.2   0.4   0.6   0.8   1.0 mm/hour⁻¹   0   10 km

Figure 2.14. The estimated actual evaporation flux (mm hour hair⁻¹) for the Sehoul study area near Rabat, Morocco for 15 September 2003 at 10.40 am, the overpass time of the Landsat 5 Thematic Mapper (taken from Van der Kwast and De Jong 2004). The actual evaporation of open water is not indicated: irrigated plots such as the golf course (in the western part of the image) have the highest evaporation, followed by cork oak forest; bare and harvested plots have the lowest evaporation – south-facing slopes are drier and have less evaporation than north-facing slopes; valley floors have a higher moisture content and evaporation rate (De Jong *et al.* 2008). The estimated values are checked (validated) using ground station data, indicating a good model performance – note that the estimated values are instantaneous values, and thus valid at that exact instance (15 September 2003 at 10.40 am)

(air pressure, temperature, relative humidity, wind speed, short- and longwave radiation), and field measurements (topsoil moisture content, pan evaporation). Important variables for evaporation that can be retrieved from satellite imagery with varying precision are land use, vegetation cover, albedo, surface temperature and, in combination with models, topsoil moisture content (De Jong *et al.* 2008). The results of such a combined approach are maps of the net radiation, soil heat transfer, sensible heat transfer, and latent heat transfer (all in MJ m⁻² hour⁻¹) or actual evaporation (mm hour⁻¹), all instantaneous values that can be related to the land use, vegetation cover, topography, and exposition of the slopes in the area (Figure 2.14). Information on the mapped variables, such as net radiation and so on, is limited to the overpass time of the satellite and cloudless days, but an important advantage of the combined approach is the spatial extent of the information and the high resolution (with pixels varying from 30 to 120 m for the Landsat Thematic Mapper sensors and from 15 to 90 m for the TERRA ASTER satellite configuration) at which information on the earth's surface energy balance is provided.

→ *Summary*

- If we take the amount of water vapour in the earth's atmosphere to equal a liquid water layer of 25 mm, the average residence time of water vapour in the atmosphere can be calculated at 9 days.

- Persistence of either wet or dry conditions on a seasonal or multi-year timescale, such as the multi-year droughts in the Sahel region of Africa, are caused by a positive soil moisture–precipitation feedback.

- The unit of both force and weight is the Newton, abbreviated as N. A Newton is defined as the amount of force required to accelerate a mass of one kilogram (kg) at a rate of one metre per second each second (m s^{-2}): 1 N = 1 kg m s^{-2}. The base unit of mass is the kilogram.

- Pressure p is defined as the magnitude of a normal force F (N) divided by the surface area A (m^2) over which this normal force acts. The unit of pressure is the pascal, abbreviated as Pa (= N m^{-2}). The air pressure, also called atmospheric pressure, is the force exerted by the weight of the molecules in the air. It is common practice to present air pressure data using the hectopascal (hPa), which is 100 times the value of a pascal (hecto = 100).

- The actual vapour pressure e_a is the actual (partial) pressure of the water vapour molecules in the air at a certain temperature. The saturation vapour pressure e_s is the pressure exerted by water vapour molecules in the air, when the air is saturated with water vapour. Vapour pressure is usually presented using the kilopascal (kPa = 1000 Pa) or the hectopascal (hPa).

- The relative humidity RH is defined as the ratio of the actual vapour pressure e_a (kPa) to the saturation vapour pressure e_s (kPa) at the same temperature.

- Condensation is the change from water vapour to liquid water at the dew point, when the relative humidity of the air is 100%. Sublimation is the change from water vapour to ice at the frost point, when the relative humidity of the air is 100% and the temperature of the air is 0°C or less.

- A cloud is a visible mass of very small water droplets (1–100 µm) and/or ice crystals (when temperatures are low) floating in the atmosphere.

- Water vapour has a higher energy than liquid water, which in turn has a higher energy than ice. Thus, with condensation, freezing, and sublimation, energy in the form of heat is released to the atmosphere: with evaporation and melting, heat energy is absorbed. Heat related to a change in phase of water is called latent heat.

- Dry air that ascends in the atmosphere expands and cools down at a fixed rate of 1°C for every 100 m it rises; this rate is called the dry adiabatic lapse rate. The saturated adiabatic lapse rate is 0.6°C for every 100 m that the air rises: this lesser cooling rate compared to dry adiabatic cooling is due to latent heat being released during condensation.

- Water vapour in the air may condense on small particles, such as dust, sea salt, or chemical substances that float in the air; these small particles act as condensation nuclei, causing clouds to form. Of particular importance as cloud condensation nuclei are atmospheric aerosols, which are solid and liquid particles, 0.001 to 10 µm in size, that occur naturally. Atmospheric aerosols both absorb and scatter solar radiation and thus play an important role in regulating the amount of solar radiation that reaches the surface of the earth.

- There are two main processes that generate precipitation: the collision–coalescence process in warm clouds, the primary mechanism in tropical regions, and the Bergeron–Findeisen process in cold clouds, a positive feedback mechanism, which is the dominant process in the middle and upper latitudes of the earth.

- There are different types of precipitation, including: convective precipitation, a result of local heating of the air at the earth's surface; orographic precipitation, when horizontal air currents are forced to rise over natural barriers such as mountain ranges; cyclonic precipitation in a cyclone (low-pressure area); and frontal precipitation, when cold and warm air masses meet. All types have in common that they result from the uplift of air.

- The Coriolis effect describes the curved deflection of a moving air parcel to the right in the Northern Hemisphere and to the left in the Southern Hemisphere. The Coriolis effect is related to the earth's rotation, one rotation per day, and acts on relatively long-lived phenomena such as mid-latitude cyclones and large vortices in the ocean.

- In large weather systems, the direction of air movement is affected by the earth's rotation (the Coriolis effect) and at or near the earth's surface by friction. From Buys Ballot's law (1857), it follows that wind travels anticlockwise near the earth's surface around areas of low pressure (100–2000 km across) in the Northern Hemisphere and clockwise in the Southern Hemisphere.

- In the centre of a low-pressure area the air rises, expands, and cools, causing the formation of clouds by condensation at a higher level and cyclonic precipitation. For high-pressure areas, the air movement is in the opposite direction. The air movement near the earth's surface in a high-pressure area (anticyclone) diverges away from the centre. This divergence near the earth's surface is due to descending air at the centre of the high-pressure area. Air that descends warms up. Since warmer air can contain more water vapour than cooler air, a high-pressure area usually brings us fair weather.

- The earth receives much heat at the equator and little at the poles, which causes transport of heat in the direction of the poles and cold air in the direction of the equator. The earth's rotation (the Coriolis effect) and the uneven distribution of the land masses and oceans creates a number of global circulation cells.

- When cold and warm air masses meet, frontal precipitation is formed. At a cold front the advancing cold air forces the warm air to rise along a steep slope, causing rapid lifting and intense rain of short duration. At a warm front, the advancing warm air, being lighter, easily rides up over the cold air. This leads to a gentle slope between the air masses, which causes a gradual lifting and cooling, and moderate rainfall of long duration.

- To estimate the areal precipitation from point precipitation data, several methods may be employed depending on the characteristics of the area under study. One may, for instance, use arithmetic averaging, Thiessen polygons, isohyets, statistical interpolation methods, radar or, for large areas, satellite images. With the latter two methods, the magnitude of the output needs to be linked (calibrated) and checked (validated) with precipitation data at the ground, the so-called ground truth.

- Potential evaporation is the maximum evaporation rate (mm day^{-1}) when the moisture content of the soil and the vegetation conditions do not limit evaporation. Actual evaporation is the evaporation rate (mm day^{-1}) under existing atmospheric, soil, and vegetation conditions. The actual evaporation is therefore always less than or equal to the potential evaporation.

- The reference crop evaporation is the potential evaporation (mm day^{-1}) of an idealized grass crop that serves as a reference value for determining the potential evaporation of other crops.

- To obtain open-water evaporation rates from pan evaporation data, the pan evaporation values need to be multiplied by a pan coefficient, a fractional value (larger than zero and smaller than one) that depends on the type of pan used and the time of use (season).

- An energy flux density (MJ m^{-2} day^{-1}; M = 10^6) is the amount of energy that traverses a small area perpendicular to the direction of the energy flow in a time interval, divided by that area and by that time interval.

- The sensible heat H is heat energy that is transported by a body that has a temperature higher than that of its surroundings. Sensible heat energy is transferred between the earth's surface and the atmosphere when there is a difference in temperature between them. If the overlying air is warmer than the surface, then heat is transferred downwards, and H has a negative value. A special case of H having a negative value is the oasis effect. This effect can occur in hot areas such as deserts. When warm, dry air in equilibrium with dry soil reaches a wet surface such as a lake (oasis), the evaporation rate increases and sensible heat is used to maintain this high rate.

- Latent heat λE_a is energy that is used for evaporation. It is called 'latent' because the energy is stored in the water molecules, to be released later during the condensation process.

- The saturation deficit (kPa) is the difference between the saturation pressure e_s at the surface and the actual vapour pressure e_a of the air above the evaporating surface: the larger the saturation deficit, the higher the evaporation rate will be.

- The aerodynamic resistance r_a (s m^{-1}) is the resistance encountered by water vapour in diffusing into the air from a vegetation surface (interception evaporation) or water body, and is reciprocal to the roughness of the earth's surface and the wind speed near the surface.

- The surface resistance r_s (s m^{-1}) is a physiological resistance imposed by vegetation stomata on the movement of water vapour by transpiration: r_s varies with water availability (soil moisture content).

- To estimate evaporation using the Penman–Monteith equation, one needs to know the air temperature T (°C), the relative humidity RH, the sun's shortwave radiation incident at the top of the earth's atmosphere S_0 (MJ m^{-2} day^{-1}), the cloudiness fraction n/N, the albedo α, the aerodynamic resistance r_a (s m^{-1}) and the surface resistance r_s (s m^{-1}).

Groundwater

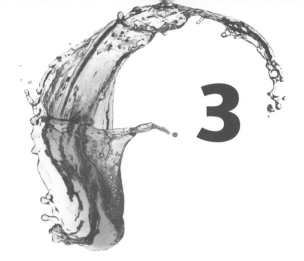

3

Introduction

Water beneath the land surface that fully saturates the pores (minute holes) in the ground is called **groundwater** (Figure 1.1). Almost all groundwater is of atmospheric origin; that is, it stems from precipitation, followed by infiltration and percolation. Only a very small part is **connate**; that is, made up of water captured during sedimentation. Groundwater can be found everywhere under the earth's surface, even under the driest deserts. Under deserts, the groundwater stems from recharge (replenishment) during an earlier wetter climate. The groundwater storage under the Sahara desert is an estimated 150 000 km^3 (4th World Water Forum 2006). The depth at which we encounter groundwater differs greatly. Groundwater is naturally filtered and purified by contact with the ground and can be used as agricultural water, industrial water, or – usually after purification (Box 3.1) – drinking water. An **aquifer** is a subsurface layer that easily stores and transmits groundwater. More water may be pumped out of an aquifer than is naturally recharged; also, groundwater may be polluted by agriculture, industry, or other human activities. Fortunately, groundwater can be artificially recharged to counter subsidence of the ground and/or to protect groundwater quality. Also, a number of types of pollution can be isolated and removed. Importantly, groundwater should be kept intact for generations to come and therefore knowledge of the groundwater system is important.

One can probably build a mental picture of surface water flow more easily than of groundwater flow. However, because the general physical laws used in hydrology are best explained for **steady groundwater flow**, groundwater flow for which the velocity components at any location do not change with time, our journey into hydrology now continues with groundwater. Steady groundwater flow can be described using model equations that all follow from combining **Darcy's law** (the flow equation) and the water balance (equation). In hydrology, the latter is also known as **continuity** or the continuity equation.

This chapter on groundwater starts by examining two generally held beliefs, which are that water flows from high to low elevations, and that water flows from high to low pressure. Two simple examples are presented that will discard these as misconceptions, before considering the proper notion of what causes water to flow. Special attention is given to the hydrology of low-lying areas, as the number of people per square kilometre and per 100 m elevation range is highest for the 0 (sea level) to 100 m elevation range (Cohen and Small 1998).

BOX 3.1 Water purification

Groundwater, surface water, and atmospheric water are all used as a source of drinking water. As a rule of thumb, when groundwater resides in a sandy aquifer for 60 days or more, it is biologically purified and free from disease-causing micro-organisms. However, to make natural water suitable for human consumption, further purification is needed. The aim of water purification is the removal from the water of dangerous substances such as bacteria, algae, viruses, fungi, toxic metals (e.g. lead and copper), and chemical pollutants. An additional aim is the improvement of the water's smell and taste.

Water purification involves a string of processes. Without going into great detail, the most familiar of these processes include coagulation, flocculation, filtration, ozonation, and softening.

Particles that are suspended in water such as clays have a negative electrical charge. This electrical charge keeps the particles suspended as they repel each other. By adding a coagulant (for instance, aluminium sulphate or iron chloride), these electrical charges are eliminated, a process called coagulation. Flocculation is the follow-up process of the joining of the particles into flocks that are heavy enough to settle at the bottom of a water reservoir. By coagulation, flocculation, and settlement, most of the suspended solids, phosphates, organic matter, bacteria, viruses, and virtually all heavy metals are removed from the water.

Three types of filtration are used at different stages in the chain of water purification: rapid sand filtration, activated carbon filtration, and slow sand filtration. The term 'activated' means that an exceptionally high surface area of the carbon is in contact with the water. Filtration over sand beds, gravel beds, and activated carbon ensures the removal of ammonia, suspended solids, organic matter, harmful bacteria, algae, manganese, and iron. A relatively new water purification technique, which is being used increasingly, is membrane filtration by polymer films with microscopic pores.

Ozonation is the addition of ozone, a gas with a high oxidizing power, to water: ozone and water should be thoroughly mixed. Ozonation is used to remove organic matter, agro-chemicals, and pathogens such as viruses and bacteria. Also, ozone is a powerful disinfectant that can often replace ill-smelling and tasting chlorine as disinfectant. As a result of ozonation, the taste, smell, and colour of the water are improved.

Softening, reducing the hardness of water, can be brought about by adding caustic soda (sodium hydroxide; NaOH) to the water. This causes calcium and magnesium hydrogen carbonates ($Ca(HCO_3)_2$ and $Mg(HCO_3)_2$) to be removed from the water.

The derivations presented in this chapter are for the purpose of providing insight into the workings of (groundwater) hydrology. One should recognize that all equations are either variations of Darcy's law (remember how to apply this one for different situations) or are derived from combining Darcy's law and continuity (also remember how to apply the equations stated in Figure 3.33!).

Groundwater 'problems' in this chapter are presented in such a way that water flow can be envisaged as taking place in one direction only; horizontally, vertically, or along the radius of a circle.

Readers interested in the mathematical derivations of the end results presented in this chapter are referred to the Mathematics Toolboxes (under M) near the end of this book; if required, mathematical grounding is provided in the Conceptual Toolkit (under C).

3.1 Misconceptions

One common misconception is that water always flows from a high to a low elevation.

For those who live in a **polder**, land below sea level reclaimed from a body of water and protected by dykes (Figure 3.1), water is pumped out regularly to ensure that groundwater levels remain at a near-constant level beneath the land surface. What would happen if the authorities were to suddenly stop dewatering these polders? Surely the groundwater would rise, and this would be a water flow from a low to a high elevation. Thus water does not flow from a high to a low elevation by law of nature.

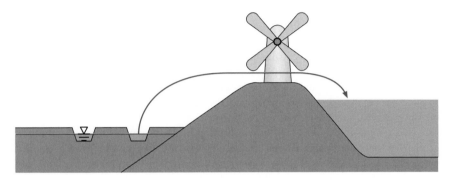

Figure 3.1. A polder, the groundwater level plus a watermill pumping water to a higher water level in a river

Another misconception is that water always flows from high to low pressure.

In Figure 3.2, the pressure at the bottom of the bucket equals the ratio of the weight (or force) of both the column of water and air above, and the cross-sectional area A (m^2). Higher up, at the air/water interface, the pressure equals the ratio of the weight (or force) of only the column of air above and the same cross-sectional area A (m^2). Thus, the pressure is higher at the bottom of the bucket than at the air/water interface. Despite this, there is no spontaneous flow from bottom to top. Thus water does not flow from high to low pressure by law of nature.

What, then, makes water flow from one location to another?

3.2 Drilling a hole ...

To measure the mechanical energy of water at a certain location, you can drill a hole in the ground to a depth below the water table and then insert a tube with a small diameter (< 30 cm) and a small screen (a few centimetres long) at the bottom. Groundwater should be able to flow through the screen easily, and especially fine materials such as clay (soil material with a size range smaller than 2 μm) and silt (2–50 μm) must be filtered out so as not to clog the screen. To prevent the latter, the borehole alongside the screen is filled with coarse-grained, poorly sorted, permeable ground material (with a diameter coarser than the screen openings). On top of this, the borehole alongside the tube is filled with bentonite, which is a clay mineral that swells with water. By this means,

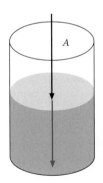

Figure 3.2. A cylinder-shaped bucket of water

Mechanical energy is the energy that an object has because of its motion or position.

precipitation, surface water, and groundwater from a higher location are prevented from entering the screen. Such a tube with a short screen at the bottom is called a **piezometer** and the elevation to which the water rises in a piezometer is called the **hydraulic head** (or the piezometric head/level or potentiometric head/level). Importantly, the hydraulic head for a piezometer at a certain location and its screen at a certain depth is a measure of the mechanical energy of the water at that location and depth.

You can also drill deeper and place a longer tube with its screen at a greater depth. As shown in Figure 3.3, the hydraulic heads in both tubes need not be the same. This may seem strange … How can this be?

3.3 Bernoulli to the aid

A **water table** that can establish itself freely is by definition the level at which the (pore) water pressure in the ground equals the air pressure of the overlying air. On average, this air pressure is the **standard atmospheric pressure** of 1013.25 hPa (1 atmosphere). Instead of using this standard value as the average water pressure at a free air/water interface (such as the water table), it is common practice to define the water pressure at a free air/water interface as zero. Because of this, **water pressure** has a value that should be interpreted with the existing air pressure at the air/water interface as a reference: a zero water pressure means that water pressure equals the existing air pressure at the air/water interface, a positive water pressure means that the water pressure is higher than this air pressure, and a negative water pressure means that the water pressure is lower than this air pressure.

A water particle moving along a **streamline** possesses three interchangeable types of mechanical energy (joules): kinetic energy, potential energy, and pressure energy. If there is no loss of mechanical energy due to friction and if we take water as incompressible, then **Bernoulli's law** for steady flow states that the sum of these three interchangeable types of energy is constant, a notion known as **conservation of energy**. Bernoulli's law is named after the Dutch/Swiss mathematician Daniel Bernoulli (1700–1782) and can be represented as follows:

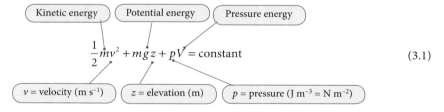

$$\frac{1}{2}mv^2 + mgz + pV = \text{constant} \qquad (3.1)$$

Steady flow has already been defined as flow for which the flow velocity components at any location do not change with time. Another term used for steady flow in the hydrological literature is **stationary flow**. On the contrary, when the magnitude or direction of flow does change with time, or when both change, the flow is called **unsteady** or **transient**.

By dividing by the volume V, we obtain an expression per unit volume of the water (mass m divided by volume V becomes density ρ):

$$\frac{1}{2}\rho v^2 + \rho g z + p = \text{constant} \qquad (3.2)$$

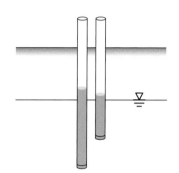

Figure 3.3. Two piezometers at the same location with their screens at different depths and different hydraulic heads

A **streamline** is a curve such that the tangent at any point coincides with the direction of the flow velocity at that point.

Kinetic energy = energy of motion; potential energy = the energy that an object (e.g. a water particle) has because of its position above a chosen reference level; pressure energy = energy related to pressure.
m = mass (kg)
g = acceleration due to gravity = 9.8 m s^{-2}
V = volume (m^3)
ρ = water density ≈ 1000 kg m^{-3}

Dividing by the near-constants ρ and g causes the mechanical energy terms to be presented per unit weight (J per N = Nm per N), and thus per length unit (metre):

$$\frac{v^2}{2g} + z + \frac{p}{\rho g} = \text{constant} \qquad (3.3)$$

As flow velocities in groundwater are slow (see section 3.7), the first term, the kinetic energy per unit weight, may be neglected, and thus:

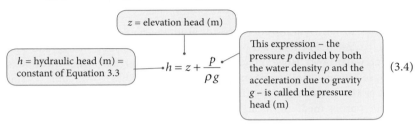

$$h = z + \frac{p}{\rho g} \qquad (3.4)$$

z = elevation head (m)

h = hydraulic head (m) = constant of Equation 3.3

This expression – the pressure p divided by both the water density ρ and the acceleration due to gravity g – is called the pressure head (m)

In words, the **hydraulic head** (the constant in Equation 3.3), which is a measure of the mechanical energy of the water at the location of the screen, equals the sum of the elevation head and the pressure head. This is shown in Figure 3.4 for a piezometer with a reference level chosen below the screen of the piezometer. The **elevation head** then equals the elevation above the chosen reference level, whereas the **pressure head**, following the common practice of defining the water pressure at a free interface between air and water as zero, equals the length of the column of water above the screen. Note that when there is no water flow, the water levels in the tube and in the ground surrounding the tube (the latter being the water table) are the same. The condition of there being no water flow is called **hydrostatic equilibrium**.

Mechanical energy as a length unit: dividing the mechanical energy (J) by the water volume (m^3) and then by both the water density (kg m^{-3}) and the acceleration due to gravity (m s^{-2}) gives J m^{-3} kg^{-1} m^3 m^{-1} s^2 = N m m^{-3} kg^{-1} m^3 m^{-1} s^2 = kg m s^{-2} m m^{-3} kg^{-1} m^3 m^{-1} s^2 = m, and thus mechanical energy as a length unit.

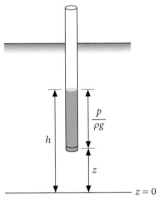

Figure 3.4. A piezometer: hydraulic head h = elevation head z + pressure head $\frac{p}{\rho g}$; reference level, $z = 0$

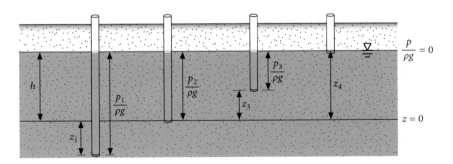

Figure 3.5. Unconfined groundwater: elevation heads and pressure heads under hydrostatic equilibrium conditions (no water flow)

$$h = z_1 + \frac{p_1}{\rho g} = 0 + \frac{p_2}{\rho g} = z_3 + \frac{p_3}{\rho g} = z_4 + 0 \qquad (3.5)$$

Unconfined groundwater is groundwater where the water table can establish itself freely; it is therefore also called **phreatic groundwater**, phreatic meaning free. Figure 3.5 shows the elevation heads and pressure heads for a number of piezometers with their screens at different depths in unconfined groundwater with no water flow (hydrostatic equilibrium). Note that the screen of the leftmost pie-zometer lies below the reference level and that the elevation head z_1 is therefore a negative number (< 0 m). Also note that the screen of the rightmost piezometer is located at the water table and that the pressure head thus equals 0 m. This is also stated in the accompanying Equation 3.5. Below the water table, the pressure heads are positive relative to the existing air pressure (> 0 m), whereas above the water table, in the unsaturated zone, the pressure heads are negative (< 0 m). Finally note, as mentioned earlier, that **under hydrostatic equilibrium conditions (no water flow), the water table and the hydraulic heads measured in the piezometers have the same level.**

Thus far, we have first considered steady groundwater flow without friction, where Bernoulli's law holds, followed by considerations concerning hydrostatic equilibrium (no water flow). However, groundwater flow through a porous medium, such as rock, sediment, or soil, is not without friction, since it has to overcome resistances caused by the water movement itself (friction among the water molecules) as well as by the porous medium through which the water flows. These resistances cause a conversion of mechanical energy into heat (energy) and thus a loss of mechanical energy or hydraulic head in the direction of the flow. In other words, water flow is in the direction of the lower mechanical energy level or hydraulic head. **This answers the question raised at the end of section 3.1, the correct concept being that water flows from a high to a low mechanical energy level or hydraulic head. Using this law of nature, one can establish the direction of groundwater flow from hydraulic head readings of piezometers at different locations and screens at different depths.**

Figure 3.6 shows two piezometers with their screens located at the same reference level, but at a horizontal distance from each other. At the level of the piezometer screens, the groundwater flow is in the direction of the lower hydraulic head h_2.

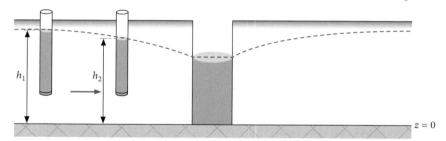

Figure 3.6. Unconfined groundwater: horizontal water flow

Figure 3.7 shows piezometers at the same location, but with their screens located at two different depths. Here also, water flow is in the direction of the lower hydraulic head h_2. For the figure on the left, the water flow is in a downward direction (**leakage** or **downward seepage**); whereas for the figure on the right, the water flow is directed upward (**seepage**). The position of the water table can be deduced from readings of piezometers with their screens near the water table. Under both leakage and seepage conditions, the water table and the hydraulic heads at different depths do not coincide. With leakage, the water table must have a higher level than the hydraulic head at some depth in order for groundwater to flow downwards; for seepage, the hydraulic head at some depth must have a higher level than the water table in order for groundwater to flow upwards. The water table could be established by using a tube that has a screen all the way down. Following the **law of communicating vessels**, the air pressure at the air/water interface in such a tube-long screen is the same as the air pressure at the water table around the tube.

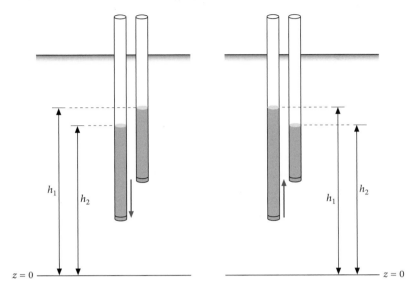

Figure 3.7. Unconfined groundwater: vertical water flow
The direction of flow can be determined by placing two piezometers with their screens at different depths at the same location: the hydraulic head is a measure of the mechanical energy at the piezometer screen and groundwater flow is always in the direction of the lower hydraulic head h_2 – in the left-hand figure, the flow is from the upper screen to the lower screen, and thus downward; in the right-hand figure, the flow is from the lower screen to the upper screen, and thus upward.

3.4 Aqui...

It is now timely to introduce hydrological terms that describe different types of sub-surface water-storing and water-confining layers.

An **aquifer** is a subsurface water-storing layer with high water permeability; **permeability** is defined for our purpose here in general terms, as a measure of the availability of a porous material (soil, sediment, or rock) to transmit water.

Unconfined groundwater (phreatic groundwater) has already been defined as groundwater where the water table can establish itself freely. Thus an aquifer with unconfined groundwater is called an **unconfined aquifer**.

A **confining layer** is a subsurface layer with little or no water permeability. Confining layers that totally block groundwater flow are impermeable layers or **aquifuges**. Confining layers with low water permeability are called semi-permeable layers, **leaky confining layers**, or **aquitards**.

Groundwater captured between two impermeable layers is **confined groundwater**, and the aquifer is a **confined aquifer**. Groundwater captured between two confining layers, one or both of these being semi-permeable, is called **semi-confined groundwater** and the aquifer is a **semi-confined aquifer** or **leaky aquifer**.

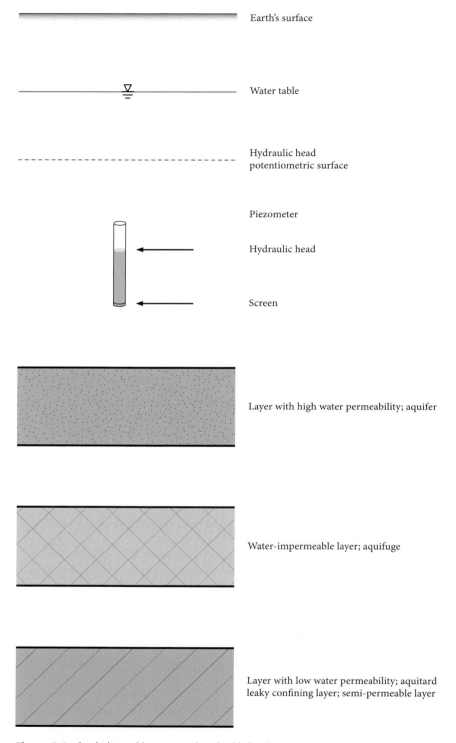

Earth's surface

Water table

Hydraulic head
potentiometric surface

Piezometer

Hydraulic head

Screen

Layer with high water permeability; aquifer

Water-impermeable layer; aquifuge

Layer with low water permeability; aquitard
leaky confining layer; semi-permeable layer

Figure 3.8. Symbols used in cross-sections in this book

Figure 3.8 shows the symbols that are used for these different types of layers in cross-sections in this book.

Figure 3.9 is a cross-section showing different types of subsurface water-storing and water-confining layers. The large hydraulic head in the semi-confined or leaky aquifer is caused by a high water table to the left in Figure 3.9 and by a high water pressure in the aquifer due to the weight of the overlying ground. The semi-confined aquifer is recharged from upslope and its hydraulic head diminishes slightly from left to right in Figure 3.9, due to the friction that the groundwater flow experiences. When the hydraulic head of a confined or semi-confined aquifer lies above the land surface, as in Figure 3.9, the groundwater is called **artesian**. Above the overlying aquitard, we observe unconfined groundwater, and to the far right we find groundwater above an impermeable layer that overlies an unsaturated zone. Unconfined groundwater underlain by an unsaturated zone is called **perched groundwater** and the water table a **perched water table**. Note that the hydraulic head is highest in the semi-confined aquifer, which means that there is seepage, groundwater flowing upwards from the confined aquifer through the aquitard that separates both aquifers.

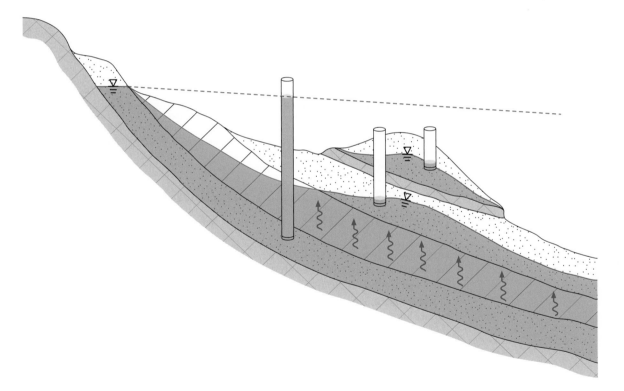

Figure 3.9. A cross-section showing different types of water-storing and water-confining layers

Figure 3.10 shows a schematic cross-section of seepage into a polder. The polder is made up of a semi-permeable clay layer and is separated from the river by dykes. Below the clay layer is an aquifer above an impermeable layer. By pumping any excess (precipitation and seepage) water into the river, the water table in the polder is artificially maintained just below the land surface level. The aquifer below the polder is provided with water from the rivers as shown in Figure 3.10. Groundwater in the aquifer is confined by the impermeable layer (aquifuge) below and the semi-permeable layer above (aquitard). The hydraulic head in the aquifer beneath the river equals the water level of the river and diminishes gradually to the centre beneath the polder, due to both the friction that groundwater flow experiences and seepage through the upper semi-permeable layer into the polder.

The hydraulic heads in the aquifer are measured in the two piezometers shown on the left in Figure 3.10. The geographical distribution of the hydraulic heads in the aquifer, the **potentiometric surface**, is presented in cross-section by the broken curve in Figure 3.10. Note that the hydraulic head in the aquifer is everywhere larger than the level of the water table in the polder, causing groundwater to seep through the semi-permeable layer. Seepage is largest just behind the dykes, where the difference in hydraulic head between the aquifer and the polder's water table is largest, and it diminishes towards the centre of the polder.

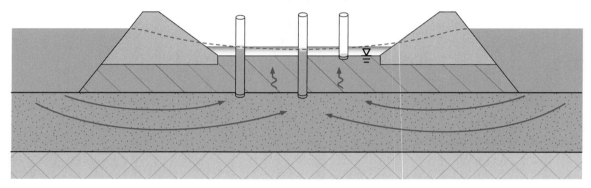

Figure 3.10. Seepage in a polder area

3.5 Effective infiltration velocity and infiltration rate

How much rain infiltrates the soil? How deeply does this rain infiltrate the soil? These are some basic questions when studying hydrological processes.

As mentioned earlier in section 2.4, it is easiest to think of rainfall depth as a layer of water over a flat surface, measured in millimetres. A uniform rainfall over a land surface of 2 mm (2×10^{-3} m), as shown in Figure 3.11, equals 2×10^{-3} m^3 for every square metre (m^2), which is 2 litres per square metre. In hydrology, it is customary to use length units for both mechanical energy terms (section 3.3) and volumes of water (section 1.4).

Figure 3.11. Rainfall depth at $t = t_1$ and infiltration depth at $t = t_2$ ($t_2 > t_1$)

Let us imagine that 2 mm of rain falls and that it infiltrates the soil in 1 hour. If we define a volume of soil and the interstices (voids) or pores that it contains as a fraction of 1 (or 100%), we can further imagine that a fraction of 0.15 (15%) of this volume contains water and a volume fraction of 0.2 (20%) contains air. Both water and air are contained in the interstices or pores of the soil; the remaining part of the soil is referred to as the **solid soil matrix**. Figure 3.12 shows different types of interstices.

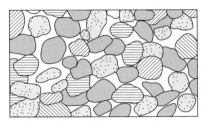

Well-sorted, rounded, porous sand;
porosity $n \approx 0.4$

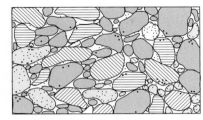

Poorly sorted, partly cemented sand with smaller grains in the interstices;
porosity $n \approx 0.2$

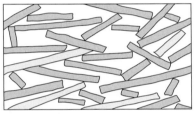

Very porous, loosely packed clay;
porosity $n \approx 0.7$

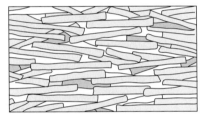

Slightly compacted, porous clay;
porosity $n \approx 0.5$

Fissures in limestone;
low to moderate primary porosity*;
high secondary porosity**

Fractures in crystalline rock;
low primary porosity*;
high secondary porosity**

*: **primary porosity**: porosity developed during the formation of the rock
: **secondary porosity: larger-scale porosity developed after the formation of the rock
   by (secondary) processes such as fracturing, cracking, and/or dissolution

Figure 3.12. Types of interstices (after De Vries and Cortel 1990)

The **volumetric water content** θ in our example equals 0.15 (15%). The **porosity** n is the maximum volume fraction of water in the soil. When air is removed from the soil and all pores are water saturated, the volume fraction of water in the soil is 0.35; the porosity n in our example thus equals 0.35 (35%). The **effective porosity** n_e is defined as the volume fraction of a rock, sediment, or soil that participates in the water flow process.

Figure 3.13 shows a dead-end pore that does not participate in the water flow process. If not all pores or parts of pores participate, then the effective porosity is less than the porosity. Of course, when all pores participate in the water flow process, effective porosity equals porosity.

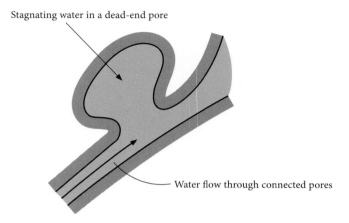

Stagnating water in a dead-end pore

Water flow through connected pores

Figure 3.13. A dead-end pore

When, in our example, all rain infiltrates, when all soil pores are interconnected and participate in the infiltration process ($n_e = n$), and when air can escape readily from these pores, then the 2 mm of rain has $n_e - \theta = 0.35 - 0.15 = 0.20$ or one fifth of the fraction of soil at its disposal to fill with water. The depth to which the rainfall infiltrates thus equals 5×2 mm = 10 mm, and the **effective infiltration velocity** v_e equals 5×2 mm hour^{-1} = 10 mm hour^{-1} = 0.24 m day^{-1} (Figure 3.11), whilst the **infiltration rate** equals the above-mentioned 2 mm hour^{-1} or 0.048 m day^{-1}. The effective infiltration velocity (0.24 m day^{-1}) and the infiltration rate (0.048 m day^{-1}) are thus different entities.

Although the units (m day^{-1}) are the same, the effective infiltration velocity is a water flow velocity, whilst the infiltration rate is (**not** a water flow velocity, but) a volume flux density. A **volume flux density** is defined as a volume flux (a volume of water travelling in a unit of time) per unit area perpendicular to the direction of flow. Note that this area perpendicular to the water flow contains both solid ground particles and pores, but that water flow is through (interconnected) pores only, **not** through the solid particles. To obtain a volume flux density, the volume flux is thus divided by an area (made up of solid ground particles and pores) that is larger than the area (of the pores) in which the actual flow takes place. **The volume flux density (m day^{-1}) in a porous medium is therefore essentially different from the water flow velocity (m day^{-1}).**

As equations:

$$\text{depth of infiltration (mm)} = \frac{P}{n_e - \theta}$$

P = precipitation depth (mm)

n_e = effective porosity (−)

θ = volumetric water content (−)

(3.6)

v_e = effective infiltration velocity (m day^{-1}) f = infiltration rate (m day^{-1})

$$v_e = \frac{f}{n_e - \theta} \qquad (3.7)$$

Again, note the important difference between the effective infiltration velocity v_e and the infiltration rate f. The effective infiltration velocity v_e is a macroscopic velocity, 'macroscopic' referring to scales encountered in the everyday world (roughly the opposite of 'microscopic'). Again, the infiltration rate is a volume flux density, a volume flux per area perpendicular to the flow; this area contains both soil pores (with water flow) and solid soil matrix (with no water flow).

3.6 The soil as a wet sponge

Percolation is vertical subsurface water flow under the influence of gravity whereby water drains from the soil water zone (unsaturated zone) to the groundwater (saturated zone). Percolation takes place when the soil is wetter than the field capacity. The **field capacity** is the maximum water content that a soil can hold against gravity. It is a slightly imprecise measure, as there is no fixed water content at which gravity drainage from a soil ceases altogether. It can therefore best be compared with a rather wet sponge; that is, just not wet enough to have water dripping from it in large quantities by gravity. The (volumetric) water content at field capacity is thus lower than the (volumetric) water content at saturation of the soil. At saturation, all the pores in the soil are filled with water; at field capacity, the largest pores in the soil may still be filled with air.

Part of the subsurface water is readily available for water flow, whilst another part is retained by surface tension forces as films around soil particles or in small capillary openings; therefore, this water is not readily available for water flow. Figure 3.14 shows clay to have a higher porosity than sand. Also note that the summation of water

clay: soil particles < 2 μm = 2×10^{-6} m in diameter

silt: soil particles ranging from 2 to 50 μm in diameter

sand: soil particles ranging from 50 μm to 2 mm in diameter

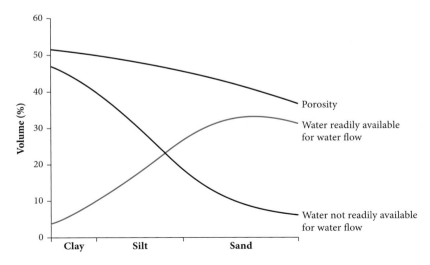

Figure 3.14. The porosity of clay, silt, and sand in relation to the volume fraction of water readily available for water flow

readily available for water flow and water not readily available for water flow equals porosity. Despite its higher porosity, because water is strongly bound to clay surfaces, the volume fraction of water readily available for water flow in clayey sediments is much less than for sandy sediments. Figure 3.14 shows silt to hold an intermediate position.

3.7 Brothers in science: Darcy and Ohm

Darcy's law

In 1856, the French engineer Henri P.G. Darcy (1803–1858) formulated the law for water flow through a porous medium that has carried his name ever since: **Darcy's law**. To improve the water supply to the city of Dijon, he carried out experiments that are schematized in Figure 3.15.

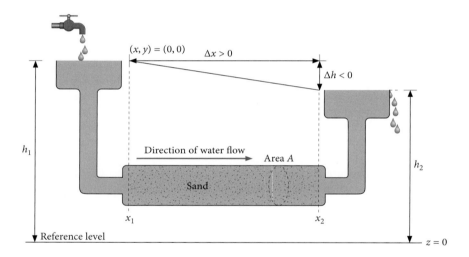

Figure 3.15. A schematic representation of Darcy's experiment

In the original experiments, Darcy used a vertical cylinder filled with water-saturated sand. Darcy's experimental set-up is represented in simplified form in Figure 3.15 by a horizontal cylinder filled with water-saturated sand. Note that the whole system is filled with water. Also note that the screens at both sides of the cylinder keep the sand in place, but that water can pass freely through the screens.

If the water levels in the two reservoirs differ, then water will flow through the sand-filled cylinder. In Figure 3.15, water flows from left to right. The water level in the left reservoir is kept at a constant elevation h_1 by letting water spill over the edge; similarly, the water in the right reservoir, which is recharged by the water moving through the cylinder, is kept at a constant elevation h_2. At every elevation along the left screen, the hydraulic head equals h_1.

In Figure 3.16, the hydraulic head h_1 at the midpoint of the left screen is shown as the summation of the elevation head z_1 and the pressure head $p_1/(\rho g)$. Similarly, at every elevation along the right screen, the hydraulic head equals h_2. Once equilibrium has been established and the water flow has become stationary, the **volume flux** Q (or **discharge** Q), which is the volume of water that moves through any vertical area A per unit of time ($m^3 s^{-1}$), can simply be measured by collecting the water that spills over the receiving reservoir during a fixed time interval. Note that the area A is perpendicular to the water flow.

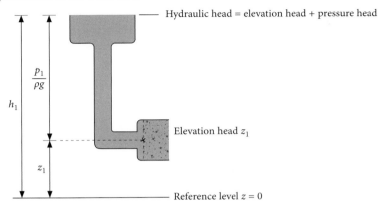

Hydraulic head = elevation head + pressure head

$\frac{p_1}{\rho g}$

h_1

Elevation head z_1

z_1

Reference level $z = 0$

Figure 3.16. The left-hand part of the schematic representation of Darcy's experiment

Porosity affects water flow. Flowing water has to overcome the resistance caused by the porous, sandy medium in the cylinder. The **difference in hydraulic head** Δh over the cylinder length equals the hydraulic head at the water-receiving end h_2 minus the hydraulic head at the water-despatching end h_1: $\Delta h = h_2 - h_1$ (and **not** the other way around). The **hydraulic gradient** i is defined as the difference in hydraulic head Δh over a porous medium divided by the distance over which this difference in hydraulic head occurs. The distance is defined as the location to which water flows minus the location from where it was despatched (and **not** the other way around). For horizontal water flow, the distance Δx thus equals $x_2 - x_1$, and

i = hydraulic gradient (–) Δh = difference in hydraulic head (m)

$$i = \frac{\Delta h}{\Delta x} = \frac{h_2 - h_1}{x_2 - x_1} \qquad (3.8)$$

Δx = distance (m)

Darcy found the volume flux Q to be directly proportional to both the hydraulic gradient (the larger i is, the larger Q becomes) and the area A perpendicular to the water flow (the larger A is, the larger Q becomes).

The volume flux Q increases when the water temperature rises. Also, the kind of material that is placed between the screens influences the volume flux Q.

The latter may have to do with differences in porosity: well-sorted or uniform sand (sand with a predominant grain size) has a higher porosity than poorly sorted sand (where smaller sand particles fit into the pores between the larger sand particles), as shown in Figure 3.12. It is easier for water to flow through sand that has a higher porosity and, as a consequence, all other circumstances being equal (i and A being unchanged), the volume flux or discharge Q will be higher.

Clay minerals have a laminated structure. Due to their mineral structure, clay minerals are negatively charged at their surface. This negative charge is neutralized by cations, positively charged ions such as Na^+ and Ca^{2+}, in the soil water. Thin films of water and neutralizing cations adhere to the clay mineral surface by electrostatic forces. Sand, on the other hand, has a granular structure, with much water held in pores between the grains by capillary forces.

An even stronger effect on the volume flux or discharge Q is caused by the way in which water is retained by the material: Figure 3.14 and Table 3.1 show clay to have a higher porosity than sand. However, in water-saturated sand the water is less strongly bound to the material than in clay. As a consequence, all other circumstances again being equal (i and A being unchanged), the volume flux or discharge Q in sand will be higher than in clay.

Table 3.1 Porosity n (–) for different types of sediment and rock

Clay	0.4–0.7
Silt	0.35–0.5
Sand	0.25–0.4
Poorly sorted sand and gravel	0.2–0.4
Gravel	0.25–0.4
Chalk	0.1–0.4
Sandstone	0.05–0.3
Limestone and dolomite	0.0–0.2
Schist	0.0–0.1
Crystalline rock	0.0–0.1

If we put the above findings into a formula, we obtain **Darcy's law** as follows:

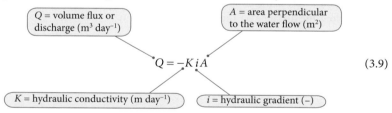

$Q =$ volume flux or discharge (m^3 day^{-1}) $A =$ area perpendicular to the water flow (m^2)

$$Q = -K\,i\,A \tag{3.9}$$

$K =$ hydraulic conductivity (m day^{-1}) $i =$ hydraulic gradient (–)

K is a proportionality factor, by means of which the characteristics of both the water and the material through which the water flows are taken into account. As the volume flux Q has units of m^3 day^{-1}, as the hydraulic gradient i is dimensionless, and as the area A perpendicular to the water flow is given in m^2, the unit of the proportionality factor K must be m day^{-1}. K is called the **(saturated) hydraulic conductivity**. The hydraulic conductivities K (m day^{-1}) of different types of sediment and rock are presented in Table 3.2. The table shows the hydraulic conductivity of sand to be much higher than for clay, by a factor of 5 to 10^{10}. Table 3.1 shows the porosity values for the same types of sediment and rocks.

The concepts of porosity and hydraulic conductivity must not be confused. The example of sand (lower porosity, higher hydraulic conductivity) versus clay (higher porosity, lower hydraulic conductivity) is evident when comparing Tables 3.1 and 3.2. Another example, involving solid rock, is that of pumice – volcanic rock with a light mass due to the many vesicular hollows formed by trapped gases during cooling and

Table 3.2 Hydraulic conductivity K (m day^{-1}) for different types of sediment and rock

Clay	0.00000001–0.2
Silt	0.1–1
Sand	1–100
Poorly sorted sand and gravel	5–100
Gravel	100–1000
Chalk	1–100
Sandstone	0.001–1
Limestone and dolomite	0.1–1000
Schist	0.0000001–0.01
Crystalline rock	0.00001–1

solidification of the rock. The porosity of these rocks is thus very high; but since the many hollows are not interconnected, the hydraulic conductivity is very low.

This takes us to yet another important matter that needs to be acknowledged for both porosity and hydraulic conductivity, which is the **sample volume**. A small sample of limestone – say, the size of a hand-held sample – may have a low porosity and a low hydraulic conductivity, whereas a large sample – say, of the order of 10 m^3 – can have a much higher porosity and hydraulic conductivity due to the presence of fissures in the limestone. When dealing with groundwater flow, we are usually interested in the

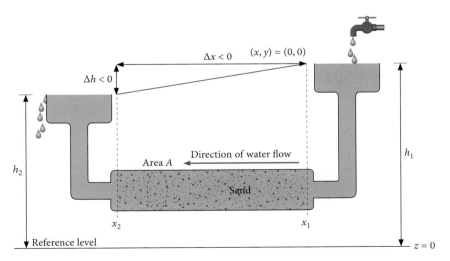

Figure 3.17. The schematic representation of Darcy's experiment, but now with groundwater flow directed from right to left

latter scale. If we enlarge the sample size, we will eventually obtain a consistent measurement of the property or parameter of interest. An important concept is therefore the **representative elementary volume** (Bear 1969), which is the minimal sample volume that must be studied to give a consistent value for the measured parameter of interest.

Groundwater flows in the direction of the lower hydraulic head. Therefore, when groundwater flows from left to right as in Figure 3.15 ($x_2 > x_1$), which is in a positive direction ($x_2 - x_1 > 0$), both the difference in hydraulic head ($\Delta h = h_2 - h_1$) and the hydraulic gradient i as defined above have a negative sign ($h_2 < h_1$; $\Delta h = h_2 - h_1 < 0$; $x_2 - x_1 > 0$; $i < 0$). Because K and A in Darcy's law have positive values, and because the volume flux Q is positive when the flow is directed from left to right, a minus sign is incorporated in Darcy's law.

Figure 3.17 shows a schematic representation of Darcy's experiment, but now with groundwater flow directed from right to left ($x_1 > x_2$), that is, in a negative direction ($x_2 - x_1 < 0$). Then, as $K > 0$, $A > 0$, $h_2 < h_1$; $\Delta h = h_2 - h_1 < 0$; $x_2 - x_1 < 0$, the hydraulic gradient i has a positive sign (> 0) and the volume flux Q has a negative sign (< 0). Notice how the hydraulic gradient always has a sign that is opposite to the sign of the volume flux Q.

Figures 3.18 and 3.19 show schematic representations of Darcy's experiment, with groundwater flow by law of nature directed towards the lower mechanical energy level or hydraulic head. Flow can be from a high to a low elevation (Figure 3.18), but also from a low to a high elevation (Figure 3.19). Furthermore, flow can be from high to low pressure (Figure 3.19), but also from low to high pressure (Figure 3.18). Logically, as flow is directed towards the lower hydraulic head, and as the hydraulic head equals the sum of the elevation head and the pressure head, the difference in elevation head must be larger than the difference in pressure head in Figure 3.18, and vice versa in Figure 3.19.

By dividing both sides of Equation 3.9 by the area A perpendicular to the water flow, Darcy's law can be presented in a simplified form:

q = volume flux density or specific discharge (m day⁻¹) ⟶ $$q = \frac{Q}{A} = -K\,i \qquad (3.10)$$

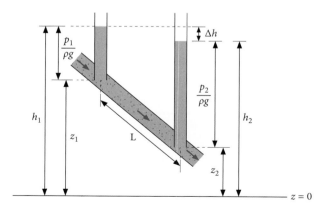

Figure 3.18. The schematic representation of Darcy's experiment with groundwater flow from low to high pressure

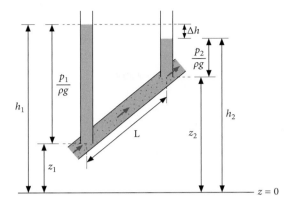

Figure 3.19. The schematic representation of Darcy's experiment with groundwater flow from low to high elevation
Note that groundwater flow in Figures 3.18 and 3.19 is parallel to the cylinder casing, that the area A is perpendicular to this water flow, and that the hydraulic gradient i, being the difference in hydraulic head Δh divided by the distance over which this difference in hydraulic head occurs, thus equals Δh divided by the distance L shown in the figures.

Note that the area A perpendicular to the water flow contains both solid ground particles and pores, and that water flow is through interconnected pores. Thus only a fraction of the area A is used for water flow, and this fraction equals the **effective porosity** as introduced in section 3.5. Dividing the volume flux or discharge Q (m^3 day^{-1}) by A (m^2) therefore yields a **volume flux density** or **specific discharge** q with units of m day^{-1}.

Note that as the volume flux Q is divided by an area (made up of solid ground particles and pores) that is larger than the area (of the pores) in which the actual flow takes place, the volume flux density or specific discharge q is not a velocity (see also section 3.5). To obtain the effective velocity v_e, one needs to divide the volume flux density or specific discharge q by the effective porosity n_e, which is analogous to Equation 3.7:

$\boxed{v_e = \text{effective velocity (m day}^{-1})}$ $\quad \bullet v_e = \dfrac{q}{n_e}$ (3.11)

Equation 3.10 can be rewritten as follows:

$$K = -\frac{q}{i} = \left| \frac{q}{i} \right|$$ (3.12)

In words, the hydraulic conductivity K (m day^{-1}; see Table 3.2) is equal to the absolute value of the volume flux density or specific discharge q when the hydraulic gradient i is one. In more technical language, K is a volume flux density (**not** a velocity) per unit hydraulic gradient.

In Exercise 3.7.1, the information presented above is to be used to determine the volume flux Q and travel time of a water particle in a tilted design of Darcy's experiment.

Exercise 3.7.1 Water flows through a sand body that has been placed between two screens in the apparatus shown in Figure E3.7.1. The water levels to the left and right in Figure E3.7.1 are kept at constant levels. The screens keep the sand in place, but allow water to pass.

$z_1 = 30$ cm $\qquad\qquad\qquad \dfrac{p_1}{\rho g} = 90$ cm

$z_2 = 60$ cm $\qquad\qquad\qquad \dfrac{p_2}{\rho g} = 40$ cm

$l = 40$ cm $\qquad\qquad\qquad\quad A = 600$ cm^2

The hydraulic conductivity K of the sand is 10 m day^{-1}; the effective porosity n_e is 0.4.

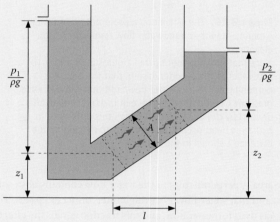

Figure E3.7.1

a. Draw the direction of water flow.

b. Determine the volume flux Q in the sand body.

c. Determine the travel time of a water particle that flows through the sand body.

Homogeneity and isotropy

Figure 3.20 explains the concepts of both homogeneity and isotropy. In a **homogeneous layer**, the hydraulic conductivity K is the same at every location. A homogeneous layer may be isotropic or anisotropic. In an **isotropic layer**, the hydraulic conductivity K at each location is the same in all directions, or, in other words, independent of direction. Figure 3.20a shows a layer that is both homogeneous and isotropic. Although aquifers are never perfectly isotropic, it may often be reasonable to assume so for calculation or modelling purposes ($K_1 \approx K_2$ and at each location $K_x \approx K_y \approx K_z$). The opposites of homogeneity and isotropy are **heterogeneity**, where K is not the same at every location, and **anisotropy**, where the value of K at a location depends on direction.

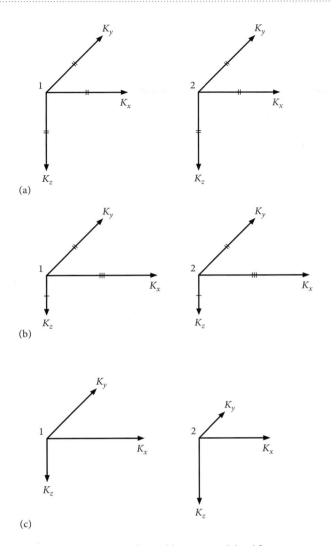

Figure 3.20. Homogeneity and isotropy explained for two locations (1 and 2) and three orthogonal directions (x, y, and z): (a) homogeneous, isotropic layer; (b) homogeneous, anisotropic layer; (c) heterogeneous, anisotropic layer

Conservative transport

Plausible values for the hydraulic conductivity K (Table 3.2), hydraulic gradient i, and effective porosity n_e (Table 3.1) in a sandy aquifer with unconfined groundwater may be as follows: $K = 10$ m day^{-1}, $i = -10^{-3}$, and $n_e = 0.4$. Using these values, we can calculate the effective velocity v_e of groundwater (Figure 3.21) from Equations 3.10 and 3.11 as follows:

$$q = -Ki = -10 \times -10^{-3} = 10^{-2} \text{ m day}^{-1};$$

$$v_e = \frac{q}{n_e} = \frac{10^{-2}}{0.4} = 2.5 \times 10^{-2} \text{ m day}^{-1} = 2.5 \text{ cm day}^{-1}$$

Note that this is very slow: 2.5 cm day^{-1} amounts to 9 m year^{-1}, which is less than 1 km in a century, or a mere stretch of 10 km in 1100 years. In Exercise 3.7.2, another example can be worked out for **conservative transport** of a polluting substance. The word 'conservative' implies that there is no interaction between the polluting substance and the groundmass, or, in other words, the polluting substance goes with the flow.

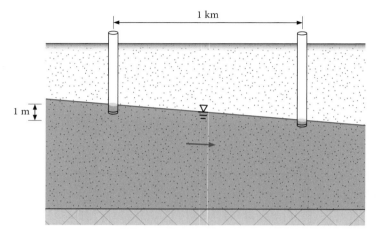

Figure 3.21. Groundwater flow in an unconfined sandy aquifer

Exercise 3.7.2 A piezometer is situated at a distance of 456.25 m from an illegal waste dump. The piezometer is placed in a homogeneous isotropic aquifer. The saturated hydraulic conductivity $K = 5$ m day^{-1}. The effective porosity $n_e = 0.4$. The water table has a slope of 1 m km^{-1} in the direction of the piezometer. At the beginning of the year 2000, the groundwater was polluted. The transport process is conservative and the polluting substance may be easily detected.

In what year will the polluting substance be detected in the piezometer?

Intrinsic permeability

The hydraulic conductivity K is a function of both the properties of the porous medium through which the water flows and the properties of the water itself, such as density and viscosity. The properties of the porous medium and groundwater can be separated (by combining the Hagen–Poiseuille law and Darcy's law through a cylindrical tube):

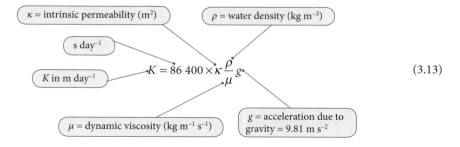

$$K = 86\ 400 \times \kappa \frac{\rho}{\mu} g \tag{3.13}$$

where $\kappa =$ intrinsic permeability (m^2), $\rho =$ water density (kg m^{-3}), s day^{-1}, K in m day^{-1}, $\mu =$ dynamic viscosity (kg m^{-1} s^{-1}), $g =$ acceleration due to gravity $= 9.81$ m s^{-2}.

Given the units of the variables on the right-hand side of the equation, 86 400 is simply a factor to directly convert the hydraulic conductivity K to units of m day^{-1}; κ is the **intrinsic permeability** (m^2) and is a function of the size of the openings through which the groundwater flows and thus of the porous medium. Both the water density ρ and the (dynamic) **viscosity** μ, being the 'thickness' or resistance of a liquid to flow, are properties of the groundwater that are temperature dependent. Water density is largest at 4°C, and viscosity decreases with increasing temperature. A hydrologist is interested in the movement of water, whilst a petroleum engineer is equally interested in the movement of subsurface carbon liquids and gases. Both these fields of expertise deal with porous subsurface media, and tribute has been paid to Henri Darcy by using the **darcy** as the unit of intrinsic permeability (1 darcy = 9.87×10^{-13} m^2).

Aquifer thermal energy storage

Figure 3.22 shows the principle of **aquifer thermal energy storage**.

The left part of Figure 3.22 shows the summer situation, where cold groundwater is pumped out of an aquifer to the left. The cold water is used for cooling a building, using its air conditioning system via a heat exchanger. Note that the groundwater that has warmed up during the heat exchange process is returned to the aquifer.

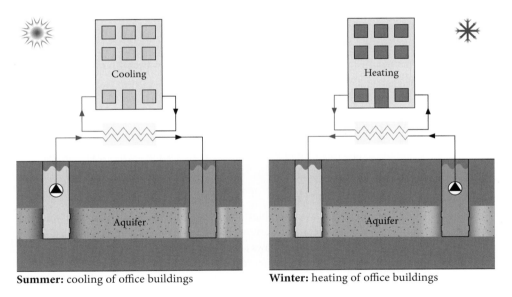

Summer: cooling of office buildings   **Winter:** heating of office buildings

Figure 3.22. Aquifer thermal energy storage (reproduced with kind permission from IF Technology B.V., Arnhem, The Netherlands)

In winter, the operation is reversed. The right part of Figure 3.22 shows the warm water, which infiltrated during the summer, now being pumped out. The warm water is used for heating the building, using its central heating system via a heat exchanger. Note that the groundwater that has cooled during the heat exchange process is again returned to the aquifer, now for later use in the summer, and so on.

Groundwater holds its temperature quite well (the thermal conductivity of aquifers is low) and groundwater flow is slow (as we just calculated), causing aquifer thermal energy storage to be an effective technique for both the cooling and heating of buildings, replacing part of the electricity demand in doing so and thus also reducing possible burning of fossil fuels and CO_2 emissions into the atmosphere.

In the 1980s, the first aquifer thermal energy storage projects started in Europe and the United States. Since then, aquifer thermal energy storage has become a much used and cost-efficient technique, with investment costs being paid back in a relatively short time span.

The hydraulic conductivity K as used in Darcy's law strongly depends on the groundwater's temperature. Therefore, calculations of the volume flux Q and the effective velocity v_e of groundwater in aquifers used for thermal energy storage should always relate to the aquifer's intrinsic permeability κ, which is independent of temperature. This is further made evident in Exercise 3.7.3, where the hydraulic conductivities of cold and warm groundwater are compared for the same aquifer.

> **Exercise 3.7.3** The intrinsic permeability κ of a geological formation equals 4.935×10^{-13} m^2 (= 0.5 darcy).
> At 5°C: $\rho_5 = 999.97$ kg m^{-3} and $\mu_5 = 1.519 \times 10^{-3}$ kg m^{-1} s^{-1}.
> At 60°C: $\rho_{60} = 983.20$ kg m^{-3} and $\mu_{60} = 0.467 \times 10^{-3}$ kg m^{-1} s^{-1}.
>
> Determine the hydraulic conductivity K in m day^{-1} for water with temperatures of 5 and 60°C: compare the answers.

Darcy and Ohm

By combining Equations 3.8 and 3.10, Darcy's law for horizontal groundwater flow can be written as follows:

$$q = -K \frac{\Delta h}{\Delta x} \tag{3.14}$$

The shape of Darcy's law resembles other laws in science, such as **Ohm's law** for the conduction of an electrical current, named after the German physicist Georg Simon Ohm (1787–1854). Ohm's laws states that the electrical current I in a continuous electrical circuit as shown in Figure 3.23 is directly proportional to the potential difference V that drives the electrical current through the circuit and inversely proportional to the resistance R of the conducting circuit. Ohm's law can be presented in the following form:

I = current (ampere) V = potential difference (volt)

R = resistance (ohm)

$$I = \frac{V}{R} \tag{3.15}$$

All variables in the above equation have absolute values (greater than or equal to zero). Comparing Equations 3.14 and 3.15, it is clear that the electrical current I is equivalent to the absolute value of the volume flux density or specific discharge q (m day^{-1}), and that the potential difference V is equivalent to the absolute value of the difference in hydraulic head Δh (m). Finally, the resistance R is equivalent to the absolute value of the reciprocal of the hydraulic conductivity K (m day^{-1}) per unit distance Δx (m). In hydrology, such a resistance (Δx over K) would thus have time (day) as unit.

Other laws resembling Darcy's law are Fourier's law for conductive heat flow and Fick's first law of diffusion. Without going into detail, one should note that Fick's, Fourier's, and Ohm's laws all exist in a material continuum, on both a microscopic and macroscopic scale level. Darcy's law, importantly, is only valid on a macroscopic scale level, and should preferably be used at the scale of the representative elementary volume (see earlier).

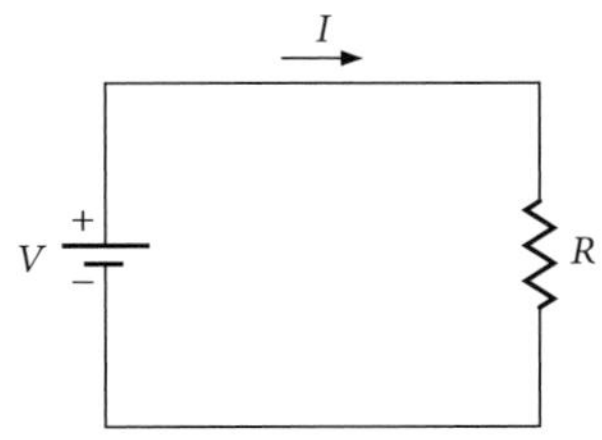

Figure 3.23. A continuous electrical circuit with voltage source V, electrical current I, and resistor R

Laminar flow

Darcy's law is limited to **laminar flow** situations; for example, situations where water can macroscopically be regarded to flow in parallel layers, with no disturbance between the layers. Thus Darcy's law works well for groundwater flow through media that resist water flow, such as sediments. The hydraulic conductivity is then a function of the sediment's particle size, shape, and sorting. The larger the grains in a well-sorted sandy sediment, a sediment having one predominant grain size, the larger the porosity and the hydraulic conductivity tend to be. The rougher the shape of the grains, the smaller the hydraulic conductivity tends to be. Poorly sorted sediments in general have lower hydraulic conductivities than well-sorted sediments, as small sediment particles may clog the pore space between the larger particles in poorly sorted sediments. Also, some 5–7% sludge in groundwater already significantly reduces hydraulic conductivity.

The opposite of laminar flow is **turbulent flow**, which is characterized by disturbances, eddies, and a chaotic flow pattern. Darcy's law cannot be applied without adaptation in, for instance, karstic limestone (Box 3.2), as fissures and fractures (**secondary porosity**), widened by the limestone's high solubility, impose little resistance upon groundwater flow.

BOX 3.2 Karst hydrology

Karst comprises terrain with distinctive hydrology and landforms that arise from a combination of high rock solubility and well-developed secondary (fracture) porosity (Ford and Williams 2007). The rock that dissolves is often limestone, which holds calcite (calcium carbonate; $CaCO_3$) as predominant mineral, but a number of other rocks/minerals such as dolomite or gypsum are also prone to dissolution. The distinctive hydrology of karst is rapid subsurface drainage, and the distinctive landscape features are sinkholes, fissures, caverns, and karstic springs, springs at the end of a water-filled cave system such as in Fontaine de Vaucluse, in the Vaucluse department of France.

Due to respiration of subsurface fauna, the carbon dioxide (CO_2) pressure of air below the land surface is higher than the CO_2 pressure of air above the land surface. When rain water (H_2O) infiltrates a limestone plateau through fissures and fractures at the surface, the calcite (calcium carbonate), carbon dioxide, and water react to form calcium ions (Ca^{2+}) and hydrogencarbonate (HCO_3^-), thus dissolving the limestone:

$$CaCO_3 + CO_2 + H_2O \rightleftharpoons Ca^{2+} + 2HCO_3^-$$

When, further downstream, this water, saturated with Ca^{2+} and HCO_3^-, surfaces in a spring, the CO_2 pressure may be reduced as plants and algae extract CO_2 from the water. Also, when air temperatures are high, part of the water will evaporate. These processes reverse the above reaction and may cause a local, secondary deposition of $CaCO_3$ at the spring outlet, made up of a hard and porous $CaCO_3$ rock/mineral known as **travertine**.

Groundwater flow through sediments is slow (Exercise 3.7.2) and can be considered as laminar, with Darcy's law applying. Groundwater flow in sediments and **karstic water flow** in limestone are both subsurface water flows. However, as karstic water flow is rapid and turbulent, karstic water flow carries all the traits of surface water flow (which will be dealt with later, in Chapter 5) and Darcy's law cannot readily be applied. One can best envisage karstic water flow as a subsurface river running through subsurface fissures, caverns, and shafts with, overall, the water level dropping in the direction of flow. Locally, however, the direction of water flow may be upwards in rising caverns or vertical subsurface shafts.

In the terrain and on a 1:25 000 topographic map, karst can be deduced by its distinctive landscape features mentioned above, but often also by the lack of a surface water drainage network over a vast area. As an example, the karstic spring of **Fontaine de Vaucluse** (average discharge 20 m^3 s^{-1}) obtains almost all its water from precipitation over an area of some 1130 km^2 that is devoid of a surface water drainage network (Emblanch *et al.* 2003).

Exercise B3.2 Determine the average yearly **precipitation surplus** (precipitation depth minus evaporation depth) in mm year^{-1} for the area upstream of Fontaine de Vaucluse.

Measuring hydraulic conductivity

The (saturated) hydraulic conductivity is an important property of the subsoil: together with the hydraulic gradient, it determines the volume flux density or specific discharge in the subsoil, as is evident from Equation 3.10. There are many techniques to estimate the hydraulic conductivity of the subsoil. The set of techniques varies from field techniques such as borehole tests, pumping tests, and tracer tests to laboratory tests involving core samples taken from the subsoil. Alternatively, the hydraulic conductivity may be indirectly established; for instance, from the grain size distribution of the subsoil. All these techniques have their advantages and disadvantages. In practice, regardless of the selected technique, it is hard to obtain a precise estimate of the hydraulic conductivity.

In this section, we will restrict ourselves to techniques using constant-head permeameters. With these techniques, a constant hydraulic head is maintained above a water-saturated core sample of the subsoil. This explains the term **constant-head permeameter**, **constant head** being an abbreviation of 'constant hydraulic head'. The hydraulic head below the core sample differs from the hydraulic head above the core sample, but is also constant. Due to this, a constant hydraulic gradient through the subsoil sample is established.

Figure 3.24 shows the principle of a constant-head permeameter used in the laboratory. As soon as the volume flux Q attains a constant value, the hydraulic conductivity

K of the subsoil sample can be determined by application of Darcy's law (Equation 3.9) as follows:

$$\boxed{\Delta h = h_2 - h_1 < 0 \text{ m}}$$

$$Q = -K A \frac{\Delta h}{L} \qquad \boxed{\begin{array}{l} L = \text{length of the sub-} \\ \text{soil core sample } (L > 0 \text{ m}) \end{array}}$$

(3.16)

Note that the water flow through the subsoil sample is upward (Q has a positive sign).

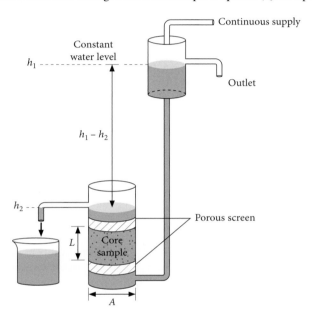

Figure 3.24. A constant-head permeameter for determining the hydraulic conductivity of a water-saturated soil core sample from the subsoil

Another measuring technique also involving the use of one or more core samples of the subsoil, but usually performed in the field or during fieldwork, is the **Kopecki field method**. A bottle is filled with water and closed by a cork with two open tubes. The bottle is then turned upside down and water is fed on to the subsoil sample as shown in Figure 3.25. The short tube lets the air in, while the larger tube lets the water out. When the water above the subsoil core sample reaches the outlet level of the larger tube, the water flow from the bottle is temporarily halted. In this way, a constant water level above the subsoil core sample is created, with water infiltrating from this water layer into the subsoil core sample. After the subsoil core sample is water saturated and as soon as the volume flux Q attains a constant value, the hydraulic conductivity K of the subsoil sample can be determined by application of Darcy's law (Equation 3.9). Note that the water flow through the subsoil sample is downward and that Q thus has to be assigned a negative sign. As outflow from the subsoil core sample is under existing air pressure, the pressure head at the water-despatching end of the subsoil sample equals 0 m. If we take the bottom of the subsoil core sample as the reference level, the hydraulic head at

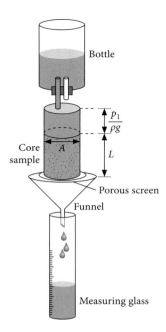

Figure 3.25. The Kopecki field method for determining the hydraulic conductivity of a water-saturated core sample from the subsoil

the water-despatching bottom end of the subsoil sample thus equals an elevation head of 0 m plus a pressure head of 0 m, totalling 0 m. On the water-receiving upper side, the hydraulic head (m) equals an elevation head of L plus a pressure head of $p_1/(\rho g)$ (both L and $p_1/(\rho g)$ being positive numbers), totalling $L + p_1/(\rho g)$. The difference in hydraulic head (m) thus equals $0 - (L + p_1/(\rho g)) = -(L + p_1/(\rho g))$.

The hydraulic gradient equals the difference in hydraulic head divided by the distance $0 - L = -L$ over which the difference in hydraulic head occurs; thus $-(L + p_1/(\rho g))/(-L) = (L + p_1/(\rho g))/L$. Application of Darcy's law (Equation 3.9) then gives:

$$Q = -KA \frac{L + \dfrac{p_1}{\rho g}}{L} (Q < 0)$$

> The pressure p_1 at the top of the subsoil core sample divided by both the water density ρ and the acceleration due to gravity g = water level above the subsoil core sample (m)

(3.17)

Exercise 3.7.4 In Figures 3.24 and 3.25, two different types of constant-head permeameters are presented: $Q = 5 \times 10^{-5}$ m³ day⁻¹, $A = 2 \times 10^{-3}$ m², $L = 5$ cm, $\Delta h = -2$ cm, and $p_1/(\rho g) = 2$ cm.
Derive equations for Q as a function of the saturated hydraulic conductivity K, L, A, and Δh or $p_1/(\rho g)$ for both permeameters, and determine the saturated hydraulic conductivity K of both soil core samples.

Exercise 3.7.5 A vertical soil core sample is taken in the vicinity of Larochette, Luxembourg. The length of the core is 5 cm. The surface area of the sample is 20 cm². The saturated hydraulic conductivity is determined using the Kopecki field method. The water level above the water-saturated soil core sample is kept at a constant height of 2 cm.
The water that flows through the soil sample is collected and has a constant discharge or volume flux Q of 140 ml in 15 minutes.

a. Draw the change in hydraulic head h for both the soil and water above it.

b. Determine the hydraulic head h at the bottom of the soil core sample.

c. Determine the hydraulic head h at the top of the soil core sample.

d. Determine the hydraulic gradient i in the soil core sample.

e. Determine the saturated hydraulic conductivity K of the soil sample.

f. Is the sample taken from a clayey weathered soil of the Keuper marls or from a sandy soil of the Luxembourg sandstone formation?

Accuracy and scale of measurement

The accuracy with which we determine the hydraulic conductivity of a (sub)soil core sample is first of all determined by the accuracy with which the core sample itself is collected. In principle, the (sub)soil core sample should be fully undisturbed, but in practice this can never be fully achieved. Collecting a good sample is an art in itself, which can best be explained in the field. In any case, determining the hydraulic conductivity from a badly collected (sub)soil sample is useless.

One should also keep in mind that the cores used for sampling are usually quite small; the cylinder-shaped cores often have a diameter of 5 cm and a length of 5 cm, establishing a volume of 100 cm³. Even if the sample is collected with great accuracy, the determined hydraulic conductivity only holds for this relatively small sample of the (sub)soil. Also, the sample volume may well be less than the representative elementary volume for the hydraulic conductivity. The collecting of samples should always be accompanied by extensive field observations stating the conditions of the (sub)soil, especially with respect to the occurrence of cracks (number of cracks; depth of the cracks), as the larger or deeper cracks cannot be sampled (in cylinder-shaped cores). One should take a sufficient number of samples to determine the distribution of the hydraulic conductivity values and from this the best statistical average and standard deviation.

Because of these problems of scale, hydraulic conductivities obtained using different methods cannot readily be compared and, depending on the problem, it is wise to stick to one method for which there is at least some confidence that the observed differences in hydraulic conductivity do carry some weight. For instance, hydraulic conductivity values obtained using the Kopecki field method (Figure 3.25) have been observed to be much smaller than when using the constant-head permeameter of Figure 3.24. Because good measures of hydraulic conductivity in the subsoil are difficult to establish, hydraulic conductivities are often calibrated in hydrological models of the subsoil involving data on input (precipitation), output (discharge), and hydraulic heads as observed from piezometers in the field.

3.8 Refracting the water

From experimental physics, we know that light is refracted at the boundary between two layers with different densities. The relationship describing this is known as **Snell's law of refraction**, after the Dutch astronomer and mathematician Willebrord Snellius (born Willebrord Snel van Royen; 1580–1626). As a simple illustration, if you try to catch a fish in shallow water, you will soon find out that the fish's underwater position is slightly off from where you see it. This is due to the refraction of light at the air/water boundary.

Groundwater flow can experimentally be shown to be subject to refraction. Similar to a light ray, we have a **streamline** (as defined in section 3.3) in hydrology. Figure 3.26 shows two streamlines when a steady groundwater flow passes from a clay layer with a low hydraulic conductivity K_1 into a sandy layer with a high hydraulic conductivity K_2. Both layers are horizontal, homogeneous, and isotropic. Darcy's law for the upper clay layer shown in Figure 3.26 states that:

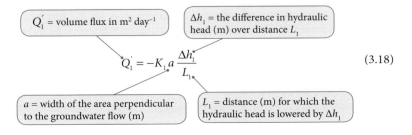

Q_1' = volume flux in m² day⁻¹

Δh_1 = the difference in hydraulic head (m) over distance L_1

$$Q_1' = -K_1 a \frac{\Delta h_1}{L_1}$$
(3.18)

a = width of the area perpendicular to the groundwater flow (m)

L_1 = distance (m) for which the hydraulic head is lowered by Δh_1

Note that the dimension (y direction) perpendicular to the drawing plane of Figure 3.26 is left out of the equation. The volume flux Q'_1 therefore is expressed in units of m^2 day^{-1}. From now on, the superscript ′ ('prime') will be used to denote that the volume flux is expressed in units of m^2 day^{-1}. Similarly, Darcy's law for the lower sandy layer can be drawn up as follows:

Q'_2 = volume flux in m^2 day^{-1}

Δh_2 = the difference in hydraulic head (m) over distance L_2

$$Q'_2 = -K_2 b \frac{\Delta h_2}{L_2}$$

(3.19)

b = width of the area perpendicular to the groundwater flow (m)

L_2 = distance (m) for which the hydraulic head is lowered by Δh_2

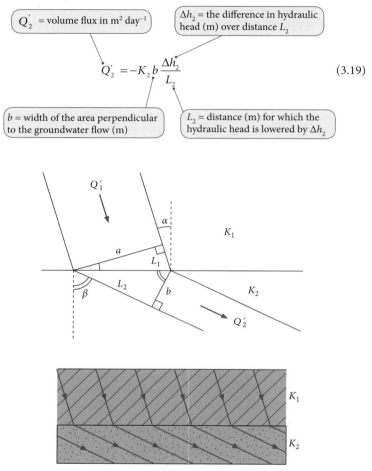

Figure 3.26. Streamline refraction for groundwater flowing from a layer with a low hydraulic conductivity K_1 to a layer with a high hydraulic conductivity K_2

As the line with width a is perpendicular to the steady groundwater flow, the hydraulic head is the same for every position along this line, or, in other words, the line denoted with a is an **equipotential**. The same goes for the line with width b, albeit that the latter hydraulic head, lying downstream, is lower by an increment Δh. This increment or difference in hydraulic head Δh is the same for both the right and left streamlines in Figure 3.26, albeit that for the right streamline Δh is established over a shorter distance L_1 and for the left streamline over a longer distance L_2. Thus, $\Delta h = \Delta h_1 = \Delta h_2$. Also, as the volume of water that leaves the upper clay layer per unit of time must be equal to the volume of water entering the sandy layer within that same unit of time, the water balance equation or **continuity** equation can be drawn up as follows:

$$Q_1' = Q_2' \tag{3.20}$$

Combining this with Darcy's law stated as Equations 3.18 and 3.19 (and substituting Δh for Δh_1 and Δh_2) gives

$$K_1 \frac{a}{L_1} = K_2 \frac{b}{L_2} \tag{3.21}$$

Figure 3.26 shows the angle of incidence of a streamline as α, and the angle of refraction of a streamline as β. By simple geometric reasoning (the sum of all corner angles in a triangle equals $180°$) other positions for α and β alongside the boundary between the two layers can be deduced. These positions are indicated in Figure 3.26 by the use of the same angle symbols.

From Figure 3.26, it can further be deduced that $\tan\alpha = L_1 / a$ and $\tan\beta = L_2 / b$. Combining this with Equation 3.21 gives

$$\frac{K_1}{K_2} = \frac{\tan\alpha}{\tan\beta} \tag{3.22}$$

Table 3.2 shows the range of hydraulic conductivity values for clay and sand. As an example, if we take the hydraulic conductivity of a clay layer, K_1, as 10^{-5} m day^{-1} and the hydraulic conductivity of an underlying sandy layer, K_2, as 10 m day^{-1}, the ratio of K_1 over K_2 equals 10^{-6}. In the situation of Figure 3.26, with groundwater flowing downwards at an angle α of approximately $20°$ and thus $\tan \alpha = 0.363...$, we can calculate from the ratio K_1 over K_2 of 10^{-6} (Equation 3.22) that $\tan \beta = 363\,970$ and thus (using the $\tan^{-1}$ function, or an equivalent function on a scientific calculator) that β is almost $90°$, meaning an almost horizontal groundwater flow in the sandy layer:

$$\frac{K_1}{K_2} = \frac{10^{-5}}{10} = 10^{-6} = \frac{\tan\alpha}{\tan\beta} = \frac{\tan 20°}{\tan\beta} = \frac{0.363...}{\tan\beta} \Rightarrow \tan\beta = \frac{0.363...}{10^{-6}} = 363\,970 \Rightarrow \beta = 90°$$

Also, a value of α of $1°$, with groundwater flow almost vertical, would give a value of β of almost $90°$:

$$\frac{K_1}{K_2} = \frac{10^{-5}}{10} = 10^{-6} = \frac{\tan\alpha}{\tan\beta} = \frac{\tan 1°}{\tan\beta} = \frac{0.017...}{\tan\beta} \Rightarrow \tan\beta = \frac{0.017...}{10^{-6}} = 17\,455 \Rightarrow \beta = 90°$$

If we start the calculation from the other end, with an almost horizontal flow through the sandy layer with, let us say, $\beta = 89.997°$, $\tan \beta = 19\,098$, and with a ratio K_1 over K_2 of 10^{-6} (Equation 3.22), $\tan \alpha = 0.019...$ and α is slightly over $1°$, meaning an almost vertical groundwater flow:

$$\frac{K_1}{K_2} = 10^{-6} = \frac{\tan\alpha}{\tan\beta} = \frac{\tan\alpha}{\tan 89.997°} = \frac{\tan\alpha}{19\,098} \Rightarrow \tan\alpha = 19\,098 \times 10^{-6} = 0.019... \Rightarrow \alpha = 1°$$

A very important consequence of streamline refraction (Equation 3.22) is therefore that in a setting of horizontally layered sediments – for instance, a semi-permeable clay layer (such as Holocene clay) overlying a sandy layer (such as Pleistocene sand) – groundwater flow in the sandy layer is almost horizontal and in the clay layer almost vertical. We can make very good use of this in our groundwater flow models, as we will see in the sections to come.

3.9 Keep it simple and confined

Steady groundwater flow in a confined aquifer

The combination of Darcy's law (the flow equation) with the continuity equation (also called the water balance equation or the mass balance equation) is a standard procedure in hydrology to derive equations that describe the distribution of the aquifer's hydrological heads (the aquifer's potentiometric surface) for specific flow cases.

In the preceding chapter, Darcy's law and continuity were combined to provide insight into the process of streamline refraction at the interface between two layers with differing hydraulic conductivities. Darcy's law and continuity will now be used for another flow case, that of steady confined groundwater flow between two canals as shown in Figure 3.27. First, the solution is obtained by physical reasoning and then the same end result is mathematically derived by the standard procedure of combining the two above-mentioned laws or equations.

In Figure 3.27, groundwater in a homogeneous aquifer is captured between two impermeable layers. The water level in both canals lies above the bottom of the upper impermeable layer, causing the aquifer to be confined (see section 3.4). The canals are incised to the top of the lower impermeable layer; for this, the expression **fully penetrating** is used. Had the canals been incised to a level above the top of the lower impermeable layer, the canals would have been called **partially penetrating**. Because the water level in the left canal is higher than in the right canal, groundwater flows from left to right through the aquifer. The water in the canals is artificially kept at the levels indicated in Figure 3.27. Because of this, groundwater flow is steady or, in other words, stationary (see section 3.3).

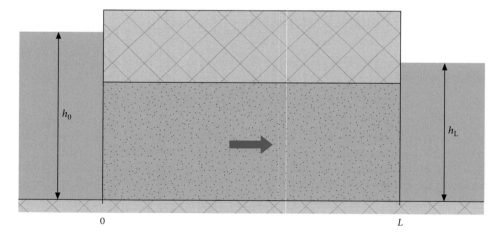

h_0

h_L

0

L

Figure 3.27. Steady groundwater flow in a confined, homogeneous aquifer between two parallel, fully penetrating canals with different water levels

The top of the lower impermeable layer will serve as the reference level. At $x = 0$ metres, the hydraulic head in the aquifer h_0 equals the water level in the left canal, and at $x = L$ metres, the hydraulic head in the aquifer h_L equals the water level in the right canal. In each point of the vertical in $x = 0$, the hydraulic head or potential has the same value h_0. Thus the vertical in $x = 0$ is an equipotential. The same reasoning goes for the vertical in $x = L$: each point of the vertical has the same hydraulic head or potential h_L and the vertical in $x = L$ is also an equipotential, albeit with a lower hydraulic head value. Because the steady flow is from left to right and is confined between two impermeable layers, flow must be horizontal and at right angles to both equipotentials in $x = 0$ and $x = L$. In other words, all verticals between $x = 0$ and $x = L$ are equipotentials, with their hydraulic heads diminishing from left to right.

According to Darcy's law:

$$Q = -KA\frac{h_L - h_0}{L} \tag{3.23}$$

For an increment Δx along a streamline and a corresponding difference in hydraulic head Δh, Equation 3.23 can be rewritten as

$$Q = -KA\frac{\Delta h}{\Delta x} \tag{3.24}$$

$\Delta h / \Delta x$ is the hydraulic gradient expressed as an (average) rate of change of h over an increment Δx. When we take Δx smaller and still smaller again, we can define dh / dx as the limit of $\Delta h / \Delta x$ when Δx approaches a zero value. In mathematical notation,

$$\frac{dh}{dx} = \lim_{\Delta x \to 0} \frac{\Delta h}{\Delta x} \tag{3.25}$$

For any point in a vertical equipotential, we can thus write

$$Q = -KA\frac{dh}{dx} \tag{3.26}$$

As the hydraulic head h varies in one direction only, here the horizontal x direction, and not in a perpendicular y or z direction, we may use the differential notation dh / dx; that is, the notation with the ordinary (flat-backed) differential ds. If the hydraulic head h were to vary in more than one direction, we would have to use partial differential equations and the notations $\partial h / \partial x, \partial h / \partial y$, and $\partial h / \partial z$ for the perpendicular directions x, y, and z, thus using partial differential operators (∂s). Due to the effect of streamline refraction (see the last part of section 3.8) and by making some assumptions, we shall manage to simplify all steady groundwater flow 'problems' tackled in this book to models using one (linear or radial) direction of flow per layer only. Metaphorically speaking, we will use a big sledgehammer to straighten all ∂s to ds, thus keeping the mathematics (in section M at the back of this book) as simple as possible. Hydrology, first of all, is the business and art of understanding how water in the landscape and ground behaves, and, fortunately, for this purpose the use of relatively simple mathematics often suffices.

As A is the saturated area perpendicular to groundwater flow inside the confined aquifer, this area A equals the (vertical z) depth of the aquifer, denoted by D, times the width W, which is the (horizontal y) direction perpendicular to the cross-section of Figure 3.27:

$$Q = -KDW\frac{dh}{dx} \tag{3.27}$$

Q' = volume flux per unit width or discharge per unit width (m² day⁻¹)

$$Q' = \frac{Q}{W} = -KD\frac{dh}{dx} \tag{3.28}$$

'Per unit width' actually means that we dispose of the width of the aquifer, so that we only deal with the horizontal x and vertical z dimensions of the cross-section of Figure 3.27. The volume flux per unit width Q' has units of m² day⁻¹, much the same units as the ones that you use when you paint your garden fence. In hydrological practice, groundwater flow problems are usually reduced to two dimensions in cross-section or plan view, for ease of calculation.

Because the aquifer is homogeneous, the hydraulic conductivity K has the same value for every horizontal position x. The saturated thickness or depth D is also constant for every horizontal position x. Because the water in both canals is artificially kept at the levels indicated in Figure 3.27, the hydraulic gradient ($dh/dx = \Delta h/\Delta x$) is also the same for every horizontal position x. It thus follows from Equation 3.29 that Q' is the same for every horizontal position x and that the hydraulic head linearly decreases with horizontal direction x; mathematically, a linear relation between h and x can be written as follows:

C_1 = constant (slope) C_2 = constant (intercept)

$$h = C_1 x + C_2 \qquad (3.29)$$

The values for the constants C_1 and C_2 can be found by inserting the boundary conditions for this flow case. First insert the $x = 0$ boundary condition and then the $x = L$ boundary condition (h_0, h_L, and L are all known values):

When $x = 0$, then $h = h_0 \Rightarrow h_0 = C_1 \times 0 + C_2$; $h_0 = C_2$; $C_2 = h_0$

When $x = L$, then $h = h_L \Rightarrow h_L = C_1 \times L + C_2 = C_1 L + h_0$; $C_1 = \dfrac{h_L - h_0}{L}$

Substituting the values found for C_1 and C_2 gives

$$h = \frac{h_L - h_0}{L} x + h_0 \qquad (3.30)$$

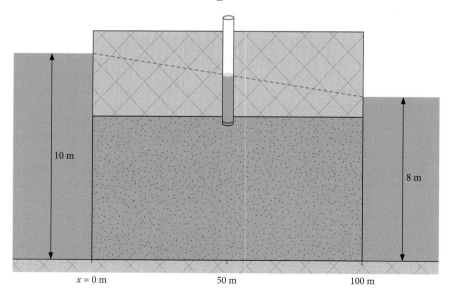

Figure 3.28. Change in hydraulic head in a confined, homogeneous aquifer between two parallel, fully penetrating canals with different water levels

Rewriting Equation 3.29 in the form of Equation 3.30 by inserting the boundary conditions as shown above offers the possibility of finding values for the aquifer's hydraulic conductivity at different distances from the left canal. For instance, if $h_0 = 10$ m, $h_L = 8$ m, and $L = 100$ m, then $h = -0.02x + 10$, and the hydraulic head midway between $x = 0$ and $x = L = 100$ m; thus at $x = 50$ m, it would then be equal to $-0.02 \times 50 + 10 = 9$ m. The above derivation shows the reader how to insert boundary conditions in an equation describing the potentiometric surface (distribution of the hydraulic heads) of an aquifer, here for Equation 3.29. Note that exactly the same method can be used when dealing with equations describing the potentiometric surface for more complicated flow problems. In our present situation, the hydraulic head can also be measured by inserting a piezometer through the upper impermeable layer into the confined aquifer, as shown in Figure 3.28.

Equation 3.28 can also be found by mathematically combining Darcy's law (Equation 3.28) and the continuity equation ($Q' =$ constant), as shown under M1 at the end of this book. An equation derived from combining Darcy's law (the flow equation) with the continuity equation for steady groundwater flow is called a **Laplacian,** after the French astronomer and mathematician Pierre-Simon (marquis de) Laplace (1749–1827).

Exercise 3.9 Thorough field research has unravelled the position of a subsurface sand layer in a Dutch polder. An impermeable dam is positioned on top of this layer, as shown in Figure E3.9. The sand layer is a homogeneous, confined aquifer. Groundwater flow is stationary. The saturated hydraulic conductivity K of the sand equals 5 m day^{-1}. The effective porosity n_e equals $\frac{1}{3}$.

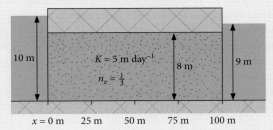

Figure E3.9

a. Determine the hydraulic head at $x = 25$, 50, and 75 m.

b. Determine the volume flux Q' in m^2 day^{-1}.

c. Determine the effective velocity v_e in m day^{-1}.

d. Assuming conservative transport how long does it take for a pollutant at $x = 0$ m (the left canal) to reach $x = 100$ m (the right canal)?

Pressure in a confined aquifer

As the water in a confined aquifer is captured between impermeable layers and because the level of the hydraulic head lies above the top of the aquifer as shown in Figure 3.28, water is stored under pressure in the pore space of the aquifer. **What components deal with pressure in a confined aquifer and how do they function?**

The **total pressure** or total stress σ_t (N m^{-2}) at the interface of the upper impermeable layer and the confined aquifer is established by the weight of the overlying layer. This weight causes the grains in the aquifer to be squeezed together, causing a pressure between the grains – an **intergranular pressure** or effective stress σ_i (N m^{-2}). Part of the overlying weight is carried by the water in the pores. This causes a (pore) **water pressure** or neutral stress p (N m^{-2}). When more water is contained under pressure in the pores, more of the weight is carried by the water.

In an equilibrium situation, the total pressure σ_t equals the sum of the intergranular pressure σ_i and water pressure p (Figure 3.29). When, under such circumstances, water is pumped out of the aquifer, the water pressure p and the level of the hydraulic head h decrease and the intergranular pressure σ_i increases, causing **compaction** of the aquifer, thereby (slightly) reducing porosity. Because groundwater can be stored in confined aquifers under high pressures, confined aquifers seem to provide a seemingly continuous volume of water to be pumped out. Conversely, when water is recharged in the aquifer, the water pressure p and the level of the hydraulic head h increase and the intergranular pressure σ_i decreases, causing **expansion** of the aquifer, thereby (slightly) increasing porosity. When, after first pumping out water and later recharging the same volume of water, the aquifer returns to its former shape with the same level(s) of hydraulic head, the confined aquifer is said to be perfectly **elastic**. In practice, however, not all confined aquifers are elastic, especially those containing a relatively high percentage of clay, and continuous withdrawal of water from such aquifers may cause permanent compaction and **subsidence** of the land surface. Because subsidence is usually a slow process and does not occur simultaneously with the lowering of the hydraulic head, early hydrologists in the 1930s failed to understand how seemingly endless volumes of water could be captured from confined aquifers. On the other hand, if large volumes of water are contained in a confined aquifer, the water pressure may become too high ($p \geq \sigma_t$), causing the aquifer to burst open and to change to **quicksand**. Also, earthquakes may act as a trigger to suddenly increase groundwater pressure, forcing groundwater upwards through (newly formed) joints and fissures, and causing earth materials to be deposited at the surface, a process called **liquefaction**.

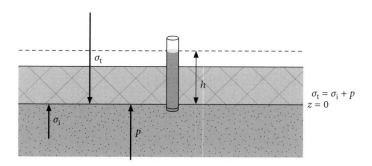

Figure 3.29. The total pressure σ_t, intergranular pressure σ_i, and (pore) water pressure p (N m^{-2}) at the interface of an upper impermeable layer and confined aquifer under equilibrium circumstances

3.10 Continuity and its consequences

Horizontal groundwater flow

Equation 3.23 (Darcy's law) for the flow case of Figure 3.27 can be rewritten for the volume flux per unit width of the aquifer in a number of ways. For instance:

T = transmissivity (m² day⁻¹) i = hydraulic gradient (−)

$$Q' = -KD\frac{h_L - h_0}{L - 0} = -T\frac{\Delta h}{L} = -Ti \qquad (3.31)$$

The **transmissivity** T is a measure of how easily an aquifer transmits water and has the same units as Q' (m² day⁻¹). For a homogeneous, confined aquifer, the transmissivity T equals the product of the hydraulic conductivity K and the saturated depth D of the aquifer:

$$T = K \times D \qquad (3.32)$$

Figure 3.30 shows the flow case of a confined aquifer with m horizontal layers, each one homogeneous, between two parallel, fully penetrating canals: m is taken as 4 in Figure 3.30. Groundwater flow is stationary and the boundary conditions are fixed. The latter means that the hydraulic gradient i is the same for each layer. Continuity (the water balance) teaches us that the volume flux or discharge through the confined aquifer equals the sum of the volume fluxes of each layer. When we express the volume fluxes per unit width:

$$Q' = Q'_1 + Q'_2 + ... + Q'_m \qquad (3.33)$$

When we insert Darcy's law in the last form of Equation 3.31, we obtain

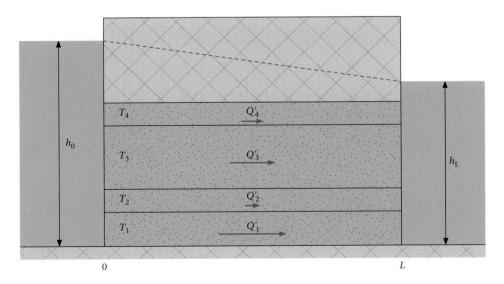

Figure 3.30. Steady groundwater flow through horizontal, homogeneous layers in a confined aquifer between two parallel, fully penetrating canals with different water levels

Exercise 3.10.1 Rewrite Darcy's law for the whole aquifer (Equation 3.31) in terms of **horizontal flow resistances** R as follows:

$$Q' = -T\frac{\Delta h}{L} = -\frac{\Delta h}{\left(\dfrac{L}{T}\right)} = -\frac{\Delta h}{R}$$

(E3.10.1.1)

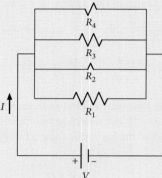

Figure E3.10.1. Parallel connected resistors in an electrical circuit

For layer 1, this gives

$$Q_1' = -T_1\frac{\Delta h}{L} = -\frac{\Delta h}{\left(\dfrac{L}{T_1}\right)} = -\frac{\Delta h}{R_1}$$

(E3.10.1.2)

The horizontal flow resistances

$$R\left(=\frac{L}{T}\ \frac{m}{m^2 day^{-1}}\right), R_1, R_2, ..., R_m,$$

have units of day m^{-1}.

Now use the continuity equation (Equation 3.33) to show that:

$$\frac{1}{R} = \frac{1}{R_1} + \frac{1}{R_2} + ... + \frac{1}{R_m}$$

(E3.10.1.3)

This relation also holds for parallel connected resistors R_1 to R_m in an electrical circuit, as shown for $m = 4$ in Figure E3.10.1. Horizontal groundwater flow through horizontal layers in a confined aquifer and an electrical current through parallel connected resistors are thus each other's analogue.

$$-Ti = -T_1 i - T_2 i - ... - T_m i$$

(3.34)

Dividing by the constant value of $-i$ gives the following end result:

$$T = T_1 + T_2 + ... + T_m$$

(3.35)

Thus, in a confined aquifer consisting of horizontal layers, the **total transmissivity** T is the sum of the transmissivities of each layer. This relation is an analogue of parallel connected resistors in an electrical circuit, and is useful when calculating the total volume flux through horizontally layered aquifers.

Consequently, we may derive the **substitute hydraulic conductivity** K of the confined aquifer from Equation 3.35 as follows:

$$K = \frac{K_1 D_1 + K_2 D_2 + ... + K_m D_m}{D}$$ (3.36)

D = total saturated aquifer depth (m) = $D_1 + D_2 + ... + D_m$

When calculating the total volume flux through horizontally layered aquifers, good use can be made of the substitute hydraulic conductivity K, as the total volume flux per unit width Q' (m² day⁻¹) is equal to the product of the substitute hydraulic conductivity K (m day⁻¹), the total saturated aquifer depth D (m), and the hydraulic gradient i (Darcy's law).

Exercise 3.10.2 Answer the following questions with regard to the cross-section and data presented in Figure E3.10.2.

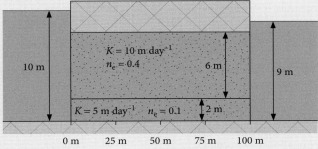

Figure E3.10.2

a. Determine the hydraulic head at x = 25, 50, and 75 m.

b. Determine the volume flux Q' in m² day⁻¹.

c. Determine the volume flux densities q in m day⁻¹.

d. Determine the effective velocities v_e in m day⁻¹.

e. Assuming conservative transport how long does it take for a pollutant at x = 0 m (the left canal) to reach x = 100 m (the right canal)?

Vertical groundwater flow

The end result of section 3.8 has taught us that the groundwater flow entering a semi-permeable horizontal layer is refracted to the vertical. Figure 3.31 shows steady vertical upward groundwater flow (seepage) through an aquitard consisting of horizontal, homogeneous semi-permeable layers (layers with a low hydraulic conductivity). To quickly distinguish this from horizontal flow (cases), the vertical (saturated) hydraulic conductivity (K_z) of the semi-permeable layers is presented as lower-case k and the saturated depth or thickness of these layers as lower-case d (instead of capital D).

As the water flows steadily upwards in the direction of the lower hydraulic head in Figure 3.31, the volume or mass of water per unit of time that is transported remains the same, as it is neither created nor destroyed on its way. Thus continuity (the water

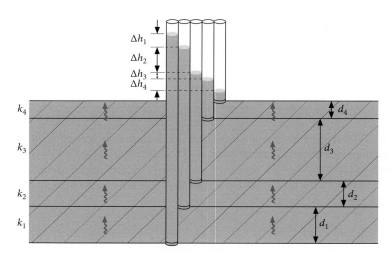

Figure 3.31. Steady seepage through an aquitard consisting of horizontal, homogeneous semi-permeable layers

balance; conservation of mass) teaches us that the vertical volume flux or discharge is the same for every homogeneous semi-permeable layer:

$$Q = Q_1 = Q_2 = ... = Q_m = \text{constant} \qquad (3.37)$$

In the example of Figure 3.31, m equals 4. For a constant area A perpendicular to this water flow, Equation 3.37 becomes

$$\frac{Q}{A} = \frac{Q_1}{A} = \frac{Q_2}{A} = ... = \frac{Q_m}{A} = \text{constant} \qquad (3.38)$$

and thus

$$q = q_1 = q_2 = ... = q_m = \text{constant} \qquad (3.39)$$

The total difference in hydraulic head over all homogeneous semi-permeable layers equals the sum of the hydraulic heads over each homogeneous semi-permeable layer:

$$\Delta h = \Delta h_1 + \Delta h_2 + ... + \Delta h_m \qquad (3.40)$$

Dividing by the constant value of $q = q_1 = q_2 = ... = q_m$ (Equation 3.39) gives

$$\frac{\Delta h}{q} = \frac{\Delta h_1}{q_1} + \frac{\Delta h_2}{q_2} + ... + \frac{\Delta h_m}{q_m} \qquad (3.41)$$

Darcy's law in combination with Equation 3.39 teaches us that

$$q = -k\frac{\Delta h}{d} = q_1 = -k_1\frac{\Delta h_1}{d_1} = q_2 = -k_2\frac{\Delta h_2}{d_2} = ... = q_m = -k_m\frac{\Delta h_m}{d_m} \quad (3.42)$$

Substituting Equation 3.42 in Equation 3.41 gives

$$\frac{d}{k} = \frac{d_1}{k_1} + \frac{d_2}{k_2} + ... + \frac{d_m}{k_m} \qquad (3.43)$$

d / k, d_1 / k_1, d_2 / k_2, ..., d_m / k_m have units of (m m^{-1} day =) days and represent the resistances to vertical groundwater flow through all layers: layers 1, 2, ... , m. The **hydraulic resistance** or **vertical flow resistance** c is defined as follows:

$$c = \frac{d}{k} \qquad (3.44)$$

As an end result, Equation 3.43 can therefore be written as

$$c = c_1 + c_2 + ... + c_m \qquad (3.45)$$

In words: the **total hydraulic resistance** c of a number of horizontal, homogeneous semi-permeable layers equals the sum of the hydraulic resistances of each layer. This relation is useful when calculating the vertical volume flux density through horizontally layered semi-permeable layers.

This is analogous to an electric current running through resistors connected in series, where the total resistance R of the electrical circuit is known to equal the sum of the resistances R_1, R_2, ... , R_m, as shown in Figure 3.32 for $m = 4$. Note that the vertical flow resistance and horizontal flow resistance used in Exercise 3.10.1 have different units.

From Equation 3.43, the **substitute hydraulic conductivity** k of the semi-permeable layers can be determined as follows:

$$d = \text{total depth (m)} = d_1 + d_2 + ... + d_m \qquad k = \frac{d}{\left(\dfrac{d_1}{k_1} + \dfrac{d_2}{k_2} + ... + \dfrac{d_m}{k_m}\right)} \qquad (3.46)$$

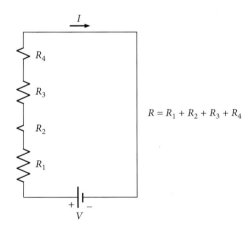

$$R = R_1 + R_2 + R_3 + R_4$$

Figure 3.32. Resistors connected in series in an electrical circuit

When calculating the volume flux through horizontally layered semi-permeable layers for steady flow, good use can be made of the substitute hydraulic conductivity k, as the vertical volume flux density q (m day^{-1}) is equal to the substitute hydraulic conductivity k (m day^{-1}) times the hydraulic gradient $i = \Delta h / d$ (Darcy's law).

What do we need to know?

The above derivations provided an insight into applying continuity and Darcy's law. Figure 3.33 now summarizes the consequences for two steady groundwater flow situations.

For steady horizontal groundwater flow, the volume fluxes per unit width Q' of horizontal, homogeneous layers may be added up (Equation 3.33); as a consequence (the hydraulic gradient i is constant for all layers), the transmissivities T may also be added up (Equation 3.35).

For steady vertical groundwater flow through horizontal, homogeneous semi-permeable layers the volume flux densities q of each layer are equal (Equation 3.39); as a consequence the hydraulic resistances c of these layers may be added up (Equation 3.45).

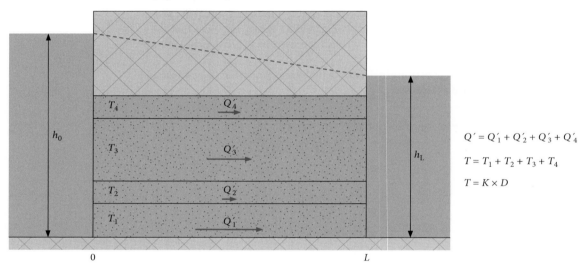

$$Q' = Q'_1 + Q'_2 + Q'_3 + Q'_4$$

$$T = T_1 + T_2 + T_3 + T_4$$

$$T = K \times D$$

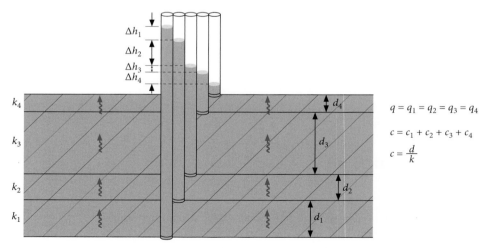

$$q = q_1 = q_2 = q_3 = q_4$$

$$c = c_1 + c_2 + c_3 + c_4$$

$$c = \frac{d}{k}$$

Figure 3.33. A summary of the consequences of continuity for two steady groundwater flow situations

Note that the hydraulic resistance c (days) is essentially different from the travel time of water particles (also in days). For the latter, of course, we need to know the effective porosity in order to first determine the effective velocity v_e of the groundwater (section 3.7; Equation 3.11). This is also evident from Exercises 3.10.4 and 3.10.5.

Exercise 3.10.3 A sandy formation consists of four layers of equal thickness. Values for the hydraulic conductivity of these layers are 1, 5, 10, and 50 m day^{-1}. Determine the substitute horizontal and vertical hydraulic conductivity of the formation.

Exercise 3.10.4 A saturated, horizontal, leaky confining (semi-permeable) layer with a thickness d of 8 m has an effective porosity n_e of $\frac{1}{3}$; the hydraulic resistance to vertical groundwater flow $c = 900$ days; the difference in hydraulic head Δh over the layer equals −0.3 m.
a. Determine the effective velocity and the residence time for vertical steady flow through the layer.
b. By digging a canal, the thickness of the leaky confining layer is halved. The difference in hydraulic head over the layer remains the same. Again, determine the effective velocity and the residence time for vertical steady flow through the layer.
c. What conclusion can be drawn when comparing the above answers?

Exercise 3.10.5 Figure E3.10.5 shows a 5 m-deep lake in a polder that receives seepage water from below. The confining layer below the lake is made up of two clay layers: the lower layer 1 is 2.5 m thick, and has a hydraulic conductivity of 5×10^{-3} m day^{-1} and an effective porosity of 0.1; the upper layer 2 is 10 m thick, and has a hydraulic conductivity of 10^{-2} m day^{-1} and an effective porosity of 0.2. Taking the bottom of the lower clay layer as the reference level, the hydraulic head in a piezometer with a short screen just below the lower clay layer is 18.1 m = 0.6 m above the free water surface of the lake.

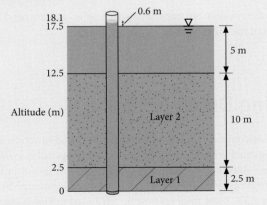

Figure E3.10.5

a. Determine the hydraulic resistance to vertical groundwater flow for the 12.5 m-thick confining layer.

b. Determine the hydraulic head at the interface between the two clay layers.

c. Determine the effective velocity and the residence time for vertical steady flow through the 12.5 m-thick confining layer.

Exercise 3.10.6 can be solved by applying the exact same principle of continuity and its consequences that were just derived for steady vertical groundwater flow through a series of homogeneous semi-permeable layers, but now for steady horizontal groundwater flow through a series of homogeneous aquifers. In effect, it is the same problem, but rotated through 90°.

Exercise 3.10.6 In a confined aquifer, steady flow conditions apply. The aquifer consists of three parts with different hydraulic conductivities. For further data, see the presented cross-section.

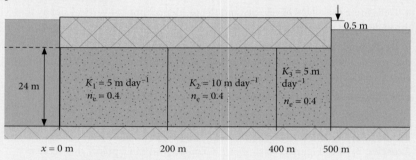

Figure E3.10.6
a. Determine the volume flux Q' of the aquifer in $m^2\ day^{-1}$.

b. Assuming conservative transport, how long does it take for a pollutant at $x = 0$ m (the left canal) to reach $x = 500$ m (the right canal)?

c. Determine the hydraulic gradient for the left, middle, and right parts of the aquifer.

d. Determine the relationship between the hydraulic gradients and hydraulic conductivities of the three parts in Figure E3.10.6.

3.11 Going Dutch

In an isotropic, unconfined aquifer, precipitation-induced groundwater flow to a partially penetrating canal (a canal not incised to the top of a possibly lower-lying impermeable layer) occurs along curved pathways (streamlines) as shown in Figure 3.34. Also, a partially penetrating canal is in part recharged by groundwater from below, also as shown in Figure 3.34. These findings may be attributed to Dutch engineer Johan M.K. Pennink (1853–1936), director of the Amsterdam Water Works at the beginning of the twentieth century, who monitored the hydraulic heads at different depths using a range of piezometers positioned perpendicular to drainage channels in the coastal dunes of

The Netherlands (De Vries 1982). For steady flow in an isotropic medium, streamlines are perpendicular to equipotentials (section 3.9) and the latter are shown as dashed lines in Figure 3.34. The water pressure at the position of the screen of a piezometer, expressed in metres water pressure (the pressure head), equals the length in metres of the column of water contained in that piezometer. Figure 3.34 thus also shows that groundwater flow is from low to high pressure (from a high to a low elevation) in the upper, recharging part of a hill slope (see also Figure 3.18), and from a low to a high elevation (from high to low pressure) in the lower, discharging part of a hill slope (see also Figure 3.19).

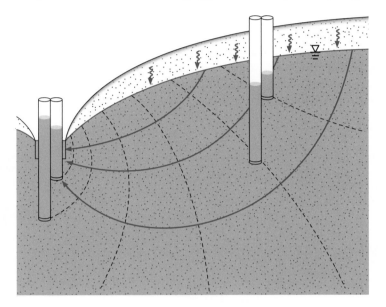

Figure 3.34. Unconfined groundwater flow through a hill slope to a partially penetrating canal; part of a local groundwater flow system

With a semi-permeable layer in the subsoil, streamlines between a recharge and seepage area are also curved. Figure 3.35 shows recharged, vertical groundwater flow through a horizontal semi-permeable layer being refracted in a horizontal direction after entering the lower aquifer, and being refracted back to the vertical in its seepage trajectory after re-entering the semi-permeable layer.

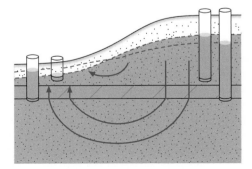

Figure 3.35. Potentiometric surfaces in cross-section and streamlines connecting a recharge and seepage area

This takes us to the situation in the western part of The Netherlands, where low-lying polders (section 3.1), made up of horizontal, leaky confining (semi-permeable) deltaic clay and peat layers of Holocene age, lie on top of a horizontal, sandy aquifer (or aquifers) of older, Pleistocene age. As shown in Figure 3.1, surface water is regularly pumped out of these polders to ensure that the connected groundwater levels in the polders remain at a near-constant level below the land surface. The deeper, confined groundwater in the Pleistocene aquifer is replenished (recharged) underground from higher locations such as the Pleistocene ice-pushed ridges in The Netherlands (e.g. Utrechtse Heuvelrug) and/or from dune areas in much the same manner as shown in Figure 3.9, and/or from lakes and rivers as shown in Figure 3.10. The groundwater in the Pleistocene aquifer is confined and as a consequence its potentiometric surface may be higher than the artificially maintained groundwater level of a polder, especially if that polder lies deep. This leads to (upward) seepage (Figure 3.7, on the right; Figure 3.35) in deep polders such as, for instance, Polder Groot Mijdrecht (–6.40 m = 6.40 m below mean sea level), as shown in Figure 3.36.

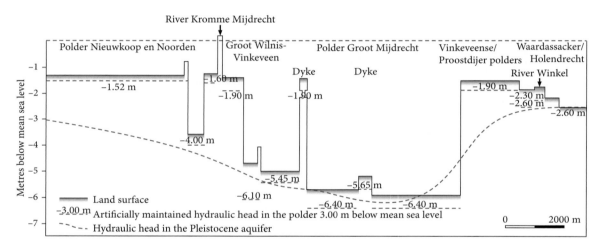

Figure 3.36. A cross-section through a polder area in The Netherlands, showing the potentiometric surfaces, in cross-section, of shallow, artificially maintained, and deeper aquifer groundwater (after De Vries 1980)

The deepest polder in The Netherlands and Europe is the Zuidplaspolder, near Rotterdam, a reclaimed lake (in Dutch: 'plas' or 'meer'), situated 6.70 m below mean sea level.

As a calculation example, for the upper, leaky confining layer of Holocene age, an average vertical hydraulic conductivity k of approximately 5×10^{-3} m day^{-1} and an average vertical thickness or depth d of 10 m may be assumed. The hydraulic or vertical resistance c then amounts to some 2000 days. For an average difference in hydraulic head Δh over this upper leaky confining layer of –0.5 to –1 m, the **seepage intensity**, defined as the volume flux density q in a vertical direction, following Darcy's law,

$$q = -k\frac{\Delta h}{d} = \frac{-\Delta h}{c}$$ (3.47)

then equals 0.25–0.5 mm day^{-1}.

This seepage intensity q is inversely related to the hydraulic resistance c of the upper, leaky confining layers. Disturbance of these layers may thus cause a reduction in hydraulic resistance and an increase in seepage intensity. The largest seepage is found in deep polders along the ice-pushed ridge Utrechtse Heuvelrug, because the Holocene peat layer in these polders had been excavated to almost the level of the Pleistocene aquifer sands to provide domestic heating by burning the peat. Polders with large seepage are called **calamity polders**. Calamity polders with freshwater seepage may be used for the provision of drinking water (Box 3.3).

BOX 3.3 Drinking water for the city of Amsterdam

Seepage intensities in the Horstermeerpolder and the Bethunepolder near the Utrechtse Heuvelrug amount to more than 20 mm day^{-1}. Seepage water in the Horstermeer has a high chloride content. Seepage water in the Bethunepolder is fresh and is used, together with water from the Loosdrechtse Plassen (Loosdrecht Lakes) and Amsterdam-Rijnkanaal (Amsterdam–Rhine Canal), to provide drinking water for the city of Amsterdam (Kosman 1988). Besides this, pre-treated water from the River Rhine is infiltrated into the coastal dunes south of Zandvoort, also as a source of drinking water for Amsterdam (Van Til and Mourik 1999).

Exercise 3.11.1 The screens of three piezometers are installed in an aquifer at the same depth below the ground surface. Piezometer B is situated 800 m south of piezometer A. Piezometer C is situated 1600 m east of piezometer B.
The hydraulic head is measured with the ground surface as zero level. The measured hydraulic heads are –0.66 m for piezometer A (= 0.66 m below the ground surface), –0.54 m for piezometer B, and –0.86 m for piezometer C.
The ground surface is flat and lies 3 m above mean sea level.

a. Determine the hydraulic gradient of the groundwater flow in the area enclosed by the piezometers.

b. Determine the direction of the groundwater flow in the area enclosed by the piezometers (north = 0°).

Exercise 3.11.2 A polder has an area of 5 km^2: 2 km^2 is open water and 3 km^2 is land. The yearly precipitation = 750 mm. The yearly open-water evaporation = 600 mm. The yearly actual evaporation on land amounts to 70% of the yearly open-water evaporation. On a yearly basis, a pumping station pumps 2×10^6 m^3 of water from the polder.

a. Determine the actual evaporation (mm) for the polder as a whole.

b. Determine the seepage flux density (mm day^{-1}) for a year in which the water table and open-water level remain unchanged.

The **storage coefficient**, defined here as the quotient of added or extracted water depth (m) and the accompanying change in water table (m), for the polder (sub)soil equals 0.4.

c. Determine the seepage flux density (mm day^{-1}) for a year in which the water table and open-water level have risen by 0.20 m.

3.12 Flow nets

Maps such as Figures 3.37 and 3.38 showing streamlines and equipotentials for two-dimensional groundwater flow are called **flow nets** or **potentiometric maps**. Flow nets show the direction of groundwater flow and are useful for estimating groundwater discharges in plan view or cross-section.

Figure 3.37 shows the directions of groundwater flow for both **effluent seepage**, seepage leaving a groundwater body, and **influent seepage**, seepage into a groundwater body, with the change in direction of seepage being caused by a river dam. Effluent seepage is what we normally encounter in humid areas, whereas influent seepage is usually found from **wadis** (Arabic), **oueds** (in North Africa), or **arroyos** (Spanish), dry river beds that contain water only during times of heavy rain, in (semi-)arid areas. Influent seepage may, however, also be encountered in the western part of The Netherlands, where rivers are located higher than the polders, as shown in Figures 3.1, 3.10, and 3.36. Water is pumped from the polders to the higher water reaches: the system of water reaches and lakes where water is pumped to for temporary storage, and for ultimate discharge to sea is called the **boezem** (Dutch).

If, for a flow net of a homogeneous, isotropic aquifer as shown in Figure 3.38, the transmissivity T is known, the volume flux Q' (m^2 day^{-1}) can be estimated from Darcy's law as the product of T (m^2 day^{-1}) and the hydraulic gradient i (the difference

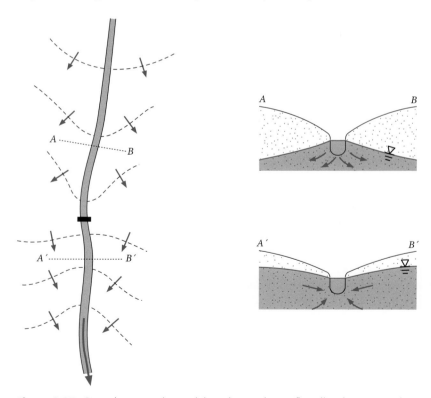

Figure 3.37. Groundwater equipotentials and groundwater flow directions near a river dam in plan and cross-section views (after De Vries and Cortel 1990)

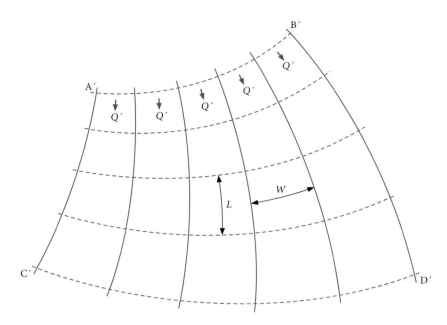

Figure 3.38. Streamlines and equipotentials for a homogeneous, isotropic aquifer in plan view

in the hydraulic head divided by the flow distance), the latter as established from the flow net. The transmissivity of an unconfined aquifer is then defined as the product of the hydraulic conductivity K and the average saturated depth of the aquifer $\bar{D}$:

$$T = K \times \bar{D} \qquad (3.48)$$

Compare this with Equation 3.32 for a confined aquifer.

Conversely, if the volume flux Q' (m² day⁻¹) is known to be constant (steady flow) in a heterogeneous aquifer, then the transmissivity at different locations can be estimated, again by applying Darcy's law using information on the hydraulic gradient provided by the flow net.

Streamlines and equipotentials in a homogeneous, isotropic aquifer can be drawn in such a way that identical rectangles are formed covering a constant difference in hydraulic head Δh, as shown in Figure 3.38. The zone between two adjacent flow lines is called a **stream tube**. If the saturated thickness or depth D of an aquifer is assumed (almost) constant (which may be reasonable for an unconfined aquifer, as the absolute value of the hydraulic gradient is of the order of 1 m per km) then Darcy's law can be written as follows:

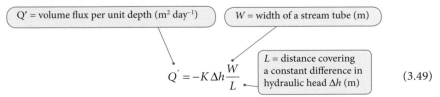

Q' = volume flux per unit depth (m² day⁻¹) W = width of a stream tube (m)

$$Q' = -K \Delta h \frac{W}{L} \qquad (3.49)$$

L = distance covering a constant difference in hydraulic head Δh (m)

For each flow net rectangle, the volume flux Q' has the same constant value, because K is constant (the aquifer is homogeneous and isotropic), Δh is taken constant, as is

W divided by L. It comes in handy to choose Δh in such a way that the rectangles become squares. As W then equals L, Equation 3.49 reduces for each square (independent of W and L) to

$$Q' = -K\Delta h \qquad (3.50)$$

In fact, Figure 3.38 has already been drawn in such a way that $W = L$, and thus we are dealing with identical squares. The volume flux Q' per unit depth D along the curve $C'D'$ can be simply determined as the sum of the contributions of each bordering square. There are five bordering squares and thus the volume flux per unit depth along $C'D'$ equals $5 \times Q'$. This method is called the **squares method** and makes for easy calculations and interpretations as shown below.

Both along $C'D'$ and $A'B'$, the volume flux per unit depth equals $5 \times Q'$. Because the width along $A'B'$ is less than along $C'D'$, the groundwater passing $A'B'$ must have a larger volume flux density q than the water passing $C'D'$. Another way of explaining this is that the squares bordering $A'B'$ are smaller than the squares bordering $C'D'$. This means that the hydraulic gradient in the vicinity of $A'B'$ is larger (Δh = constant, but L is smaller) than near $C'D'$. According to Darcy's law, the volume flux density q at $A'B'$ must then be larger than at $C'D'$. When the effective porosities n_e near $A'B'$ and $C'D'$ are the same, the **effective groundwater velocity** v_e at $A'B'$ is larger than at $C'D'$, or, in other words, the groundwater at $A'B'$ will flow faster than at $C'D'$.

Exercise 3.12 After a long period without rain, the discharge of a river at measuring station A equals 50 m³ s⁻¹. Discharge 100 km downstream at station B equals 60 m³ s⁻¹. Between the stations, the river is recharged from both sides by an unconfined aquifer. The equipotentials of the groundwater are parallel to the river. The average slope of the groundwater surface may be taken as $\frac{1}{1000}$.

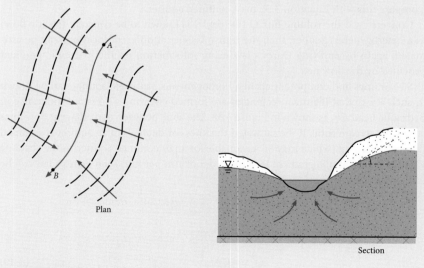

Plan

Section

Figure E3.12

Determine the transmissivity of the aquifer in m² day⁻¹.

From the foregoing, it is also evident that continuity in a homogeneous, isotropic medium is to be translated as the volume flux Q being constant (or $Q' =$ constant), but that continuity as such does not imply that the volume flux density q is constant.

3.13 Groundwater flow regimes and systems

If after today there was no more precipitation, the water table on a local and regional scale would ultimately become a flat surface. In reality, however, such a manifestation is countered by the ongoing hydrological cycle and recharge of the groundwater by precipitation, infiltration, and percolation. Because of this continued replenishment (recharge), the water table in an unconfined aquifer is at some scale related to the topography or land surface. As an example, the average altitude of a water table under a high land surface is higher than the water table in the lowlands surrounding this high land surface.

Two major groundwater flow regimes can be distinguished, as shown in Figure 3.39. From piezometer readings, one can observe a **topography-controlled water table**: the water table closely follows the local topography as a subdued replica (Figure 3.39a). Alternatively, with a **recharge-controlled water table**, the water table does not follow the local topography, but does follow the regional topography. In other words, it

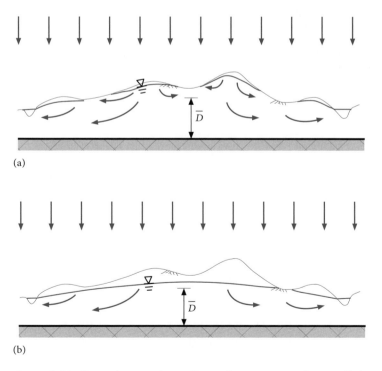

Figure 3.39. Two major groundwater flow regimes: a topography-controlled water table (a) and a recharge-controlled water table (b) for the same land surface (adapted from Haitjema 1995)

follows the topography at a cruder scale (Figure 3.39b). The degree to which the shape of the water table is topography controlled or recharge controlled depends on the recharge rate, the aquifer transmissivity, the aquifer geometry, and the topography itself (Haitjema and Mitchell-Bruker 2005).

Haitjema and Mitchell-Bruker (2005) conducted a number of modelling studies and compared their results with the earlier conceptual theories of Tóth (1963), concerning a topography-controlled water table with local cells (Figures 3.39a, 3.40, and 3.41), and both Dupuit (1863) and Forchheimer (1886), concerning what we now label as a recharge-controlled water table, dominated by horizontal flow and mostly regional flow cells (Figure 3.39b). They concluded that a high average annual recharge rate in comparison with the aquifer hydraulic conductivity (such as, for instance, in wetlands), a relatively flat land surface, and a shallow aquifer with low transmissivity are factors that are favourable for the establishment of a topography-controlled water table. Conversely, a low average annual recharge rate in comparison with the aquifer hydraulic conductivity, a relatively uneven land surface, and a deep aquifer with high transmissivity, typical of a 'usable' aquifer – that is, an aquifer containing substantial and easily pumped-out volumes of water – are factors that are favourable for the establishment of a recharge-controlled water table. From their findings, Haitjema and Mitchell-Bruker (2005) produced a dimensionless decision criterion to assess the likelihood for either topography-controlled or recharge-controlled water tables:

> Average distance between surface waters $\bar{L}$ over the average aquifer thickness $\bar{D}$ (–)

> Average distance between surface waters $\bar{L}$ over the maximum distance between the surface water levels and the terrain elevation z_{max} (–)

> Average annual recharge rate $\bar{R}$ over the aquifer's hydraulic conductivity K (–)

$$\frac{\bar{R_r}}{K} \times \frac{\bar{L}}{\bar{D}} \times \frac{\bar{L}}{z_{max}} > m \quad \text{topography controlled} \tag{3.51}$$

> $m = 8$ for an aquifer that is strip-like in shape (one-dimensional groundwater flow)
> $m = 16$ for an aquifer that is circular in shape (radial-symmetric groundwater flow)

$$\frac{\bar{R_r}}{K} \times \frac{\bar{L}}{\bar{D}} \times \frac{\bar{L}}{z_{max}} < m \quad \text{recharge controlled} \tag{3.52}$$

Thus, when the conditions are right, local, intermediate, and regional groundwater flow systems may develop in a drainage basin, as shown in Figure 3.40.

The time of contact of groundwater with the subsoil influences the chemical composition of the groundwater. The groundwater's chemical composition at different locations may be established from a chemistry toolkit that is relatively simple to use in the field, and that is most useful in establishing the groundwater's origin and whether the groundwater at these locations and at a specific time is part of a local, intermediate, or regional flow system. The latter activity is part of **hydrological systems analysis**, an analysis of the close relationships between land use, groundwater recharge, vegetation patterns, groundwater flow systems, water quality, and surface water networks, which

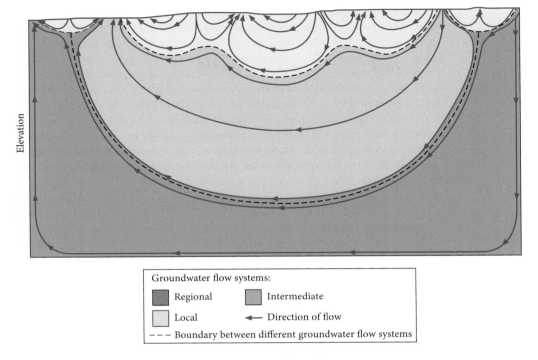

Figure 3.40. Groundwater flow systems according to Tóth (1963)

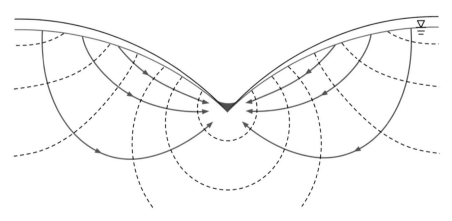

Figure 3.41. The concept of a local groundwater flow system or cell (adapted from Hubbert 1940)

can be applied to solve a wide spectrum of water resources inventory and management problems (Engelen and Kloosterman 1996).

Groundwater that has just infiltrated, **atmocline water**, still carries the traits of atmospheric water, such as a low ion concentration, measurable as a low **electrical conductivity** (*EC*) (Box 3.4) of the order of 10 μS cm^{-1}. Groundwater that has been in contact with the subsoil for a long time, **lithocline water**, has a higher ion concentration (*EC*s are usually smaller than 1000 μS cm^{-1}) and can have a high **hardness** (high mineral content) when it contains substantial amounts of Ca^{2+} due to the subsoil dissolution of lime ($CaCO_3$) in sediments or rocks through which the groundwater travels (see also Box 3.2). When fresh and salt water are, or have been, in contact in the subsoil, the sodium (Na^+) and/or chloride (Cl^-) levels in the groundwater are high. Groundwater that contains much salt, **thalassocline water** (as in seawater, which has an EC of approximately 50 000 µS cm^{-1}), can have such a high ion concentration that the EC measuring device becomes damaged. Also, the **pH**, a measure inversely related to the concentration of hydrogen (H^+) ions in water (Box 3.4), and the **alkalinity**, the capacity to neutralize acid (Box 3.4), are important variables to be measured instantaneously in the field. Both the values of the above-mentioned variables and the chemical composition of the groundwater at different locations and times are important findings for unravelling the dynamic framework of groundwater flow over time.

BOX 3.4 Alkalinity, electrical conductivity (*EC*), and *pH*

Alkalinity is the capacity of a solution to neutralize acid; in practice, it is the concentration of both HCO_3^- and CO_3^{2-} ions in a watery solution.

Electrical conductivity (*EC*) is a measure of the water's ability to conduct electricity, and therefore a measure of the water's ion concentration. The *EC* can be measured in the field with a simple small stick-like apparatus attached by means of a wire to a measuring device. The *EC* is measured in µS cm^{-1} (micro-siemens per centimetre), which is the same as µmho cm^{-1} (micro-mho per centimetre). As a **siemens**, named after the German inventor and industrialist Ernst Werner von Siemens (1816–1892), is the reciprocal of an ohm, the term 'mho' is simply derived (as a sort of word joke) from spelling the word 'ohm' backwards. As the *EC* is temperature dependent, it is standard practice to recalculate its value to the *EC* that would occur at a temperature of 20°C.

pH is a measure of the concentration of hydrogen ions (H^+) in a solution:

$[H^+]$ = the concentration of hydrogen ions (H^+) in mol litre^{-1}

$$pH = -\log[H^+] \qquad\qquad (B3.4)$$

pH < 7: the solution is acidic;
pH = 7: the solution is neutral;
pH > 7: the solution is alkaline.

After determining the *EC*, *pH*, alkalinity, and chemical composition of the water, there are many ways to display the differences in water chemistry, such as by means of Piper (1944), Stiff (1951), or Schoeller (1955) diagrams, all of which involve reworking concentrations to units of milli-equivalents per litre (Box 3.5). As an example, Figure 3.42 shows adapted **Stiff diagrams** of the Tollbaach drainage basin (catchment) near Haller in the Gutland of east Luxembourg. This forested drainage basin lies within the Luxembourg sandstone formation, which is the most important aquifer of the Grand Duchy of Luxembourg, as 90% of its water supply is obtained from it (Von Hoyer 1971). Figure 3.42 shows the effect that manure dumping on arable land in the western part of the drainage basin had on the water quality of springs 2, 3, and 5 in 1983. These springs show high nitrate (NO_3^-) levels (Box 3.6) because of local groundwater flow over a slightly tilting impermeable clay layer within the Luxembourg sandstone, directed away from the arable land.

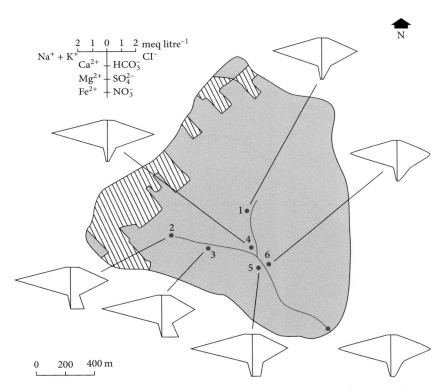

Figure 3.42. Adapted Stiff diagrams for stream and spring waters in the forested Tollbaach drainage basin, Luxembourg (17 November 1983); shaded areas are arable land (Hendriks 1990)

nitrate-polluted water may cause their skin to turn blue, the blue baby syndrome, indicating a shortage of oxygen in blood and brains, thus constituting a serious threat to their health and life. Public awareness of the worst-case effects of drinking nitrate-polluted water may still be considered appropriate in certain areas.

3.14 Fresh and saline: Ghijben–Herzberg

Fresh water from rain that falls on a circular-shaped island in the sea causes the development of a freshwater hydraulic head h_f relative to the mean sea level, as shown in Figure 3.43. As the infiltrating and percolating fresh water replaces the saline water originally present below the surface, and as fresh water has a slightly lower density than saline water, a freshwater lens will develop that floats on the denser saline water. Figure 3.43 shows the situation for an unconfined aquifer in a circular island with a precipitation surplus (precipitation exceeds evaporation on a yearly basis) after equilibrium has been reached. The situation drawn in Figure 3.43 is for a large part comparable to what we will encounter below coastal dunes, such as along the Dutch coast. It is interesting to compute the depth of the freshwater lens, as this provides a source of drinking water.

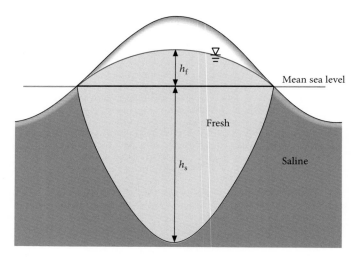

Figure 3.43. Fresh groundwater floating on saline groundwater in an unconfined aquifer in a circular island in the sea

Fresh water has a density ρ_f of 1000 kg m^{-3}, whereas saline groundwater has a density ρ_s of approximately 1025 kg m^{-3}. If we assume hydrostatic conditions (no groundwater flow) at the interface of the fresh and saline groundwater, a sharp interface, and no mixing of the fresh and saline water, we can compute the depth of the freshwater lens h_s relative to the mean sea level (see Figure 3.43) as follows.

The hydraulic head equals the sum of the elevation head and pressure head (section 3.3). This is true for both the fresh and saline groundwater at the sharp interface. For the hydraulic head at the interface in fresh water, we can write

$$h_s + h_f = z + \frac{p}{\rho_f g} \qquad (3.53)$$

Equally, for the hydraulic head at the interface in saline water, we can write

$$h_s = z + \frac{p}{\rho_s g} \qquad (3.54)$$

We may take the reference level of the elevation head, where $z = 0$, anywhere along the interface of the fresh and saline groundwater. When $z = 0$, the above Equations 3.53 and 3.54 reduce to

$$h_s + h_f = \frac{p}{\rho_f g} \Rightarrow p = \rho_f g(h_s + h_f) \qquad (3.55)$$

$$h_s = \frac{p}{\rho_s g} \Rightarrow p = \rho_s g h_s \qquad (3.56)$$

There can only be one pressure p at any point in the groundwater. Consequently, setting Equations 3.55 and 3.56 equal and solving for h_s yields

$$h_s = \frac{\rho_f}{\rho_s - \rho_f} h_f \qquad (3.57)$$

This relation is true at every location along the interface of the fresh and saline groundwater, and is known as the **Ghijben–Herzberg relation** after the Dutch engineer–officer Willem Badon Ghijben (1845–1907) and the German engineer A. Herzberg, who slightly later independently reached the same conclusion. Subsequently, it was discovered that the freshwater/saltwater interface equilibrium had been formulated earlier by Joseph DuCommon in the United States in 1818, but Badon Ghijben was not aware of this at the time (De Vries 1994).

If we take $\rho_f = 1000$ kg m^{-3} and $\rho_s = 1025$ kg m^{-3}, Equation 3.57 reduces to

$$h_s \approx 40 h_f \Rightarrow h_s + h_f \approx 41 h_f \qquad (3.58)$$

The thickness or depth of a freshwater aquifer under hydrostatic equilibrium conditions thus equals approximately 41 times the **convexity** (h_f), which is defined as the difference in altitude between the top of the water table and the open-water surface (acting as boundary condition): the convexity is also called the **differential head**.

The actual situation encountered beneath coastal dunes differs from the above as, in reality, the hydrostatic conditions upon which the Ghijben–Herberg relation is based do not exist. Also, in reality, the interface between fresh and saline water is not sharp. Due to **molecular diffusion**, molecules that move randomly through water – the so-called **Brownian motion** – and chemicals that move from regions of high concentration to regions of low concentration (Van der Perk 2006), the interface will consist of a transition zone of brackish water, water that is intermediate between fresh and saline water (Box 3.7).

BOX 3.7 The coastal dunes of the Netherlands

In the coastal dunes of the Netherlands, the maximum thickness of the freshwater lens before the extraction of drinking water was approximately 160 m (De Vries 1980). The thickness of the brackish water zone was 5–10 m, with chloride contents increasing with depth from 50 mg litre^{-1} up to 16000 mg litre^{-1} (Van den Akker 2007). Since the second half of the nineteenth century, groundwater has been pumped from the freshwater lens below the dunes to be used as drinking water; in 1903, the extraction of groundwater by means of deep wells began (De Vries 1994). To avoid **desiccation** of the dunes as well as the unwanted **upwelling** of saline or brackish water, surface water – for instance,

Fresh water, < 300 mg Cl$^-$ per litre; weakly brackish water, 300–3000 mg Cl$^-$ per litre; brackish water, 3000–10000 mg Cl$^-$ per litre; strongly brackish to saline water, 10000–16000 mg Cl$^-$ per litre; seawater, > 16000 mg Cl$^-$ per litre.

pre-treated water from the River Rhine – is nowadays infiltrated from both infiltration canals and infiltration wells to replenish the fresh groundwater below the dunes (Van Til and Mourik 1999). Since the extraction started, the thickness of the brackish-water transition zone has extended to some 30–50 m (Van den Akker, personal communication).

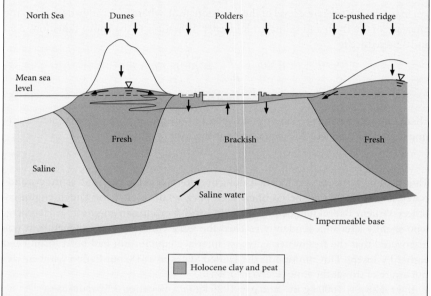

Figure B3.7. A geohydrological profile through the western part of The Netherlands (after De Vries 1980)

Figure B3.7 is a geohydrological profile through the western part of the Netherlands, showing the distribution of fresh, brackish, and saline water. Fresh groundwater can be seen to be found both below the coastal dunes and under a Pleistocene ice-pushed ridge some 50 km to the east.

3.15 Groundwater hydraulics

As the basics of steady groundwater flow have been discussed in the preceding sections, and to enhance the story of steady groundwater flow, this section presents a selection of (solved) groundwater flow cases. The reader will see a world of simple solutions to a number of groundwater flow problems opening up from the use of only relatively simple mathematics. (However, should one decide to skip section 3.15 and move on to Chapter 4 on soil water, then there is no harm done to the build-up of the story of this book as a whole.)

The general build-up for each presented groundwater flow case in this section is that Darcy's law is combined with the continuity equation to obtain the potentiometric surface, which is the spatial distribution of the hydraulic heads with linear distance x (for one-dimensional flow cases), or with radial distance r (for radial-symmetric flow cases). Such a build-up has already been introduced in section 3.9 for the (simplest) flow case of a confined groundwater between two canals. All aquifers and aquitards (semi-permeable layers or leaky confining layers) in the following groundwater flow cases are assumed to be homogeneous and isotropic.

Readers interested in the complete mathematical derivations of the end results presented in this section are referred to the Mathematics Toolboxes (under M) near the end of this book; if required, the necessary mathematical grounding is provided in the Conceptual Toolkit (under C).

Further, to obtain a real grasp of the benefits of groundwater hydraulics when studying steady groundwater flow, a sufficiently large number of exercises are included in this section.

A summary of the equations that are useful as a reference and starting point for solving the exercises on groundwater hydraulics is given in Table 3.3; in fact, it is good practice to solve these exercises by only using Table 3.3 as a memory aid! The Answers section at the end of this book provides answers to intermediate steps, to highlight the best strategy for solving the exercises.

Unconfined flow: Dupuit–Forchheimer

Darcy's law for steady groundwater flow in an unconfined aquifer between two parallel, fully penetrating canals with different water levels (see Figure 3.44) can be written as follows:

$$Q' = -Kh\frac{dh}{dx} \tag{3.59}$$

Assuming the hydraulic gradient to be equal to the slope of the water table, plus for small water-table gradients, streamlines to be horizontal and equipotentials to be vertical (the **Dupuit–Forchheimer assumptions**), Darcy's law and the continuity equation ($Q' = $ constant) can be combined to yield the result that the potentiometric surface for steady groundwater flow in an unconfined aquifer between two parallel, fully penetrating canals with different water levels can be described as follows:

$$h^2 = C_1 x + C_2 \tag{3.60}$$

The derivation of Equation 3.60 is presented under M2 at the end of this book. Compare the above Equation 3.60 for unconfined groundwater with the linear relation of Equation 3.29 for confined groundwater ($h = C_1 x + C_2$). **Because h is to the power of two in Equation 3.60, the equation is a parabolic one, as the reader may have noticed from the shape of the water tables drawn in Figures 3.34 and 3.41;**

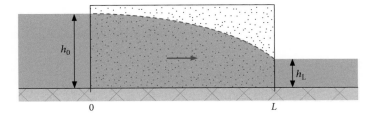

Figure 3.44. Steady groundwater flow in an unconfined aquifer between two parallel, fully penetrating canals with different water levels

in Figure 3.44, the potentiometric surface has already been drawn as a parabolic curve.

As with confined groundwater (see section 3.9), the values for the constants C_1 and C_2 can be found by inserting the boundary conditions for this flow case. First, insert the $x = 0$ boundary condition and then the $x = L$ boundary condition (h_0, h_L, and L are all known values). This then yields

$$h^2 = \frac{h_L^2 - h_0^2}{L}x + h_0^2 \tag{3.61}$$

From Equation 3.60, it can be derived that (see section C2.3)

$$h = \left(C_1 x + C_2\right)^{\frac{1}{2}} \tag{3.62}$$

and, consequently,

$$\frac{dh}{dx} = \tfrac{1}{2}C_1 \left(C_1 x + C_2\right)^{-\frac{1}{2}} \tag{3.63}$$

Combining Equations 3.62 and 3.63 yields

$$h\frac{dh}{dx} = \left(C_1 x + C_2\right)^{\frac{1}{2}} \times \tfrac{1}{2}C_1 \left(C_1 x + C_2\right)^{-\frac{1}{2}} = \tfrac{1}{2}C_1 \tag{3.64}$$

Substituting Equation 3.64 in Equation 3.59 gives

$$Q' = -\tfrac{1}{2}C_1 K \tag{3.65}$$

or (when using $C_1 = (h_L^2 - h_0^2)/L$)

$$Q' = -K\frac{h_L^2 - h_0^2}{2L} = -K\frac{h_L + h_0}{2}\frac{h_L - h_0}{L} \approx -K\bar{D}i \tag{3.66}$$

The first part of Equation 3.66,

$$Q' = -\tfrac{1}{2}K\frac{h_L^2 - h_0^2}{L} \tag{3.67}$$

is known as the **Dupuit–Forchheimer equation** after the French civil engineer Arsène J.E.J. Dupuit (1804–1866) and the Austrian engineer Philipp Forchheimer (1852–1933).

Because Q' is constant (continuity) and as $\bar{D}$ decreases in the direction of flow ($Q' = q \times \bar{D}$), or because K is constant (homogeneous aquifer) and as i increases in the direction of flow ($q = -K \times i$), this must mean that the volume flux density q increases in the direction of flow. This is another example of how continuity stands for $Q' =$ constant, but that continuity as such in no way implies that the volume flux density q is constant (see also the end of section 3.12).

Exercise 3.15.1.1 Figure E3.15.1.1 shows a cross-section of a sand dam with vertical sides and open water at both sides: $h_1 = 6$ m; $h_2 = 3$ m. The horizontal hydraulic conductivity of the sand equals 1 m day^{-1}. Water flow is in the direction of the x-axis only.

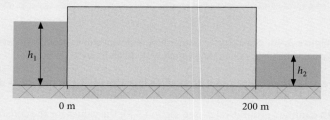

Figure E3.15.1.1

a. Determine the hydraulic head at $x = 40, 80, 120,$ and 160 m.

b. Make a drawing of the change in hydraulic head with x.

c. Determine the volume flux Q' of the aquifer in m^2 day^{-1}.

For the Dupuit–Forchheimer assumptions to hold, as a rule of thumb, L in an isotropic aquifer must be at least five times larger than $\bar{D}$ (Haitjema and Mitchell-Bruker 2005).

In summary,

$$h^2 = C_1 x + C_2 \qquad (3.68)$$

All other equations in this section on steady, unconfined groundwater flow can be derived from this one equation by merely using Darcy's law and relatively simple mathematics.

All exercises 3.15.1 on unconfined flow can be solved by using Equation 3.68, Darcy's law, and continuity.

Exercise 3.15.1.2 Figure E3.15.1.2 shows a cross-section of an unconfined aquifer overlying an impermeable layer. To the left and right of the aquifer, parallel, fully penetrating canals with a different water level overlie the impermeable layer. Ten metres to the right of the left canal, the hydraulic head is determined with piezometer 1: the hydraulic head is 6.75 m above the impermeable layer. The hydraulic head in piezometer 2 at 75 m to the right of the left canal is 6.25 m. The aquifer is homogeneous and isotropic, and has a hydraulic conductivity of 10 m day^{-1}.

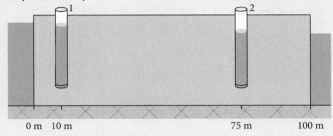

Figure E3.15.1.2

Determine the water levels in the left and right canals.

Exercise 3.15.1.3 Figure E3.15.1.3 shows a cross-section of a sand dam with vertical sides and open water at both sides: $h_1 = 6$ m; $h_2 = 3$ m. Between $x = 0$ and $x = 40$ m, the hydraulic conductivity K_1 of the sand is 1 m day^{-1}; between $x = 40$ and $x = 200$ m, the hydraulic conductivity K_2 of the sand is 10 m day^{-1}. Water flow is in the direction of the x-axis only.

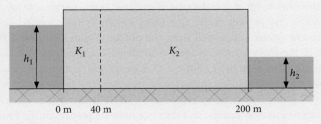

Figure E3.15.1.3

a. Determine the hydraulic head at $x = 40$ m.

b. Make a drawing of the change in hydraulic head with x.

c. Determine the volume flux Q' of the aquifer in m² day⁻¹.

d. Does the hydraulic head change if we change the hydraulic conductivities K_1 and K_2, and if so, in what way?

Exercise 3.15.1.4 An 8 m-thick aquifer between two parallel canals has a steady groundwater flow. The aquifer is homogeneous and isotropic, and overlain and underlain by impermeable layers. The water level in the left canal is 10 m above the impermeable layer (fully penetrating); the water level in the right canal is 6 m above the impermeable layer. The distance between the left and right canals is 100 m.

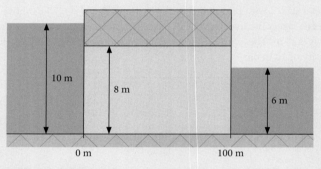

Figure E3.15.1.4

a. At what distance from the left canal is the hydraulic head above the impermeable layer equal to the thickness of the aquifer?

b. Determine the hydraulic conductivity K of the aquifer for a volume flux Q' of 1 m² day⁻¹.

'Hollands profiel'

Figure 3.45 shows the '**Hollands profiel**', the profile for steady groundwater flow in a Dutch polder made up of a leaky confining layer (semi-permeable layer) consisting of Holocene clay or peat overlying a leaky aquifer (semi-confined aquifer) with Pleistocene sand. Water is pumped from the polder (Figures 3.1 and 3.10) and the water table in the leaky confining layer is artificially maintained at a level h_a (m). The dyke in the cross-section of Figure 3.45 is simplified to an impermeable vertical screen above the leaky aquifer. Importantly, however, the following approximation is of use for all groundwater flow cases where a water-holding semi-permeable layer overlies a water-saturated aquifer, and thus not only for polders in the Province(s) of Holland in The Netherlands (as will be evident from Exercises 3.15.2.3 and 3.15.2.4).

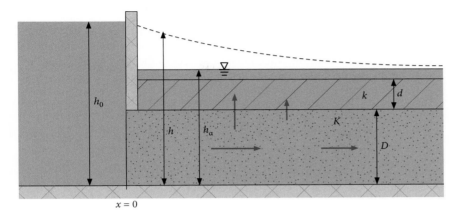

Figure 3.45. Steady groundwater flow in a leaky aquifer
h_a is taken above the land surface for reasons of didactic convenience; in reality, h_a would lie slightly below the land surface, but this would cause the saturated thickness d of the leaky confining layer (semi-permeable layer) to be smaller than the thickness of the layer itself; this in turn would lead to a smaller value of the vertical flow resistance c

Darcy's law for the leaky aquifer can be written as follows:

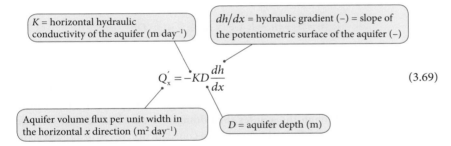

$$Q'_x = -KD\frac{dh}{dx} \qquad (3.69)$$

K = horizontal hydraulic conductivity of the aquifer (m day^{-1})

dh/dx = hydraulic gradient (−) = slope of the potentiometric surface of the aquifer (−)

Aquifer volume flux per unit width in the horizontal x direction (m^2 day^{-1})

D = aquifer depth (m)

The subscript x in Q'_x denotes that flow is approximated as occurring in the horizontal x direction. Due to seepage (q_z in m day^{-1}) through the semi-permeable layer, the hydraulic gradient, which is the slope of the potentiometric surface of the water held in the aquifer, in Figure 3.45 diminishes to the right.

Darcy's law for the leaky confining layer (semi-permeable layer) can be written as follows:

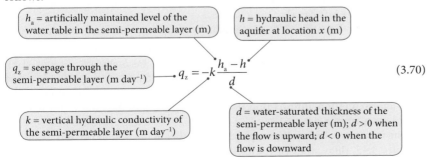

$$q_z = -k\frac{h_a - h}{d} \qquad (3.70)$$

h_a = artificially maintained level of the water table in the semi-permeable layer (m)

h = hydraulic head in the aquifer at location x (m)

q_z = seepage through the semi-permeable layer (m day^{-1})

k = vertical hydraulic conductivity of the semi-permeable layer (m day^{-1})

d = water-saturated thickness of the semi-permeable layer (m); $d > 0$ when the flow is upward; $d < 0$ when the flow is downward

The subscript z in q_z denotes that flow is approximated as occurring in the vertical z direction. By incorporating Darcy's law for both the leaky aquifer and leaky confining layer in the water balance (continuity equation) of a vertical column in the subsoil the following equation for the potentiometric surface of the leaky aquifer is derived (see M3 at the end of this book for the derivation):

λ = leakage factor (m)

$c = d/k$ = hydraulic resistance (day); $c > 0$ when the flow is upward; $c < 0$ when the flow is downward

$$ h = h_a + C_1 e^{\frac{x}{\lambda}} + C_2 e^{\frac{-x}{\lambda}} \text{ with } \lambda = \sqrt{KDc} \qquad (3.71) $$

The values for the constants C_1 and C_2 can be found by inserting the boundary conditions. For Figure 3.45, the boundary conditions are as follows:

$x = 0$, then $h = h_0 : h_0 = h_a + C_1 e^0 + C_2 e^0 = h_a + C_1 + C_2$; thus $C_1 + C_2 = h_0 - h_a$

$x = \infty$, then $h = h_a : h_a = h_a + C_1 e^\infty + C_2 e^{-\infty}$; thus $C_1 e^\infty + C_2 e^{-\infty} = h_a - h_a = 0$

$C_1 e^\infty + C_2 e^{-\infty} = (C_1 \times \infty) + (C_2 \times 0) = 0 \Rightarrow C_1 = 0$ and thus $C_2 = h_0 - h_a$

Therefore for an **infinite polder** ($x \to \infty$) as in Figure 3.45, Equation 3.71 becomes

$$ h = h_a + \left(h_0 - h_a\right) e^{\frac{-x}{\lambda}} \text{ with } \lambda = \sqrt{KDc} \qquad (3.72) $$

The **leakage factor** has the metre as unit (the square root of $m^2 \ day^{-1} \ day$).

The leakage factor is large when the leaky aquifer has a high transmissivity $T = KD$ and when the absolute value of the hydraulic resistance c of the leaky confining layer (semi-permeable layer) is also large. In Figure 3.45, horizontal groundwater flow through the aquifer then dominates over the vertical seepage flow through the leaky confining layer, and as a consequence, the hydraulic head in the aquifer h diminishes relatively slowly with the distance x from the left canal.

Conversely, the leakage factor is small when the leaky aquifer has a low transmissivity $T = KD$ and when the absolute value of the hydraulic resistance c of the leaky confining layer (semi-permeable layer) is small. In Figure 3.45, vertical seepage flow through the leaky confining layer then dominates over horizontal groundwater flow through the aquifer, and as a consequence, the hydraulic head in the aquifer h then decreases relatively quickly with the distance x from the left canal.

The shape of Equation 3.72 is a gradually declining potentiometric curve, as shown in Figure 3.45. One can reason that $e^{-x/\lambda}$ in Equation 3.72 for $x = 3\lambda$ becomes $e^{-3\lambda/\lambda} = e^{-3} = 0.05$. Likewise, for $x = 4\lambda$, the factor $e^{-x/\lambda}$ becomes $e^{-4\lambda/\lambda} = e^{-4} = 0.02$. This means that the hydraulic head h in the leaky aquifer equals h_a plus only 5% of $(h_0 - h_a)$ at $x = 3\lambda$ and likewise h_a plus only 2% of $(h_0 - h_a)$ at $x = 4\lambda$. **In other words, in this approximation, infinity thus lies three to four times the value of the leakage factor λ away from $x = 0$.**

If one wants to calculate the seepage volume flux to the polder ($m^2 \ day^{-1}$) shown in Figure 3.45, there are two routes that one may take. The simplest route is to

first recognize that as groundwater flow is steady (stationary), the volume flux or vertical water seepage Q_z' must be equal to the horizontal volume flux Q_x' entering the leaky aquifer at $x = 0$:

$$Q_z' = Q_{x=0}' = -KD\left(\frac{dh}{dx}\right)_{x=0} \tag{3.73}$$

dh / dx can be obtained by mathematical differentiation of Equation 3.71:

$$\frac{dh}{dx} = \frac{C_1}{\lambda}e^{\frac{x}{\lambda}} + \frac{C_2}{-\lambda}e^{\frac{-x}{\lambda}} \tag{3.74}$$

At $x = 0$,

$$\left(\frac{dh}{dx}\right)_{x=0} = \frac{C_1}{\lambda}e^0 + \frac{C_2}{-\lambda}e^0 = \frac{C_1}{\lambda} - \frac{C_2}{\lambda} = \frac{C_1 - C_2}{\lambda} \tag{3.75}$$

Combining this with Equation 3.73 for an infinite folder ($x \to \infty$; $C_1 = 0$) gives:

$$Q_z' = Q_{x=0}' = -KD\frac{-C_2}{\lambda} \tag{3.76}$$

Equation 3.76 holds for an infinite polder ($x \to \infty$), as shown in Figure 3.45. For such a groundwater flow case, we know that $C_2 = h_0 - h_a$. Therefore,

$$Q_z' = Q_{x=0}' = -KD\frac{h_a - h_0}{\lambda} \tag{3.77}$$

The other way to determine Q_z' for an infinite polder ($x \to \infty$) as in Figure 3.45 would be to step up from q_z (m day^{-1}) to Q_z' (m^2 day^{-1}), which is by mathematical integration:

$$Q_z' = \int_0^\infty q_z\, dx \tag{3.78}$$

Combining this with Equation 3.70 gives

$$Q_z' = \int_0^\infty -k\frac{h_a - h}{d}\, dx \tag{3.79}$$

h can be derived at every location of x from Equation 3.71 as follows:

$$h = h_a + C_1 e^{\frac{x}{\lambda}} + C_2 e^{\frac{-x}{\lambda}} \tag{3.80}$$

Combining Equations 3.79 and 3.80 for an infinite folder ($x \to \infty$; $C_1 = 0$) gives:

$$Q_z' = \int_0^\infty -k\frac{-C_1 e^{\frac{x}{\lambda}} - C_2 e^{\frac{-x}{\lambda}}}{d}\, dx = \frac{k}{d}\int_0^\infty C_1 e^{\frac{x}{\lambda}} + C_2 e^{\frac{-x}{\lambda}}\, dx = \frac{1}{c}\left[\lambda C_1 e^{\frac{x}{\lambda}} - \lambda C_2 e^{\frac{-x}{\lambda}}\right]_0^\infty$$

$$= \frac{\lambda}{c}(C_1 e^\infty - C_2 e^{-\infty} - C_1 e^0 + C_2 e^0) = \frac{\lambda}{c}C_2$$

Thus, again for $C_2 = h_0 - h_a$:

$$Q'_z = \frac{\lambda}{c}C_2 = -\frac{\lambda}{c}(h_a - h_0)$$

(3.81)

$\lambda = \sqrt{KDc} \Rightarrow \lambda^2 = KDc \Rightarrow \lambda/c = KD/\lambda$; Equation 3.81 can thus also be written as

$$Q'_z = -\frac{KD}{\lambda}(h_a - h_0)$$

(3.82)

Equation 3.82 is the same as Equation 3.77, as, of course, it should be!

Note that Equations 3.81 and 3.82 hold for an infinite polder $(x \rightarrow \infty)$, as shown in Figure 3.45, as they are based upon groundwater horizontally entering the aquifer at $x = 0$ and vertically moving out of the aquifer between $x = 0$ and $x = \infty$.

A solution for a **finite polder** (bounded at $x = 0$ and $x = L$) or other finite landforms where a water-holding semi-permeable layer overlies a water-saturated aquifer could also have been derived from Equation 3.71. Use $x = L$ as right boundary condition (instead of $x = \infty$), and let the water enter (or move out of) the aquifer from two sides (at $x = 0$ and $x = L$). You may check that the solution is as follows:

Exercise 3.15.2.1 Figure E3.15.2.1 shows that the water table in a polder behind a dyke is at surface level and 2 m lower than the open-water level at the other side of the dyke.

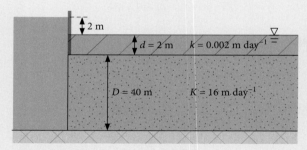

Figure E3.15.2.1

a. Determine the hydraulic head at $x = 10, 100, 250, 500, 1000,$ and 1500 m.

b. Make a drawing of the change in hydraulic head with x.

c. Determine the seepage in mm day^{-1} at $x = 10, 100, 250, 500, 1000,$ and 1500 m.

d. Determine the seepage per 1000 m length of the dyke in m^3 year^{-1} using two methods.

e. Determine the seepage between $x = 0$ and 100 m using two methods.

$$Q'_z = -\frac{\lambda}{c}\left(C_1 - C_1 e^{\frac{L}{\lambda}} - C_2 + C_2 e^{\frac{-L}{\lambda}}\right)$$

$$= -\frac{KD}{\lambda}\left(C_1 - C_1 e^{\frac{L}{\lambda}} - C_2 + C_2 e^{\frac{-L}{\lambda}}\right) \tag{3.83}$$

All exercises 3.15.2 can be solved by using Equation 3.71, Darcy's law for both the leaky aquifer and the leaky confining layer, and continuity. The settings of the exercises 3.15.2.3 and 3.15.2.4, with $h_a > h$, may be taken as representing **paddy fields**, flooded parcels of arable land used for growing rice.

Exercise 3.15.2.2 Figure E3.15.2.2 shows an aquifer and two parallel, fully penetrating canals. The aquifer has a thickness of 10 m and a hydraulic conductivity of 24 m day^{-1}. The leaky confining layer has a thickness of 3 m and a hydraulic conductivity of 2 mm day^{-1}. The distance between both canals is 1200 m. The water level in the left canal is 16 m; the water level in the right canal is 15 m. The water table in the polder in between the canals may be taken at surface level, 13 m. Steady groundwater flow occurs in the x and vertical (z) directions only.

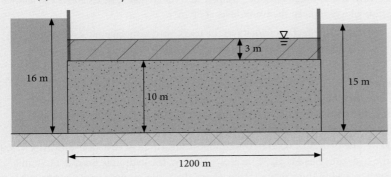

Figure E3.15.2.2

a. Determine the hydraulic head with distance x in the aquifer.

b. Determine the seepage into the polder in m^2 day^{-1}.

c. At what distance x from the left canal is the water divide in the aquifer situated?

Exercise 3.15.2.3 Figure E3.15.2.3 shows an aquifer and two parallel, fully penetrating canals with water levels of 26 m. The aquifer has a transmissivity of 500 m^2 day^{-1}. The thickness of the leaky confining layer is 1 m and its hydraulic conductivity is 2 mm day^{-1}. The distance between the canals is 500 m. The water level in the irrigated fields above the leaky confining layer is 27 m. Steady groundwater flow occurs in the x and vertical (z) directions only.

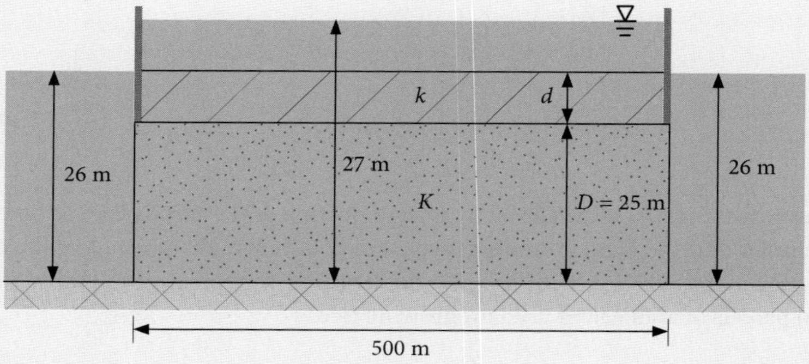

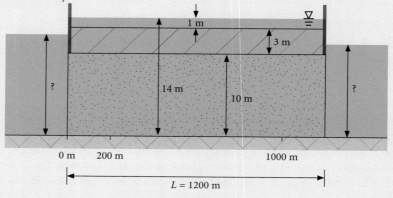

Figure E3.15.2.3

a. Determine the hydraulic head with distance x in the aquifer.

b. Determine the hydraulic head at x = 50, 100, 150, 200, and 250 m.

c. Determine the vertical volume flux density at x = 250 m in the aquifer.

d. Determine the leakage through the leaky confining layer in m^2 day^{-1} using two methods.

e. Given that the replenishment (recharge) by precipitation excess equals 1 mm day^{-1}, how much water must be pumped into the irrigated fields above the leaky confining layer to keep the water level at 27 m?

f. Determine the effective velocity in the leaky confining layer at x = 250 m: choose a value for the missing parameter.

Exercise 3.15.2.4 Figure E3.15.2.4 shows an aquifer and two parallel, fully penetrating canals. The aquifer has a thickness of 10 m and a hydraulic conductivity of 24 m day^{-1}. The thickness of the leaky confining layer is 3 m and its hydraulic conductivity is 2 mm day^{-1}. The distance between the canals is 1200 m. The hydraulic head in the aquifer is measured at two locations. At x = 200 m, the hydraulic head is 12.31 m; at x = 1000 m, the hydraulic head is 11.70 m. The water level in the irrigated fields above the leaky confining layers is 1 m above the leaky confining layer. Steady groundwater flow occurs in the x and vertical (z) directions only.

Figure E3.15.2.4

a. Determine the water level in the left and right canals.

b. Determine the leakage through the leaky confining layer in m² day⁻¹ using two methods.

c. At what distance x from the left canal is the water divide in the aquifer situated?

Exercise 3.15.2.5 Figure E3.15.2.5 shows a typically Dutch inverse landscape profile ('Hollands profiel'), with a high river, a dyke, and a low polder acting as a leaky confining layer. The water level of the fully penetrating river is situated 30 m above an impermeable layer. The aquifer overlies an impermeable layer, and has a thickness of 20 m and a hydraulic conductivity of 25 m day⁻¹. The leaky confining layer has a thickness of 6 m and a hydraulic resistance to vertical groundwater flow c of 80 days. From $x = 0$ to 10 m behind the dyke, the leaky confining layer has been made impermeable by sealing with clay. From $x = 10$ m onwards, the water table in the polder may be taken at surface level. All layers and aquifers are homogeneous and isotropic. Steady groundwater flow occurs in the x and vertical (z) directions only.

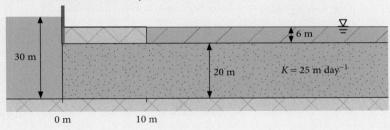

Figure E3.15.2.5

a. Determine the hydraulic head with distance x in the aquifer.

b. Determine the hydraulic head at $x = 5$, 15, and 25 m behind the dyke.

c. Determine the seepage into the polder in m² day⁻¹ using two methods.

d. Determine the seepage into the polder in mm day⁻¹ at $x = 5$, 15, and 25 m behind the dyke.

Adding recharge

For steady groundwater flow in a recharged, unconfined aquifer bordered by two parallel, fully penetrating canals with equal water levels, we may combine Darcy's law (Equation 3.59: $Q' = -Kh\,(dh/dx)$) with the continuity equation for a recharge situation:

$$N = \text{recharge rate (m day}^{-1}) \qquad x = \text{horizontal distance (m)}$$

$$Q' = Nx \tag{3.84}$$

to give the following equation for the potentiometric surface (see M4 at the end of this book):

$$h^2 = -\frac{N}{K}x^2 + C \qquad (3.85)$$

Note that because the present flow case is symmetrical, the location of $x = 0$ is taken in the middle of Figure 3.46 where h (as h_0) is largest due to recharge and groundwater flow in opposite directions to both canals. C in Equation 3.85 can then be found by inserting the boundary conditions for this flow case: first insert the $x = 0$ mid-boundary condition and then the $x = L$ boundary condition (h_0, h_L, and L are all known values). This then yields

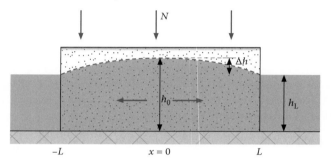

Figure 3.46. Steady groundwater flow in a recharged, unconfined aquifer bordered by two parallel, fully penetrating canals with equal water levels

$$N = \frac{\Delta h}{\left(\dfrac{L^2}{2K\overline{D}}\right)} \qquad (3.86)$$

Again, $\overline{D} = (h_0 + h_L)/2$. Equation 3.86 is known as the **Hooghoudt equation**, after the Dutch chemist and hydrologist Symen B. Hooghoudt (1901–1953). Equation 3.86 resembles Ohm's law and $L^2/(2K\overline{D})$ can be interpreted as a resistance with day as unit (m^2 per (m^2 day^{-1}) = day). Equation 3.86 can be rewritten as follows:

$$\boxed{2L = \text{drain spacing (m)}} \quad 2L = 2\sqrt{\frac{2K\overline{D}\Delta h}{N}} \quad \boxed{\Delta h = \text{convexity (m)}} \qquad (3.87)$$

Equation 3.87 gives the drain spacing $2L$ (in metres) as a function of the recharge rate N (m day^{-1}) and the convexity Δh (m) for an aquifer with transmissivity $T = K\overline{D}$ m^2 day^{-1}. **As the land surface in polders should not be too wet in order to still allow for farming practices (which restricts the values for the convexity Δh), this is a very useful equation for the determination of the optimal distance between the ditches that drain these polders.** Use Equation 3.85 to try to solve Exercise 3.15.3.1.

Equation 3.85 for canals with equal water levels – as, for instance, in polders – of course, can be interpreted as a special case of steady groundwater flow in a recharged, unconfined aquifer bordered by two parallel, fully penetrating canals, but now with different water levels – as, for instance, in mountainous areas and as shown in Figure 3.47. Note that as this flow case is not symmetrical, the location of $x = 0$ is taken to the left. From Darcy's law and continuity, the following equation for the potentiometric surface is derived (see M5 at the end of this book):

$$h^2 = -\frac{N}{K}x^2 + C_1 x + C_2 \qquad (3.88)$$

Note that for $N = 0$, and thus for unconfined groundwater flow with no recharge, Equation 3.88 reduces to $h^2 = C_1 x + C_2$, an equation that we already know from the section on steady, unconfined groundwater flow.

Further note that Equation 3.88 describes the potentiometric surface of a recharge-controlled water table, as the derivation under M5 assumes horizontal streamlines and vertical equipotentials (the Dupuit–Forchheimer assumptions). From section 3.13, we know (after Haitjema and Mitchell-Bruker 2005) that a low average annual recharge rate in comparison with the aquifer hydraulic conductivity, a relatively uneven land surface, and a deep aquifer with high transmissivity are factors that are favourable for the establishment of such a recharge-controlled water table.

Exercise 3.15.3.1 An unconfined aquifer overlying an impermeable layer is drained by two parallel, fully penetrating canals (the canals have equal water levels). The replenishment (recharge) by precipitation excess equals 10 mm day^{-1} and the differential head, the difference between the water table at the water divide and the water level in the canals, is 9 cm. The transmissivity of the unconfined aquifer is 50 m^2 day^{-1}.
Determine the distance between the two canals.

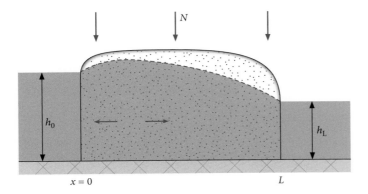

Figure 3.47. Steady groundwater flow in a recharged, unconfined aquifer bordered by two parallel, fully penetrating canals with different water levels

Exercise 3.15.3.2 From Equation 3.88, determine the equations for Q', the distance from $x = 0$ to the water divide, and the elevation of the water table at the divide.

Pump and treat

The stand-alone well (without a regional groundwater flow field)

Pumping wells (withdrawal wells; discharge wells) are functional for the withdrawal of groundwater from aquifers for use as drinking water (Box 3.1) or for other uses (in agriculture or industry). Because it takes a long time to replenish groundwater, authorities and/or drinking water companies should take care not to deplete all of the water stored underground in an aquifer. To counter depletion, water can be infiltrated from infiltration canals or recharge wells (infiltration wells or injection wells; Boxes 3.3 and 3.7). Good use can be made of a recharge well and a pumping well to isolate and treat polluted subsoil, a process generally known as 'pump and treat', as explained at the end of this section. First, let us describe the hydrological flow processes using groundwater hydraulics equations for radial-symmetric, steady groundwater flow as shown in Figure 3.48.

Confined aquifers may contain very large volumes of water, as water is stored under pressure in the pore space of the aquifer (see section 3.9). A **fully penetrating well** is a well that is incised to the top of a lower impermeable layer. The volume flux or discharge Q_0 from such a well in a confined aquifer is as follows:

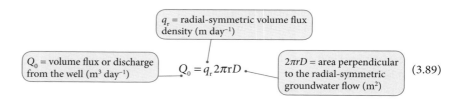

$$Q_0 = q_r 2\pi r D$$

q_r = radial-symmetric volume flux density (m day^{-1})

Q_0 = volume flux or discharge from the well (m^3 day^{-1})

$2\pi r D$ = area perpendicular to the radial-symmetric groundwater flow (m^2)

(3.89)

Combining this with Darcy's law (see M6) yields the following potentiometric drawdown surface as a function of the radial distance from the pumping well r:

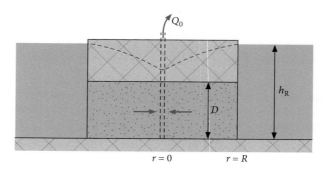

$r = 0$ $r = R$

Groundwater flow in Figures 3.48 and 3.49 is towards the centre line of a cylinder.

Figure 3.48. Radial-symmetric, steady groundwater flow to a fully penetrating pumping well in a confined aquifer

$$h = h_R + \frac{Q_0}{2\pi KD}\ln\frac{r}{R} \quad \text{for } r_w \le r \le R \qquad (3.90)$$

where h_R = pre-pumping hydraulic head (m), R = radius of influence of the pumped well (m), r_w = radius of the pumped well (m).

The volume flux or discharge Q_0 from a pumping well is positive ($r \le R$: $r/R \le 1$; $\ln(r/R) \le 0$; $2\pi KD > 0$; $h < h_R \Rightarrow h - h_R < 0$; $Q_0 > 0$), and Q_0 is negative ($Q_0 < 0$) for a recharge or injection well.

Similar to Equation 3.89, the volume flux Q_0 to a fully penetrating well in an unconfined aquifer as shown in Figure 3.49 can be described as follows:

$$Q_0 = q_r 2\pi r h \qquad (3.91)$$

Note that in Equation 3.91 the hydraulic head h, which for a pumping well diminishes in the direction of the pumping well itself, replaces the thickness or depth D of the aquifer of Equation 3.89. Again, combining Equation 3.91 with Darcy's law yields the following equation for the potentiometric drawdown surface in an unconfined aquifer:

$$h^2 = h_R^2 + \frac{Q_0}{\pi K}\ln\frac{r}{R} \quad \text{for } r_w \le r \le R \qquad (3.92)$$

For unconfined conditions, where the absolute value of the drawdown, $|h - h_R|$, is much smaller than the hydraulic head h itself, we may rewrite Equation 3.92 as follows:

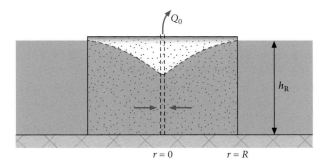

Figure 3.49. Radial-symmetric, steady groundwater flow to a fully penetrating pumping well in an unconfined aquifer

$$h^2 - h_R^2 = (h + h_R)(h - h_R) = 2\bar{D}(h - h_R) \Rightarrow h - h_R = \frac{Q_0}{2\pi K\bar{D}}\ln\frac{r}{R} \Rightarrow$$

$$h = h_R + \frac{Q_0}{2\pi K\bar{D}}\ln\frac{r}{R} \quad \text{for } r_w \le r \le R \qquad (3.93)$$

$\bar{D}$ equals the average saturated depth of the aquifer between $r = r_w$ and $r = R$:

$$\bar{D} = \frac{1}{2}(h_{r_w} + h_R) \tag{3.94}$$

Equation 3.93 for unconfined conditions is similar to Equation 3.90 for confined conditions. Using the symbol T for transmissivity ($T = K \times D$ for confined conditions; $T = K \times \bar{D}$ for unconfined conditions), the general equation for the potentiometric drawdown surface of a well under confined and unconfined conditions, where $|h - h_R| \ll h$, can be written as follows:

$$h = h_R + \frac{Q_0}{2\pi T} \ln\frac{r}{R} \text{ for } r_w \le r \le R \tag{3.95}$$

Q_0 for a pumping well is positive; Q_0 for a recharge well is negative. Equation 3.95 could also have been written as (see M6 for the derivation):

h_2 = hydraulic head at radial distance r_2 from the well (m)

h_1 = hydraulic head at radial distance r_1 from the well (m)

$$h_2 - h_1 = \frac{Q_0}{2\pi T}\ln\frac{r_2}{r_1} \text{ for } r_w \le r \le R \tag{3.96}$$

The latter equation is known as the **Thiem equation**, after the German scientist Adolph Thiem (1836–1908), and can be used to determine the aquifer transmissivity T for a confined aquifer from pumping test data provided that a number of assumptions are (largely) met, such as the existence of steady state conditions, a fully penetrating well, and a homogeneous and isotropic aquifer, uniform in thickness and infinite in extent (Kruseman and De Ridder 1994; Schwartz and Zhang 2003).

Using Equation 3.93, the lowering of the hydraulic head ($h_R - h$) can be determined for every radial position r between $r = r_w$ and $r = R$. From Equation 3.93, it is clear that the drawdown defined as $h - h_R$ has a negative value for a pumping well ($Q_0 > 0$).

A well in a regional groundwater flow field

Figure 3.50a shows a pumping cone in cross-section (drawdown curve) of a well in the absence of a regional groundwater flow field. The hydraulic head h_r of the pumping cone is (as described by Equation 3.95):

h_r = hydraulic head caused by pumping (m)

$$h_r = h_R + \frac{Q_0}{2\pi T} \ln\frac{r}{R} \text{ for } r_w \le r \le R \tag{3.97}$$

Exercise 3.15.4.1 Figure E3.15.4.1 shows four observation wells or piezometers in an aquifer in plan view; the observed hydraulic heads are stated. The aquifer has a thickness of 20 m and a hydraulic conductivity of 10 m day^{-1}. The effective porosity n_e of the aquifer equals 0.4. A waste dump is located between $x = 0$ and 250 m and $y = 500$ and 750 m.

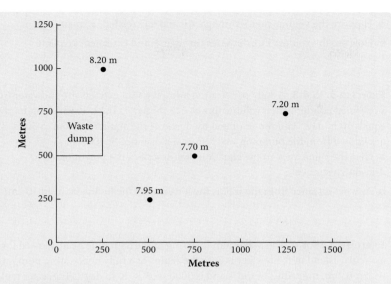

Figure E3.15.4.1

a. Draw the equipotentials of 8 and 7 m.

b. Determine the volume flux density of the natural groundwater flow.

c. Determine the effective velocity of the groundwater as well as its direction of flow.

d. On what line should one place a pumping well to capture polluted groundwater from the observation well?

e. What should be the minimum pumping discharge of the well?

Exercise 3.15.4.2 Figure E3.15.4.2 shows part of the steady groundwater flow to a pumping well in a 10 m thick homogeneous, isotropic, and confined aquifer. There is no regional groundwater flow and the hydraulic conductivity of the aquifer is not known. The pumping discharge of the well equals 314 m^3 day^{-1}. A and B are located at a distance of 200 m from the well, and C and D at a distance of 100 m. AB and CD are arcs of a circle with the pumping well as centre. At 500 m radial distance from the well, the lowering of the hydraulic head due to pumping is reduced to zero.

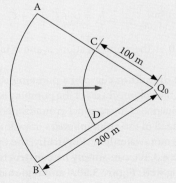

Figure E3.15.4.2

a. How are the volume fluxes through AB and CD related? Explain.

b. How are the volume flux densities through A and C related? Explain.

Exercise 3.15.4.3 A fully penetrating pumping well extracts groundwater from a confined aquifer with a volume flux of 628 m³ day⁻¹. The hydraulic conductivity equals 10 m day⁻¹ and the saturated thickness of the aquifer is equal to 50 m. The drawdown from the pumping well ceases at a distance of 1000 m.
a. Give the value for R in the equation that describes the potentiometric drawdown surface.

b. At what distance from the well is the lowering of the hydraulic head 10 cm?

Exercise 3.15.4.4 The maximum lowering of the hydraulic head allowed for a circular island in fresh water is 3 m; the radius of a fully penetrating pumping well is 0.2 m; the radius of the island is 5000 m; $K = 20$ m day⁻¹. The water table at the circumference of the island lies 25 m above a flat, impermeable layer that extends all the way below the island ($h_R = 25$ m).
a. Use Equation 3.92 to determine the maximum volume flux for the pumping well in the unconfined aquifer.

b. Determine the maximum volume flux for the pumping well had we assumed a confined aquifer with saturated depth $D = 25$ m.

c. Explain the difference in the answers for a and b.

Figure 3.50b shows the potentiometric surface of a regional groundwater flow field in cross-section. The groundwater in the regional field flows from right to left in the direction of the lower hydraulic head. The shape of the potentiometric surface of a regional, unconfined groundwater flow system in cross-section is parabolic (Equation 3.60). However, as the parabola is quite flat, we may approximate its shape as a linear function, with the hydraulic gradient i as slope and C as a constant. We may assume the hydraulic gradient i for unconfined groundwater flow to be of the order of $\frac{1}{1000}$ (a positive gradient as flow is to the left, as explained in section 3.7). Thus, for both confined and unconfined conditions:

$$h_x = ix + C = ir + C \qquad (3.98)$$

h_x = hydraulic head of the original, regional groundwater flow field (m)

Taking $x = 0$ and $r = 0$ at the same location, x and r in cross-section are interchangeable.

Figure 3.50c shows the drawdown curve of a pumping well in a regional groundwater flow field. Note that the volume fluxes of the pumping well in Figures 3.50a and c are taken as equal. The influence of the regional groundwater flow field in Figure 3.50c is to lower the hydraulic head of the pumping well's drawdown curve in the direction of the original, regional groundwater flow. **Physically, the hydraulic head h_{tot} in the drawdown curve of Figure 3.50c can simply be constructed by summation of the influences of the pumping well (Figure 3.50a) and regional groundwater flow field (Figure 3.50b).** Graphically, the hydraulic head h_{tot} in Figure 3.50c is thus derived by

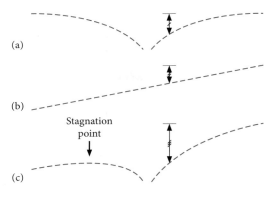

(a)

(b)

Stagnation
point

(c)

Figure 3.50. A potentiometric surface due to
groundwater extraction by a pumping well in cross-
section (a) without a regional groundwater flow field,
and (c) in a regional groundwater flow field; (b) shows
the potentiometric surface of the regional groundwater
flow field itself – the curve of (c) is obtained by
summation of the curves for (a) and (b)
The reference level for the hydraulic head has been
taken to the utmost right in the figures; the hydraulic
head is therefore negative, and equals the drawdown
for curves (a) and (c)

adding together h_r of Figure 3.50a and h_x of Figure 3.50b. Mathematically, h_{tot} is the
summation of h_r (Equation 3.97) and h_x (Equation 3.98):

h_{tot} = hydraulic head caused by pumping in
a regional groundwater flow field (m)

$$h_{tot} = h_r + h_x = h_R + \frac{Q_0}{2\pi T} \ln\frac{r}{R} + ir + C = h_R + \frac{Q_0}{2\pi T}(\ln r) - \frac{Q_0}{2\pi T}(\ln R) + ir + C \quad (3.99)$$

Stagnation point

At a certain point to the left in Figure 3.50c, the hydraulic head ceases to be influenced by
flow to the pumping well located to the right. The regional groundwater flow is to the left
and, thus, exactly at the point where flow to the right to the pumping well is replaced by
regional groundwater flow to the left, we must encounter a point of no groundwater flow.
This point is called the **stagnation point**. Figure 3.51 shows the plan view of the streamlines caused by a pumping well in a regional groundwater flow field. Note the position
of the stagnation point in this figure. The stagnation point is a local maximum in the
potentiometric surface h_{tot}. Mathematically, a local maximum can be derived as follows:

$$\frac{dh_{tot}}{dr} = 0 \quad (3.100)$$

Differentiation of Equation 3.99 gives

$$\frac{dh_{tot}}{dr} = \frac{Q_0}{2\pi T}\left(\frac{1}{r}\right) + i \quad (3.101)$$

Combining Equations 3.100 and 3.101 yields

$$\frac{Q_0}{2\pi T}\left(\frac{1}{r}\right) + i = 0 \Rightarrow \frac{Q_0}{2\pi Tr} + i = 0 \Rightarrow \frac{Q_0}{2\pi Tr} = -i \Rightarrow 2\pi Tr = \frac{-Q_0}{i}$$

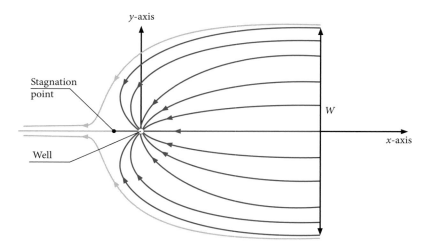

Figure 3.51. Streamlines due to groundwater extraction by a pumping well in a regional flow field in plan view (o, pumping well)

From this, the position of the stagnation point relative to the centre of the pumping well is thus found as

$$r = \frac{-Q_0}{2\pi Ti} \tag{3.102}$$

r is negative for a pumping well ($Q_0 > 0$) in a regional groundwater flow field from right to left ($i > 0$), as in Figures 3.50b and c. This is correct, as the stagnation point then is located to the left of the centre of the pumping well (where $r = 0$).

Exercise 3.15.4.5 As shown in Figure 3.51, a fully penetrating well (coordinates 0,0) pumps water from a confined aquifer with a thickness of 50 m. Before pumping, there existed a uniform groundwater flow parallel to the x-axis in a negative direction. A **uniform groundwater flow** is one in which the groundwater flow velocity has the same magnitude and direction at every point. The hydraulic gradient of this uniform groundwater flow is $\frac{1}{1000}$. The aquifer has a hydraulic conductivity of 10 m day^{-1}. The well pumps water with a discharge or volume flux of 314 m^3 day^{-1}.

a. Determine the maximum width of the area from which water is extracted by the well.

b. Determine the location of the stagnation point in the groundwater flow field.

Hydrological isolation of polluted subsoil

The left part of Figure 3.52 is equivalent to Figure 3.51, and thus shows streamlines in a regional groundwater flow field as influenced by a pumping well Q_1 ($Q_1 > 0$). **If the subsoil to the right of the pumping well is polluted, the polluted water can be drawn to the pumping well and treated, after which clean(ed) water is injected (recharged) upstream, at a location symmetrical to the centre of the polluted area, as shown in Figure 3.52. When the injection rate Q_2 of well 2 is taken equal to the pumping rate Q_1 of well 1 ($Q_2 = -Q_1$), and with a correct set-up (concerning the distance between the wells and the actual pumping rate Q_1), the polluted area is**

captured in a symmetrical water lens (largely) isolated from the regional ground-water flow system, as evident from the streamlines in Figure 3.52. Because of the continued pumping and recharge, the subsoil will get cleaner and cleaner still with the passing of time. This **hydrological isolation** is an example of a technique more generally called **pump and treat**.

Hydrological isolation of polluted subsoil is a very cost-effective measure, as the costs of pumping and injection (recharge) are very low, especially when compared to

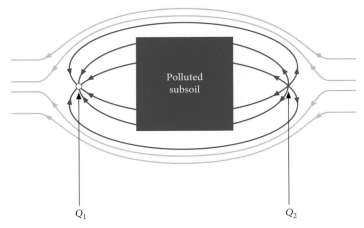

Figure 3.52. Pump and treat: hydrological isolation of polluted subsoil by a pumping well (o) with discharge Q_1 and a recharge well (x) with recharge Q_2 $(Q_2 = -Q_1)$

measures such as excavation and/or isolation of polluted soil by clay dams or concrete structures.

Superposition, image wells, and boundaries

Superposition

In general, for a fully penetrating well, numbered as n, be it a pumping well $(Q_n > 0)$ or recharge well $(Q_n < 0)$, we may rewrite Equation 3.95 as follows:

s_n = drawdown due to well n (m)
$s_n < 0$ for a pumping well $(Q_n > 0)$
$s_n > 0$ for a recharge well $(Q_n < 0)$

$$s_n = h_{r_n} - h_{R_n} = \frac{Q_n}{2\pi T} \ln \frac{r_n}{R_n} \text{ for } r_{w_n} \leq r_n \leq R_n \qquad (3.103)$$

Figure 3.53 shows the plan view of a piezometer and four wells, pumping and/or recharge wells in a flat area. Each well in Figure 3.53 has its own volume flux Q and circle of influence R, and the drawdown at distances r from a specific well can be calculated by inserting the correct values for Q, r, and R in Equation 3.103. There is no regional groundwater flow. The piezometer measures the hydraulic head as influenced by all four wells. The principle of **superposition** states that the total drawdown

s_{tot} at any location within the circles of influence of all wells (as determined by their values of R) can be calculated as the sum of the drawdowns of each well itself. Thus:

$$\boxed{s_{tot} = \text{total drawdown (m)}} \qquad s_{tot} = s_1 + s_2 + \dots + s_n \qquad (3.104)$$

In Figure 3.53, there are four wells ($n = 4$), and the hydraulic head in the piezometer h after steady conditions have set in can be found by summation of the

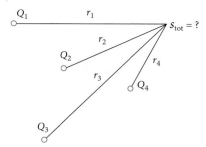

Figure 3.53 A plan view of a flat area with a piezometer and four fully penetrating wells (pumping and/ or recharge wells); the principle of superposition applies

pre-pumping hydraulic head h_R and the total drawdown s_{tot} (negative when the hydraulic head has lowered, and positive when the hydraulic head has increased) at the location of the piezometer:

$$h = h_R + s_{tot} \qquad (3.105)$$

What should a water company or the authorities do when they encounter a polluting substance in one of the water pumping wells in a pumping field where

Exercise 3.15.5.1 Figure E3.15.1 shows the location of three pumping wells and the location of a piezometer in a flat area. The aquifer from which water is pumped is confined. The hydraulic head measured in the piezometer prior to pumping was 10 m above mean sea level. $Q_1 = 314$ m³ day⁻¹; $Q_2 = 471$ m³ day⁻¹; $Q_3 = 628$ m³ day⁻¹. The transmissivity $T = 100$ m² day⁻¹. $R_1 = R_2 = R_3 = 500$ m.

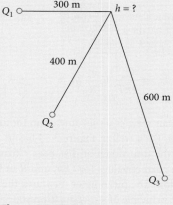

Figure E3.15.5.1

a. Calculate the hydraulic head in the piezometer after steady groundwater flow to the pumping wells has started.

b. Change Q_2 to a recharge well with the same absolute volume flux as earlier. Again, calculate the hydraulic head after steady groundwater flow has started.

drinking water is produced from groundwater? Of all feasible measures, do not close the well! It is best practice to increase the volume flux Q of the polluted well to try to draw all of the subsoil pollution to this one well. Of course, this renders the water from this well unsuitable for consumption, and the water should be treated in accordance with the type of pollutant found. Also, all other wells in the pumping field should continuously be monitored for traces of the polluting substance. In Box 3.8, two dangerous pollutants in groundwater – trichloroethylene and arsenic – are briefly discussed.

Box 3.8 Trichloroethylene and arsenic

A major threat to the production of drinking water in (former) industrial countries is caused by the subsoil presence of DNAPLs, dense non-aqueous phase liquids, such as trichloroethylene (trichloroethene), a chlorinated hydrocarbon commonly used as an industrial solvent. Trichloroethylene is found in soil and groundwater at many waste sites. Its solid (non-aqueous) phase is heavier (denser) than water and percolates to rest on impermeable and semi-permeable layers in the subsoil; it is a co-carcinogen substance – acting together with other substances to promote the formation of tumours – and it cannot easily be removed from the subsoil. This is an example of a man-made contaminant in groundwater. Countries such as Bangladesh (the Ganges Delta) and India (West Bengal), on the other hand, suffer severe problems due to high levels of naturally occurring arsenic in groundwater used for human consumption (Appelo 2008): arsenic compounds are both toxic and carcinogenic.

Open-water linear boundary

Figure 3.54 shows streamlines in plan view and the potentiometric surface in cross-section (let us say west–east) between a recharge well and a pumping well. There is no regional groundwater flow in Figure 3.54. The total drawdown s_{tot} in the cross-section can – similar to the procedure followed for Figure 3.50c – be established by adding together the drawdown s_1 caused by the recharge well ($s_1 > 0$) and the drawdown s_2 due to the pumping well ($s_2 < 0$). Thus:

$$s_{tot} = s_1 + s_2 \qquad (3.106)$$

Interestingly, this causes a constant hydraulic head equal to the pre-pumping hydraulic head h_R exactly at the midpoint between the two wells, as can be seen in the cross-section of Figure 3.54. **From the plan view of Figure 3.54, it is clear that this midpoint is part of a line with a constant hydraulic head equal to the pre-pumping hydraulic head h_R, and that this line extends in a (north–south) direction perpendicular to the cross-section.**

Now imagine a situation with only a pumping well and, to the left (west) of it, open water (a lake or canal) with a straight right (eastern) boundary coinciding with the line with constant hydraulic head h_R of Figure 3.54 just mentioned. In reality, as the

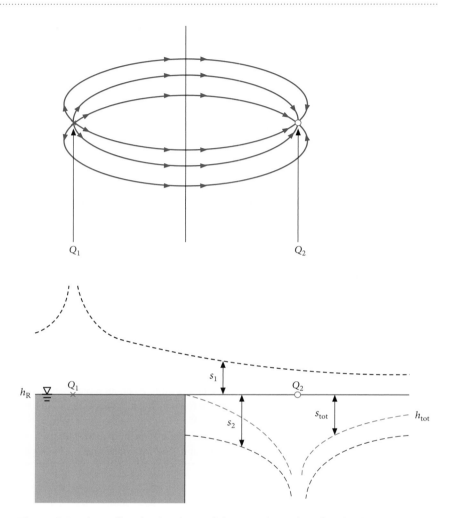

Figure 3.54. Streamlines in plan view and the potentiometric surface h_{tot} in cross-section between a recharge (x) and pumping well (o) ($Q_1 = -Q_2$); simulation of an open-water linear boundary (see the text for explanation)

open water and groundwater meet at this straight boundary (and because the open water is continuously replenished from upstream), the water level at the open-water boundary would remain at this constant level h_R all the time, before pumping and also after pumping commenced. In fact, if one had wanted to insert an open-water linear boundary with a constant hydraulic head h_R to the left (west) of the pumping well, mirroring the pumping well to the other side of the open-water boundary, converting it into a recharge well with the same absolute volume flux ($Q_1 = -Q_2$), and then adding s_1 and s_2 together to give s_{tot} would be just the way to go about it. The mirrored well is called an **image well**. In reality, this well does not exist, but is solely incorporated into the model for the wanted effect of creating an open-water linear boundary with constant hydraulic head h_R to the left (west) of the pumping well. The part of the model that exists in reality is the part to the right (east) of the open-water boundary. A known constant hydraulic head at the boundary of a water flow region is called a **Dirichlet boundary condition**, after the German mathematician Johann Peter Gustav Lejeune Dirichlet (1805–1859).

Pumping groundwater from a sandy aquifer bordering a river to provide for drinking water has as advantages that the groundwater is abundantly replenished from its connection with the surface water, and also that groundwater moving through sandy subsoil for some 60 days is biologically purified and freed from disease-causing micro-organisms (see Box 3.1).

No-flow linear boundary

Figure 3.55 shows streamlines in plan view and the potentiometric surface in cross-section (west–east), again between two wells, but now both are pumping wells

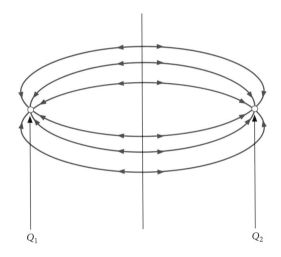

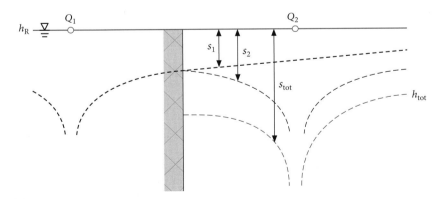

Figure 3.55. Streamlines in plan view and the potentiometric surface h_{tot} in cross-section between two pumping wells (o) ($Q_1 = Q_2$); simulation of a no-flow linear boundary

that have equal volume fluxes ($Q_1 = Q_2$). **It is clear from the cross-section that this set-up creates a flat potentiometric surface at the midpoint and midline between the two wells.** A flat potentiometric surface means no hydraulic gradient and thus no groundwater flow at the midline between the two wells, and thus the creation of a no-flow linear boundary. A known constant volume flux at the boundary of a water flow region is called a **Neumann boundary condition**, after the German mathematician Carl Gottfried Neumann (1832–1925). In this particular case, the constant volume flux is a zero volume flux.

Calculation of the drawdown near a linear boundary

Figure 3.56 explains how to calculate the drawdown at a location in the vicinity of a pumping well near (a) an open-water linear boundary and (b) a no-flow linear boundary (impermeable area). One should first mirror the pumping well as explained above, calculate the drawdown (negative value) using the distance r_1 to the pumping well, then calculate the drawdown (positive or negative) using the distance r_2 to the image well, and then apply the principle of superposition by adding the two drawdowns together to give the total drawdown s_{tot}.
Had the original well been a recharge well (instead of the pumping well of Figure 3.56), then modelling an open-water linear boundary would have involved an image pumping well, and modelling a no-flow linear boundary would have involved an image recharge well.

Exercises 3.15.5.2 and 3.15.5.3 provide some training in the principles outlined above.

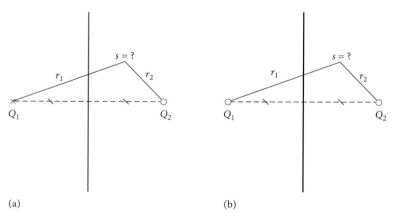

(a)   (b)

Figure 3.56. Calculation of the drawdown at a location in the vicinity of a pumping well (o) with discharge Q_2 near an open-water linear boundary (a) and a no-flow linear boundary (b)

Exercise 3.15.5.2 On behalf of the drinking water supply, groundwater is pumped from an aquifer. The pumping discharge is 1257 m³ day⁻¹. The saturated depth of the aquifer is 50 m. The aquifer consists of two layers. The lower layer has a thickness of 20 m and a hydraulic conductivity of 10 m day⁻¹. The upper layer has a thickness of 30 m and a hydraulic conductivity of 30 m day⁻¹. The aquifer may be interpreted as confined.

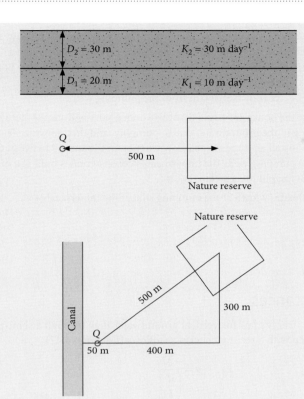

Figure E3.15.5.2

a. Determine the total transmissivity of the aquifer.

The midpoint of a nature reserve is located at a distance of 500 m from the pumping well, as shown in the middle part of Figure E3.15.5.2. There is a fear that the nature reserve will dry out or suffer from desiccation. At a radial distance of 2 km from the well, there is no lowering of the hydraulic head due to pumping.

b. Determine the lowering of the hydraulic head for the midpoint of the nature reserve.

The above situation is extended with a canal and waste dump. The canal is located at a distance of 50 m from the pumping well. The waste dump is located at a distance of 500 m from the pumping well. The lower part of Figure E3.15.5.2 shows this new situation.

c. Explain qualitatively what the effect will be of the existence of the canal on the lowering of the hydraulic head in the midpoint of the nature reserve in comparison to the earlier situation under b.

d. Quantitatively determine the lowering of the hydraulic head in the midpoint of the nature reserve for the new situation. Is this answer in agreement with your qualitative assessment under c?

e. How are the volume fluxes through both layers of the aquifer related?

The effective porosity for both layers equals 0.4.

f. How are the effective velocities in both layers of the aquifer related?

Exercise 3.15.5.3 A confined aquifer has a transmissivity of 50 m² day⁻¹. A house on wooden, supporting piles is located 1 km from a pumping well with a pumping rate (volume flux) of 314 m³ day⁻¹; the drawdown below the house due to pumping is –10 cm.

a. Calculate the drawdown at 400 m from the pumping well.

The owners of the house are afraid that the wooden piles will start to decay due to the lowering of the groundwater due to pumping, and they convince friendly authorities to install an impermeable construction in the subsoil between their house and the pumping well. The impermeable construction is built at a distance of 400 m from the pumping well.

b. Calculate the drawdown at the pumping side of the impermeable construction.

Extra resistances

In section 3.11, Darcy's law for vertical groundwater flow through a semi-permeable layer was written with a resistance term as follows (Equation 3.47):

$$q = \frac{-\Delta h}{\left(\dfrac{d}{k}\right)} \tag{3.107}$$

In section 3.10, Darcy's law for horizontal groundwater flow under confined conditions was written as follows (Equation E3.10.1.1):

$$Q' = \frac{-\Delta h}{\left(\dfrac{L}{KD}\right)} \tag{3.108}$$

In this section 3.15, the potentiometric surface in a recharged, unconfined aquifer bordered by two fully penetrating canals with equal water levels was written as follows (Equation 3.86):

$$N = \frac{\Delta h}{\left(\dfrac{L^2}{2K\bar{D}}\right)} \tag{3.109}$$

All of the analytical equations derived thus far are based on simplifying assumptions; for instance, the Dupuit–Forchheimer assumption of horizontal groundwater flow in an unconfined aquifer (see Figure 3.57a). **In reality, canals may not be fully penetrating, and streamlines in an unconfined aquifer are not horizontal but slightly bent, slightly longer, and contracting near the canal, causing an extra flow resistance (Figure 3.57b). Also, the canals may be lined with less permeable material, causing another extra resistance**, leading to a seepage face, which is a steeper hydraulic gradient near the outflow boundary of an aquifer to open water, as shown in Figure 3.57c. To account for deviations from reality, extra resistances

Ω may be built into the equations as follows (note the different units of Ω to be in accordance with the original resistance):

$$q = \frac{-\Delta h}{\left(\dfrac{d}{k} + \Omega_v\right)}$$ ⎯ Ω_v = extra vertical flow resistance (days)   (3.110)

$$Q' = \frac{-\Delta h}{\left(\dfrac{L}{KD} + \Omega_h\right)}$$ ⎯ Ω_h = extra horizontal flow resistance (day m^{-1})   (3.111)

$$N = \frac{\Delta h}{\left(\dfrac{L^2}{2K\overline{D}} + \Omega_N\right)}$$ ⎯ Ω_N = extra horizontal flow resistance (days)   (3.112)

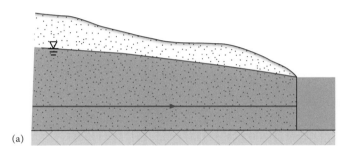

(a)

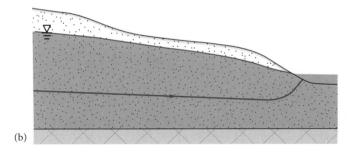

(b)

Seepage face

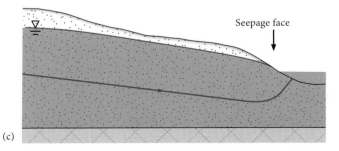

(c)

Figure 3.57. Unconfined groundwater flow: (a) the modelled situation; (b) with an extra flow resistance; and (c) with yet another extra resistance due to less permeable material along the outflow boundary, causing a seepage face

Table 3.3. A summary sheet for groundwater hydraulics

One-dimensional steady groundwater flow

Confined

$$h = C_1 x + C_2$$

(3.29)

Unconfined

$$h^2 = C_1 x + C_2$$

(3.68)

Leaky

$$h = h_a + C_1 e^{\frac{x}{\lambda}} + C_2 e^{\frac{-x}{\lambda}} \quad \text{with } \lambda = \sqrt{KDc}$$

(3.71)

Recharge; equal
water levels

$$h^2 = -\frac{N}{K}x^2 + C$$

(3.85)

Recharge; different
water levels

$$h^2 = -\frac{N}{K}x^2 + C_1 x + C_2$$

(3.88)

Radial-symmetric steady groundwater flow

Confined

$$h = h_R + \frac{Q_0}{2\pi KD} \ln\frac{r}{R} \quad \text{for } r_w \leq r \leq R$$

(3.90)

Unconfined

$$h^2 = h_R^2 + \frac{Q_0}{\pi K} \ln\frac{r}{R} \quad \text{for } r_w \leq r \leq R$$

(3.92)

Rationale

Table 3.3 summarizes the equations for the different potentiometric surfaces needed to solve the exercises throughout this section 3.15. The equations for the potentiometric surfaces are derived after combining Darcy's law and continuity for steady groundwater flow in a homogeneous and isotropic medium, usually from Laplacians such as (see section M):

$$\frac{d^2h}{dx^2} = 0 \text{ for confined aquifers} \qquad (3.113)$$

or

$$\frac{d^2h^2}{dx^2} = 0 \text{ for unconfined aquifers} \qquad (3.114)$$

These are special cases of the three-dimensional partial differential equation for steady groundwater flow in a homogeneous and isotropic medium known as the **Laplace equation**, after Pierre-Simon (marquis de) Laplace (1749–1827):

$$\frac{\partial^2h}{\partial x^2} + \frac{\partial^2h}{\partial y^2} + \frac{\partial^2h}{\partial z^2} = 0 \qquad (3.115)$$

Equipped with basic knowledge of the workings of Bernoulli's law, Darcy's law, continuity, and the principle of superposition, the reader should – after some effort, of course – be able to solve the exercises in this section using only Table 3.3 as a memory aid.

Analytical equations have been established for both steady and transient groundwater flow cases. Examples of transient groundwater flow are when pumping starts or ends, or when an open-water level is suddenly lowered or raised; thus when there is no steady state. For instance, the general flow equation for radial-symmetric unconfined groundwater flow in a homogeneous and isotropic medium is the **Boussinesq equation** (1904), named after French mathematician and physicist Joseph Valentin Boussinesq (1842–1929). This equation includes both the change of hydraulic head in time $\partial h / \partial t$ and a storage term S_y, known as the **specific yield**, which is measure of the volume of water per volume of porous material released by the forces of gravity in response to a decline of the water table. If the drawdown in the aquifer is very small compared with the saturated thickness or depth of the aquifer, and without a recharge term, the Boussinesq equation can be linearized to the form (Fetter 2001):

$$\frac{\partial^2h}{\partial x^2} + \frac{\partial^2h}{\partial y^2} = \frac{S_y}{K\overline{D}}\frac{\partial h}{\partial t} \qquad (3.116)$$

There are analytical solutions to many groundwater flow problems. Dr Gijs Bruggeman, renowned in The Netherlands for his knowledge of analytical solutions to groundwater flow problems, presented analytical solutions to some 1100 geohydrological problems (Bruggeman 1999) – for most of which, however, the reader and/or user needs quite a firm understanding of mathematics. As an introduction to analytical groundwater hydrology (groundwater hydraulics), and thus for the purpose of this book, the seven equations of Table 3.3 will suffice. **Importantly, analytical solutions to groundwater problems are especially useful to trained hydrologists, as they provide a quick first estimate to many groundwater flow problems.**

For conceptually complex hydrological problems, more complex mathematical models are needed, and for this hydrologists use numerical computational methods that run on a computer. Basically, three types of numerical approaches can be distinguished that use finite differences, finite elements, or analytical elements.

The **finite difference method** (FDM) partitions the groundwater flow domain of interest into a grid of rectangular blocks and for each block solves the finite difference equations based on Darcy's law and continuity. Finite difference equations relate each node in a grid to the hydraulic head of its neighbouring nodes and, in transient calculations, to the hydraulic head at the node at the previous time step (Fitts 2002).

A well-known example of a finite difference modelling code is **MODFLOW** (Modular Three-dimensional Finite-difference Groundwater Flow Model; McDonald and Harbaugh 1988; Anderson and Woessner 1992).

The **finite element method** (FEM) has the advantage (over a FDM) of using more flexible elements to partition (discretize) the domain of interest, such as differently shaped triangular and trapezoidal prisms. However, the mathematics involved is inherently more difficult to understand (than for the FDM). A finite element modelling code much used in The Netherlands is **MicroFEM** (Nienhuis and Hemker 2009).

The last-mentioned approach, **the analytical element method** (AEM), is a numerical method based on the superposition of a large number of analytical solutions, as discussed in this section. The use of a computer to superpose large numbers of analytical solutions in groundwater flow models was an idea pioneered by Professor Otto D.L. Strack of the University of Minnesota. A brief presentation of the principles on which the technique is based is presented in Strack (1999); a full coverage is provided in the textbooks by Strack (1989) and Haitjema (1995). The names of the model codes used end with AEM – **SLAEM**, for instance, is a single-layer (SL) version.

Interesting as they may be, to dwell further on the numerical modelling methods mentioned above is beyond the scope of this book. There is an increasing number of excellent textbooks on numerical modelling for the more advanced student or graduate student. For the undergraduate student or any other reader starting off with hydrology, it is the author's hope that this chapter has provided the reader with valuable insights into the workings of both the basic and essential hydrological laws.

→ *Summary*

- An aquifer is a subsurface layer that easily stores and transmits groundwater.

- Steady groundwater flow (stationary groundwater flow) is groundwater flow for which the velocity components at any location do not change with time: when the magnitude or direction of flow does change with time, or when both change, the flow is called unsteady or transient.

- Mechanical energy is the energy that an object has because of its motion or position.

- A piezometer is used for measuring the hydraulic head, the elevation to which the water rises in a piezometer. One can establish the direction of groundwater flow from hydraulic head readings of piezometers at different locations and screens at different depths. The potentiometric surface of an aquifer is the imaginary surface that one obtains when connecting the hydraulic heads of an aquifer.

- The hydraulic head h equals the sum of the elevation head z and the pressure head: $h = z + \dfrac{p}{\rho g}$

- Unconfined groundwater (or phreatic groundwater) is groundwater where the water table can establish itself freely; an aquifer with unconfined groundwater is called an unconfined aquifer.

- A confining layer is a subsurface layer with little or no water permeability. Confining layers that totally block groundwater flow are impermeable layers or aquifuges. Confining layers with low water permeability are called semi-permeable layers, leaky confining layers, or aquitards.

- Groundwater captured between two impermeable layers is called confined groundwater and the aquifer a confined aquifer.

- Groundwater captured between two confining layers, one or both of these being semi-permeable, is called semi-confined groundwater and the aquifer a semi-confined aquifer or leaky aquifer.

- When the hydraulic head of a confined or semi-confined aquifer lies above the land surface, the groundwater is called artesian.

- Unconfined groundwater underlain by an unsaturated zone is called perched groundwater and its water table a perched water table.

- Hydrostatic equilibrium is the condition of there being no water flow. In unconfined groundwater under hydrostatic equilibrium conditions, the water table and hydraulic heads have the same level.

- Leakage (downward seepage) is groundwater flow through a (sub)surface layer in a downward direction; seepage is groundwater flow through a (sub)surface layer in an upward direction.

- Field capacity is the maximum water content that a soil can hold against gravity. The (volumetric) water content at field capacity is normally lower than the (volumetric) water content at saturation of the soil. At saturation, all the pores in the soil are filled with water; at field capacity, the largest pores in the soil may still be filled with air. Percolation starts when the soil becomes wetter than field capacity.

- Q = volume flux or discharge (m^3 day^{-1}). Darcy's law for horizontal groundwater flow is as follows:
$$Q = -KA\frac{\Delta h}{\Delta x}$$

- Q' = volume flux per unit width or discharge per unit width (m^2 day^{-1}). Darcy's law for horizontal groundwater flow in a confined aquifer is as follows:
$$Q' = -KD\frac{\Delta h}{\Delta x}$$

- q = volume flux density or specific discharge (m day^{-1}). Darcy's law for vertical groundwater flow through a semi-permeable layer is as follows: $q = -k(\Delta h / \Delta z)$ (k = vertical hydraulic conductivity).

- The volume flux density q (m day^{-1}) is defined as a volume flux Q (m^3 day^{-1}) per unit area A (m^2) perpendicular to the direction of flow. Note that the area A is perpendicular to the water flow and that it contains both solid ground particles and pores: water flow is through (interconnected) pores only, **not** through the solid particles, which is why the volume flux density (m day^{-1}) is **not** a velocity.

- The (saturated) hydraulic conductivity K (m day^{-1}) is a proportionality factor in which the characteristics of both the water and the material through which the water flows are taken into account. Sand, for instance, has a high hydraulic conductivity, whilst clay has a low hydraulic conductivity. The intrinsic permeability κ (m^2) is only a function of properties of the porous medium through which the water flows, and thus, importantly, not of the properties of the water itself.

- The difference in hydraulic head Δh (m) equals the hydraulic head at the water-receiving end h_2 (m) minus the hydraulic head at the water-despatching end h_1 (m): $\Delta h = h_2 - h_1$.

- The hydraulic gradient i (dimensionless) is defined as the difference in hydraulic head Δh (m) over a porous medium divided by the distance (m) over which this difference in hydraulic head occurs. The distance (m) is defined as the location to which water flows minus the location from which it is despatched. For horizontal water flow, the distance Δx thus equals $x_2 - x_1$, and for vertical groundwater flow $z_2 - z_1$. The hydraulic gradient always has a sign that is opposite to the sign of the volume flux Q (or Q') and the volume flux density q.

- Darcy's law is limited to laminar flow situations; for example, situations where water can macroscopically be regarded to flow in parallel layers, with no disturbance between the layers. The opposite of laminar flow is turbulent flow, which is characterized by disturbances, eddies, and a chaotic flow pattern.

- To obtain the effective velocity v_e (m day^{-1}) of groundwater, one needs to divide the volume flux density or specific discharge q (m day^{-1}) by the effective porosity n_e (dimensionless): $v_e = q / n_e$; the effective porosity n_e is defined as the volume fraction of a rock, sediment, or soil that participates in the water flow process.

- The water balance (equation) in hydrology is also known as continuity or the continuity equation: continuity for steady groundwater flow implies that the volume flux or discharge Q (m^3 day^{-1}) is constant, as no volume or mass of water is created or destroyed during the flow process; continuity does not necessarily imply that the volume flux density q (m day^{-1}) is constant!

- The representative elementary volume is the minimal sample volume that must be studied to give a consistent value for a measured parameter of interest; for instance, the hydraulic conductivity.

- In a homogeneous layer, the hydraulic conductivity K is the same at every location. A homogeneous layer may be isotropic or anisotropic. In an isotropic layer, the hydraulic conductivity K at each location is the same in all directions, or, in other words, independent of direction. The opposites of homogeneity and isotropy are heterogeneity, where K is not the same at every location, and anisotropy, where the value of K at a location depends on direction.

- Aquifer thermal energy storage is a much used, effective, and cost-efficient technique for both cooling and heating buildings, replacing part of the electricity demand in doing so and thus also reducing the possible burning of fossil fuels and CO_2 emissions into the atmosphere.

- A very important consequence of streamline refraction is that in a setting of horizontally layered sediments, groundwater flow can be modelled as horizontal in the sandy aquifer and as vertical in the semi-permeable clay layer.

- In an equilibrium situation, the total pressure or total stress σ_t (N m^{-2}) at the interface of a confined aquifer and overlying impermeable layer equals the sum of the intergranular pressure or effective stress σ_i (N m^{-2}) and the water pressure or neutral stress p (N m^{-2}): $\sigma_t = \sigma_i + p$.

- The transmissivity T is a measure of how easily an aquifer transmits water and has the same units as Q' (m^2 day^{-1}). For a homogeneous, confined aquifer, the transmissivity T equals the product of the hydraulic conductivity K and the saturated depth D of the aquifer: $T = K \times D$.

- In a confined aquifer consisting of horizontal layers, the total transmissivity T is the sum of the transmissivities of each layer: $T = T_1 + T_2 + ...$; the substitute hydraulic conductivity K may be derived from this relation written as: $KD = K_1 D_1 + K_2 D_2 + ...$

- The hydraulic resistance or vertical flow resistance c (days) is defined as d / k, where d is the depth or thickness of the layer (m) and k is the vertical hydraulic conductivity (m day^{-1}); for ease of recognition, this book uses the characters d and k in vertical groundwater flow equations to distinguish this from horizontal groundwater flow (where the capitals D and K are used in the equations).

- The total hydraulic resistance c of a number of horizontal, homogeneous semi-permeable layers equals the sum of the hydraulic resistances of each layer: $c = c_1 + c_2 + ...$ This relation is useful when calculating the vertical volume flux density q through horizontally layered semi-permeable layers: the hydraulic resistance c (days) is essentially different from the travel time of water particles (also in days).

- A topography-controlled water table closely follows the local topography as a subdued replica, whereas a recharge-controlled water table does not follow the local topography, but does follow the regional topography. The degree to which the shape of the water table is topography controlled or recharge controlled depends on the recharge rate, the aquifer transmissivity, the aquifer geometry, and the topography itself.

- Electrical conductivity (*EC*) is a measure of the water's ability to conduct electricity, and therefore a measure of the water's ion concentration. The *EC* can be measured in the field with a simple small stick-like apparatus attached with a wire to a measuring device. The *EC* is measured in μS cm^{-1} (micro-siemens per centimetre). As the *EC* is temperature dependent, it is standard practice to recalculate its value to the *EC* that would occur at a temperature of 20°C.

- The time of contact of groundwater with the subsoil influences the chemical composition of the groundwater. Atmocline water is groundwater that has just infiltrated and therefore still carries the traits of atmospheric water, such as a low ion concentration (*EC* of the order of 10 μS cm^{-1}). Lithocline water is groundwater that has been in contact with the subsoil for a long time and therefore usually has a higher ion concentration (*EC*s are usually smaller than 1000 μS cm^{-1}): lithocline water can have a high hardness (substantial amounts of Ca^{2+}). Thalassocline water is groundwater that contains much salt; it can have such a high ion concentration that the *EC* measuring device becomes damaged.

- There are many ways to display the differences in water chemistry; for instance, by using Stiff diagrams. All of the methods involve reworking concentrations to units of milli-equivalents per litre (Box 3.5).

- From the Ghijben–Herzberg relation (hydrostatic equilibrium) it can be deduced that the thickness or depth of a freshwater lens in saline groundwater below coastal dunes equals approximately 41 times the convexity or differential head; the latter is the difference in altitude between the top of the freshwater table and the open-water surface that acts as a boundary condition.

- Hydrological isolation of polluted subsoil, pump and treat, is a very cost-effective measure to clean polluted groundwater.

Soil water

4

Introduction

Above the water table, pores in the soil may contain both air and water. This zone is known as the **unsaturated zone** or **vadose zone**, sometimes the **zone of aeration**, and the water stored there is called **soil water**.

Soil water can be considered as important in many ways. The relation between soil water, plants, and the atmosphere is important for agricultural practices and in relation to climate change effects. Also, the soil is important with regard to the recharge (replenishment) of the underlying groundwater. Importantly, soil water provides a first line of defence against groundwater pollution, as explained below.

Adsorption of soil water is a process whereby soil water accumulates on solid soil surfaces, forming a thin molecular film of water. Adsorption of soil water is mainly due to electrostatic forces binding dipole water molecules to the electrically charged surfaces of solid soil particles. As these forces are only effective very close to the surfaces of soil particles, only very thin films of water are formed around the soil particles.

Let us presume that percolating water is polluted by a nearby waste dump and that, because of this, the water contains pathogenic bacteria and viruses. Torkzaban *et al.* (2006) conclude that adsorption of viruses to both solid/water and air/water interfaces in the soil pores increases as the degree of water saturation in a soil decreases. An increased adsorption prolongs the stay of viruses in a soil. Forces acting on solid/water and air/water interfaces may rip bacteria and viruses at these interfaces apart, thereby inactivating (killing) these bacteria and viruses. Additionally, as bacteria (0.3–2.0 μm in length) are much larger than viruses (0.01–0.3 μm in length), bacteria may become obstructed in soil pores during water transport. A prolonged stay of pathogenic bacteria and viruses in soil pores at solid/water and air/water interfaces, together with the forces acting on these interfaces, causes bacteria and viruses to become inactivated. **The mechanisms of adsorption plus inactivation in the unsaturated zone thus are very effective in protecting groundwater (the saturated zone) from pathogenic bacteria and viruses**. Thus soil water provides a first line of defence against pollution of groundwater and drinking water that may be derived from it.

The term 'dipole' indicates that a water molecule has both a partial negative (electrical) charge and a partial positive charge, due to the asymmetry of the water molecule.

4.1 Negative water pressures

From section 3.3 of Chapter 3, we know that for the saturated zone, the zone where all pores in the ground are filled with water, the hydraulic head equals the sum of the elevation head and the pressure head:

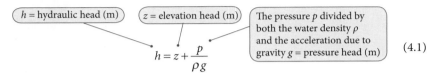

$$h = z + \frac{p}{\rho g}$$ (4.1)

- h = hydraulic head (m)
- z = elevation head (m)
- The pressure p divided by both the water density ρ and the acceleration due to gravity g = pressure head (m)

The hydraulic head is a measure of the mechanical energy of the groundwater. Also remember that the pressure head at a free air/water interface in the ground is defined as zero (section 3.3) and thus the pressure head at a water table that can establish itself freely is zero. When we apply this for groundwater under hydrostatic conditions (no water flow) and take the water table as the reference level, we can draw Figure 4.1.

In Figure 4.1 the elevation of the groundwater (in length units) is given along the vertical axis; the energy levels of the hydraulic head, elevation head, and pressure head (also in length units) are given along the horizontal axis. As the water table is taken as the reference level, the elevation head at the water table is zero metres. As mentioned above, the pressure head at the water table is taken as zero metres. Then, according to Equation 4.1, the hydraulic head at the water table, being the sum of the elevation head and the pressure head, equals zero metres.

When we descend into the groundwater, the elevation head decreases by the same amount (in length units) by which the elevation (in length units) decreases. As the water table is taken as the reference level, elevation heads below the water table are negative, and the relation between elevation head and elevation can be drawn as a line under an angle of 45° (for equally scaled axes), as shown in Figure 4.1.

When we descend into the groundwater, the pressure head (in length units) at different elevations (in length units) is linearly related to the length of the water

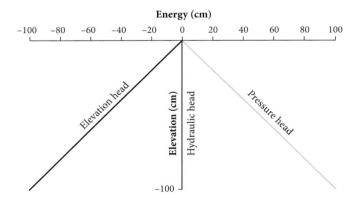

Figure 4.1. Hydraulic head, elevation head, and pressure head in the saturated zone under hydrostatic equilibrium conditions (no vertical groundwater flow)

column above the elevation of interest. Pressure heads below the water table thus are positive, and the relation between pressure head and elevation can now also be drawn as a line under an angle of 45° (for equally scaled axes, but in a different direction as for the elevation head), as also shown in Figure 4.1.

As the hydraulic head everywhere equals the sum of the elevation head and the pressure head, the hydraulic head in the saturated zone at all elevations equals zero, and thus is independent of elevation. As there are no differences in hydraulic head with elevation, there is no groundwater flow in the vertical direction, which is in agreement with our starting point, which was hydrostatic equilibrium (no vertical groundwater flow).

What would a similar graph under hydrostatic equilibrium conditions (no water flow) look like in the unsaturated zone, the zone above the water table? The soil pores then generally contain both water and air. In the unsaturated zone, water also flows from a high to a low mechanical energy level, and again this mechanical energy equals the sum of the elevation head and the pressure head (as a consequence of Bernoulli's law). When we continue to take the water table as the reference level, the elevation head at the water table remains defined as zero metres.

In accordance with the above text, when we move upwards from the water table, the elevation head increases by the same amount (in length units) by which the elevation (in length units) above the water table increases. In order for there to be no water flow, the mechanical energy in the unsaturated zone should remain zero. This can only be achieved if the pressure head above the water table decreases by the same amount by which the elevation above the water table increases, as shown in Figure 4.2. Thus, by defining the pressure head at the water table as zero, water pressures in the unsaturated zone are negative; that is, less than the existing air pressure.

Because pores in the unsaturated zone usually contain air, the mechanical energy cannot be as easily determined as below the water table, in the saturated zone, where it may simply be determined as the hydraulic head in a piezometer. In part because of this, a different term is used for the mechanical energy in the unsaturated zone: the mechanical energy in the unsaturated zone is called the **total (water) potential**.

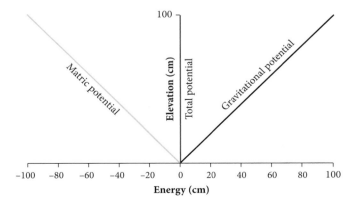

Figure 4.2. Total potential, gravitational potential, and matric potential in the unsaturated zone under hydrostatic equilibrium conditions (no water flow)

From an energy point of view, this term is fully equivalent to the term 'hydraulic head' for the saturated zone. Also, for the elevation head and the pressure head, other terms are used in the unsaturated zone: the **gravitational potential** is the equivalent of the elevation head and the **matric potential** is the equivalent of the pressure head. Note that **the matric potential is the pressure head with a negative sign**. In hydrology, the term **suction** is often used: **suction is the absolute value of the matric potential, and thus the matric potential without the negative sign. Low suction thus means a slightly negative matric potential, whereas high suction indicates a strongly negative matric potential.**

Analogous to Equation 4.1, Bernoulli's law can be written for the unsaturated zone as follows:

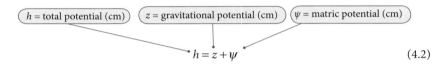

$$h = z + \psi \tag{4.2}$$

where h = total potential (cm), z = gravitational potential (cm), ψ = matric potential (cm).

Figure 4.3 combines hydrostatic equilibrium in the saturated zone with hydrostatic equilibrium in the unsaturated zone and summarizes the terms used in both zones.

In Figure 4.3 (as in Figures 4.1 and 4.2), the water table is taken as the reference, or zero, level. If another level – for instance, the land surface – had been taken as the reference level, this would only have affected the position of the elevation head

> In the hydrological literature, the terms soil water potential and hydraulic potential are sometimes used as synonyms of total (water) potential.

> In the hydrological literature dealing with physical processes at the pore scale, the term capillary pressure, being the difference between air pressure and water pressure in a capillary pore, is used as a synonym for suction.

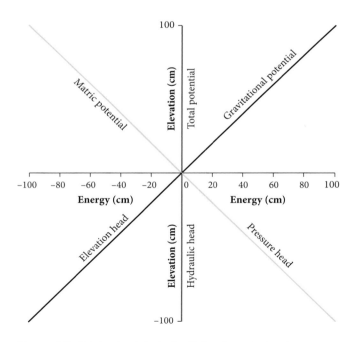

Figure 4.3. Total potential = hydraulic head, gravitational potential = elevation head, and matric potential = pressure head in both the unsaturated and the saturated zone under hydrostatic equilibrium conditions (no water flow)

gravitational potential line and the position of the hydraulic head total potential line, in the sense that both lines would shift to positions parallel to their current positions. Because the pressure head matric potential line remains locked in its present position, the choice of reference level does not influence the shape of a hydraulic head total potential line or curve. One may therefore always simply choose the reference level that comes in most handy; of course, once a level is selected for describing the energy levels at a specific location, one should stick to it.

Terms and units for water energy (water potential)

In this book we have become acquainted with dealing with **energy** (J) terms **per unit of weight** (N), which yields a length unit ($J\ N^{-1} = N\ m\ N^{-1} = m$). Another common unit is **energy** (J) **per unit volume** (m^3), which yields a unit of pressure ($J\ m^{-3} = N\ m\ m^{-3} = N\ m^{-2} = Pa$). Yet another unit may be **energy per unit mass** ($J\ kg^{-1}$).

In the soil water zone, the centimetre (cm) is selected as the unit of water potential, and thus for the total potential h, the gravitational potential z, and the matric potential ψ. To avoid large numbers at high suctions (strongly negative matric potentials), the pF has been introduced, which is the logarithm (base 10) of the suction ($-\psi$) in centimetres:

$$pF = \log(-\psi) \quad \longleftarrow \boxed{-\psi = \text{suction (cm)}} \tag{4.3}$$

Exercise 4.1.1 Take 1 atm (atmosphere) $\approx 1000\ mbar = 1000\ hPa = 10^3 \times 10^2\ Pa = 10^5\ N\ m^{-2}$ and take g = acceleration due to gravity $\approx 10\ m\ s^{-2}$.

a. Give the water pressure at a water surface in atmospheres.

b. Give the average air pressure at a water surface in atmospheres.

c. By how much does the pressure increase with every 10 m that a scuba diver descends further into fresh water?

d. Determine the water pressure in fresh water at 20 m depth in atmospheres.

e. Determine the total pressure in fresh water at 20 m depth in atmospheres.

f. 1 hPa water pressure corresponds to how much water pressure in centimetres?

Exercise 4.1.2 Figure E4.1.2 shows a water-saturated 2 m-deep clay layer within a sandy formation; the saturated clay layer carries 1 m of unconfined groundwater on top. Below the clay layer is an unsaturated zone with a pF of 2. The vertical hydraulic conductivity of the clay layer equals $2 \times 10^{-3}\ m\ day^{-1}$.

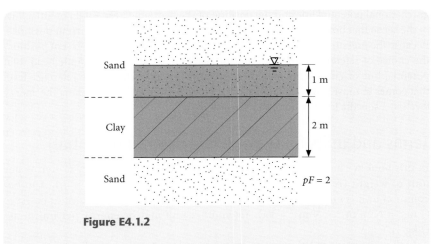

Figure E4.1.2

Determine the vertical volume flux density in the water-saturated clay layer.

4.2 Determining the total potential

The total potential h of water in the unsaturated zone cannot be measured directly in the same manner as in the saturated zone, where it can simply be measured as the hydraulic head h in a piezometer.

The elevation of groundwater in a piezometer, and thus the hydraulic head, can be determined with a hollow weight that is moved downwards in a piezometer and that makes a plopping noise when it hits the water. Alternatively, and to obtain a record of the fluctuations of the hydraulic head in a piezometer, one may install a **pressure sensor** at the bottom of a piezometer, as well as one higher up above the water level: the latter sensor measures the air pressure. By subtracting the air pressure from the total pressure (water pressure + air pressure) measured at the bottom of the piezometer, one obtains a precise estimate of only the water pressure; the latter pressure is linearly related to the length of the column of water and thus the hydraulic head. The altitude of the sensor measuring the total pressure is then the zero, or reference, level.

Because the total potential h, which is the unsaturated zone equivalent of the hydraulic head h, cannot be measured in this direct manner, the matric potential ψ is measured instead, and the total potential deduced from this. This is done with a so-called tensiometer. Figure 4.5 shows that a **tensiometer** consists of a permeable cup (plaster or ceramic) of height 5 cm or less underneath a pipe, which is fully filled with water. At the top of the pipe, the water pressure is measured using a manometer (a pressure-measuring device). A tensiometer can be inserted vertically in the topsoil or, after digging a pit, horizontally at different depths. Care should be taken to ensure that the tensiometer is fitted tightly into the soil and that the tensiometer's permeable cup is in good contact with the soil. The matric potential can be determined after equilibrium has set in between the water in the tensiometer and the soil moisture. The time to initial equilibrium after placing a tensiometer depends on the soil type and the size of the tensiometer's permeable cup; for heavy clay soils and

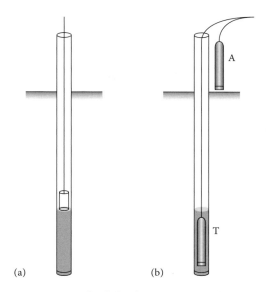

ρ = water density $\approx$ 1000 kg m^{-3}; g = acceleration due to gravity = 9.8 m s^{-2}; Δz = difference in altitude (m); $\rho g \Delta z$ thus has units of kg m^{-3} m s^{-2} m = kg m^{-1} s^{-2} = (kg m s^{-2}) m^{-2} = N m^{-2}.

$\psi = p / (\rho g)$ = has units of N m^{-2} kg^{-1} m^3 m^{-1} s^2 = (kg m s^{-2}) m^{-2} kg^{-1} m^3 m^{-1} s^2 = m.

Figure 4.4. Hydraulic head measurements using a hollow weight (a) and two pressure sensors (b): pressure sensor T measures total pressure, whereas pressure sensor A measures air pressure; stored pressure fluctuations over time can be read by connecting a laptop computer to the pressure sensors

a larger permeable cup size, the time to initial equilibrium may be of the order of weeks; therefore, it is advisable to use tensiometers with a small permeable cup in such soils.

In Figure 4.5, the water pressure p_C at the level of the permeable cup (level C) equals the sum of the water pressure p_M at the level of the manometer (level M) and the water pressure exerted by the column of water between C and M:

p_C = water pressure at C (N m^{-2})

p_M = water pressure at M (N m^{-2})

$$p_C = p_M + \rho g \Delta z \qquad (4.4)$$

$\rho g \Delta z$ = water pressure exerted by the column of water between C and M (N m^{-2})

Dividing the water pressures p (N m^{-2}) by ρ (kg m^{-3}) and g (m s^{-2}) yields the matric potential (= $p/(\rho g)$) in metres, which can be converted to centimetres. Thus, from Equation 4.4, it follows that

ψ_M = matric potential at M (cm)

ψ_C = matric potential at C (cm)

$$\psi_C = \psi_M + \Delta z \qquad (4.5)$$

Δz = difference in altitude (cm)

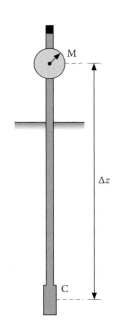

Figure 4.5. A tensiometer

As an example, if the manometer readings show that $\psi_M = -110$ cm and $\Delta z = 80$ cm, as in Figure 4.5, then the matric potential ψ_C of the soil at the porous cup level equals $-110 + 80 = -30$ cm, and the suction at this level thus equals 30 cm. With the porous cup located 60 cm below the surface and taking the surface as the reference level (z at the porous cup thus equals -60 cm), we can simply determine the total potential by applying Bernoulli's law for soil water (Equation 4.2): the total potential h then equals $-60 + (-30) = -90$ cm.

Exercise 4.2 Figure E4.2 shows two tensiometers at the same location, but with their porous cups at different depths. Hydrostatic equilibrium conditions exist. Tensiometer readings ψ_M for both tensiometers are -90 cm. The land surface is taken as the reference level.

a. Draw the potential diagram (gravitational potential, matric potential, and total potential) between -60 and -80 cm, taking the land surface as the reference level.

b. Determine the depth of the water table.

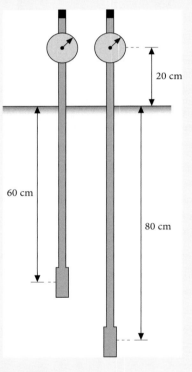

Figure E4.2

4.3 The soil as a dry filter paper or a wet sponge

A dry soil has a strongly negative matric potential or, in other words, a large suction.

When a dry soil is wetted, the added water will first of all be sucked into the smaller soil pores. One may visualize this as what happens when a dry filter paper (as an analogue of a soil) is placed upon a wet worktop: the water from the worktop will be sucked into the small pores of the dry filter paper. If, on the contrary, we were to place a dry sponge, which consisted of larger pores, upon a wet worktop, hardly any water would be sucked into the sponge. Thus, **small pores have a larger suction power than large pores**, and this too holds for a soil. A simple analogue for capillary pores is shown in Figure 4.6, and in Box 4.1 the following relation between the suction and diameter of a capillary pore is established:

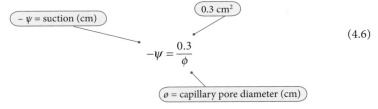

$$-\psi = \frac{0.3}{\phi} \tag{4.6}$$

- $-\psi$ = suction (cm)

0.3 cm^2

ϕ = capillary pore diameter (cm)

When a wet soil drains, the water will first of all drain from the larger pores. One may visualize this as a wet sponge (as an analogue of a soil) from which water readily drips. From a wet filter paper hardly any water would drain, as the small pores of the filter paper would hold on to the water tightly; that is, with a higher suction power (or a more negative matric potential).

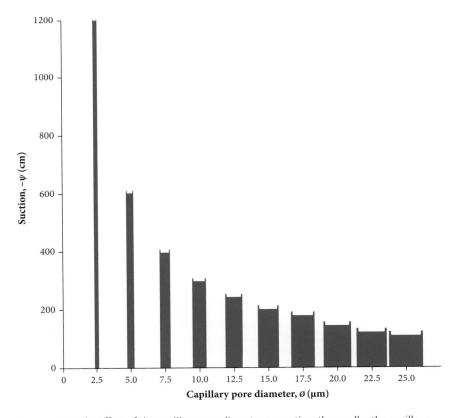

Figure 4.6. The effect of the capillary pore diameter on suction: the smaller the capillary-shaped pore, the larger is the suction

BOX 4.1 Relation between the suction and diameter of a capillary pore

Cohesion is a force of attraction between like molecules, whereas adhesion is a force of attraction between unlike molecules. Cohesion at exposed water surfaces is especially strong and strives to reduce the exposed water surface to a minimum area: because water molecules at a water surface only have like neighbours in the water, but not in the air above, they exhibit a stronger force upon their nearest neighbour on the surface – this enhanced intermolecular attraction force is the surface tension of water. The surface tension of water causes the water surface to behave as an elastic sheet, enabling small insects such as the pond skater (water strider) to walk on water, and for oil, or even small metal objects such as needles and razor blades, to float on water.

A sufficiently narrow capillary pore can suck up water in the same way as dry filter paper sucks up water from a wet worktop, a phenomenon known as capillary action. Figure B4.1 shows a capillary pore and water sucked into it. Adhesive forces between water molecules and the capillary pore wall turn the water meniscus, the air/water interface, upwards at the capillary pore wall:

this curvature is more pronounced, or, likewise, the contact angle is smaller, when the capillary pore is narrower – this is because the magnitude of adhesive forces as part of all acting forces (cohesive and adhesive forces) is larger when the pore is narrower. The contact angle may be defined as the angle at which a liquid/vapour interface meets the solid surface as measured from inside the liquid; in our example, the liquid is of course water.

The surface tension is also the cause of capillary action: when water is sucked into a capillary pore, equilibrium is established between the force lifting the column of water into the capillary pore and the weight of the column of water. The force lifting the column of water into the capillary pore is the surface tension of water, which acts on the circumference of the air/water interface (the water/vapour interface) that is in contact with the solid surface of the capillary pore wall. This force acts in an upward direction, but is offset by the contact angle, as shown in Figure B4.1. The surface tension of water σ, also called the air/water interfacial tension, is quite large in comparison with other liquids and equals 72.75×10^{-3} newton per metre ($N\,m^{-1}$) at 20°C.

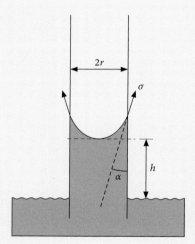

Figure B4.1. A capillary pore (see the text for an explanation of symbols)

For the lifting force, we can write

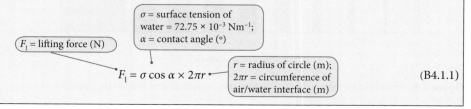

F_l = lifting force (N)

σ = surface tension of water = $72.75 \times 10^{-3}\ Nm^{-1}$;
α = contact angle (°)

$$F_l = \sigma \cos \alpha \times 2\pi r$$

r = radius of circle (m);
$2\pi r$ = circumference of air/water interface (m)

(B4.1.1)

For the weight of the column of water, we can write

F_w = weight of column of water (N)

h = height of column of water (m);
$\pi r^2 h$ = volume of column of water (m³)

ρ = water density = 1000 kg m⁻³; g = acceleration due to gravity = 9.8 m s⁻²

$$F_w = \pi r^2 h \rho g \qquad (B4.1.2)$$

Because there is equilibrium between the force lifting the column of water into the capillary pore (or holding it there) (Equation B4.1.1) and the weight of the column of water (Equation B4.1.2),

$$F_1 = F_w \qquad (B4.1.3)$$

Substituting Equations B4.1.1 and B4.1.2 in Equation B4.1.3 gives

$$\sigma \cos \alpha \times 2\pi r = \pi r^2 h \rho g \qquad (B4.1.4)$$

For small capillary pores, the contact angle α will tend towards zero, and $\cos \alpha = \cos 0° = 1$. Inserting all given constants in Equation B4.1.4 as well as substituting $r = \frac{1}{2}\phi$, where ϕ is the capillary pore diameter, gives

$$h \times \phi \approx 3 \times 10^{-5} \, \mathrm{m^2} \qquad (B4.1.5)$$

Reworking from m² to cm² gives

$$h \times \phi \approx 3 \times 10^{-1} \, \mathrm{cm^2} \qquad (B4.1.6)$$

Note that in Equation B4.1.6, h, the height of the water column, now in centimetres, can be replaced by the suction $-\psi$, also in centimetres, to give

$$-\psi \times \phi \approx 3 \times 10^{-1} \, \mathrm{cm^2} \qquad (B4.1.7)$$

Equation B4.1.7 can also be written as Equation 4.6 in the main text.

The reader should be aware that Equations B4.1.7 and 4.6 give the relation between suction and the diameter of small, perfectly shaped capillary pores. In reality, soil water flow occurs through films of water in irregularly shaped pores of different sizes. Equations B4.1.7 and 4.6 aid in a first understanding of the relation between suction and pore size, but may only be used as a rough approximation.

In summary, when water is added to a dry soil (a soil with a low volumetric moisture content), the smaller pores will be the first to suck in the water (at high suctions), and only after the soil has become quite wet will the larger pores start to fill (at low suctions). When water is drained from a wet soil (a soil with a high volumetric moisture content), the larger pores will be the first to empty (at low suctions) and only when the soil has become quite dry will the smaller pores, where water is held at high suction (powers), start losing water.

4.4 The soil moisture characteristic

The **volumetric moisture content** θ (see also section 3.5) of a soil may be defined as the volume fraction ($0 < \theta < 1$) or volume percentage ($0 < \theta < 100\%$) of water-filled pores in a soil:

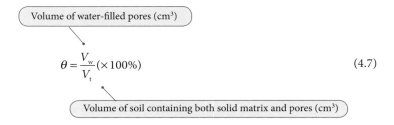

$$\theta = \frac{V_w}{V_t}(\times 100\%) \tag{4.7}$$

Sand consists of grains and the forces at work in the pore space are mainly the capillary forces that are presented in a simplified fashion in Figure 4.6. Clays have a laminated structure and negatively charged surfaces that attract cations, positively charged ions such as Na^+ and Ca^{2+}, in the soil water. The cations are bound to the clay sheets by electrostatic forces (see also section 3.7), and as a consequence the soil water that holds these cations is also bound to the clay. Suction is a unifying concept for all (types of) forces by which water is attached to solid soil particles, irrespective of the specific nature of the forces, and thus irrespective of water being held by capillary forces between the soil particles or by adsorption as a thin film of water around the soil particles.

Because the pore size distribution differs for different soils and because different forces may be at work, as outlined above, the relation between suction and volumetric moisture content differs per soil type. To fully understand the physics involved in soil water flow, it is important to know the shape of this relation.

A **soil moisture characteristic (soil moisture retention curve; pF curve)** is the relation between the suction – usually on the vertical axis – and the volumetric moisture content θ, usually on the horizontal axis. There are a number of ways to measure the moisture content in the field or from samples collected in the field: gravimetric by oven-drying as explained in Box 4.2, or by a number of indirect methods using electrical resistance blocks, a neutron probe, a gamma-ray scanner, a capacitance probe, a time domain reflectometer (TDR), or a frequency domain reflectometer (FDR). The suction is normally presented as pF, which, as stated earlier, is the logarithm (base 10) of the suction ($-\psi$) in centimetres (Equation 4.3). **All points on a soil moisture characteristic curve describe equilibrium situations between suction and moisture content.** When determining the shape of a soil moisture characteristic (Box 4.2), it may take quite some time to reach equilibrium. Figure 4.7 shows soil moisture characteristics for a sand and clay soil.

When we compare the shape of the soil moisture characteristics for a sand and clay soil, we notice – by comparing the curves along a vertical axis – that for the same volumetric moisture content, a higher suction exists in the clay soil, or that – when comparing the curves along a horizontal axis – for the same suction, a higher volumetric moisture content exists in the clay soil. This is due to the different nature of the water binding forces, and because clay has a higher porosity and a larger variety of pore sizes than (well-sorted) sand.

Many other curves could have been drawn in Figure 4.7 for other textural classes (silt, gravel) or soils (poorly sorted soils; well-sorted soils). For instance, a curve for the silt textural class, which is intermediate in size between clay and sand (or for a loam soil, which largely consists of silt) would be found in an intermediate position to the sand and clay curve.

When a saturated soil starts to drain, a certain critical suction must be exceeded for air to enter the largest pores, causing water to be released from these pores.

This critical suction is called the **air-entry suction** $-\psi_{ae}$ ($-\psi_{ae} > 0$ cm), and may be visible as the length of a vertical line at the right-hand side of the soil moisture characteristic curve, as shown for sand in Figure 4.7. Because it is harder for air to enter a small pore than a large pore, the air-entry suction is larger for soils with small pores – for example, fine-textured soils (clays) – than for soils with large pores – for example, coarse-textured soils (sands). However, as coarse-textured soils, especially when well sorted, have a predominant pore size, the phenomenon of air entry is more distinctly visible in the soil moisture characteristic curve of coarse-textured soils (sands). Also, water held in these predominant pores will drain instantaneously, with only a slight increase in suction as the soil dries out. The near-horizontal part of the soil moisture characteristic for sand in Figure 4.7 is illustrative of such a sudden drop in moisture content with only a slight increase in suction. Box 4.2 provides some insight into how a soil moisture characteristic is determined in practice.

The **field capacity** has been defined earlier as the maximum water content (the maximum volumetric moisture content) that a soil can hold against gravity (section 3.6), and the water content at which a wet sponge stops dripping provides an equally imprecise analogue. The **wilting point** may be defined as the water content (volumetric moisture content) at which a plant starts to wilt and die when a soil dries out or desiccates; the plant can then simply no longer extract water from the soil, as the little soil moisture that is left in the soil is held there by too large a suction power. In practice, both field capacity and wilting point are linked to certain pF values; for field capacity, the moisture content at $pF = 2.0$ is usually taken, whilst the moisture content at wilting point is generally taken at $pF = 4.2$.

BOX 4.2 Determination of the soil moisture characteristic

1. Take a soil core sample in the field and seal it in such a way that moisture is well-contained in the sample; the ring is usually designed to hold a soil volume of 100 cm^3. Also take loose samples of the soil and put these into well-sealed plastic bags.

2. In the laboratory, weigh the soil sample in its enveloping ring. This yields a value for $M_s + M_{wf} + M_r$; the variable notations are explained in Table B4.2.

3. Place the soil core sample (on filter paper) in a bucket filled at the bottom with water-saturated sand. Very gradually increase the level of the water in the bucket until the water is just below the top of the ring; this gradual increase may be spread over days (for sands) or weeks (for heavy clays). By means of this method, care is taken to ensure that air is not trapped in the soil sample.

4. Weigh the water-saturated soil sample in its enveloping ring. This yields a value for $M_s + M_{w0} + M_r$. Note that the mass of water in the soil sample when saturated is taken to represent the mass of water at $pF = 0$ ($\psi = -1$ cm ≈ 0 cm (saturation); see the main text of section 4.4).

5. Place the soil core sample (on filter paper) on very fine sand in a sandbox, as shown in Figure B4.2. Set the water table in the sandbox at a desired level; for instance, at 10 cm below the centre of the soil sample, as shown in Figure B4.2. After hydrostatic equilibrium (no water flow) has set in (which again may take days for sands or weeks for heavy clays), the matric potential ψ in the (centre of the) soil sample equals -10 cm, which is equivalent to $pF = 1$.

6. After equilibrium has set in, weigh the soil sample in its enveloping ring. This yields a value for $M_s + M_{w1} + M_r$.

7. Repeat steps 5 and 6 for different pF values. For instance, setting the simulated water table at 20 cm below the centre of the soil sample means that the matric potential ψ in the (centre of the) soil sample equals -20 cm, which is equivalent at hydrostatic equilibrium to a pF of 1.3; $\psi = -40$ cm is equivalent at hydrostatic equilibrium to a pF of 1.6, and $\psi = -100$ cm to a pF of 2. $\psi = -100$ cm is about as far as one can go using fine sand in the sandbox. For $pF = 2.0-2.7$ a sandbox filled with kaolinite clay is used,

Table B4.2 Variables of mass and volume for soil core samples

M_r	=	mass (g) of the ring enveloping the soil sample
M_s	=	mass (g) of the solid matrix of the soil sample
M_w	=	mass (g) of water in the soil sample
M_{w0}	=	mass (g) of water in the soil sample when saturated ($pF = 0$)
M_{w1}	=	mass (g) of water in the soil sample at $pF = 1$
M_{w2}	=	mass (g) of water in the soil sample at $pF = 2$
M_{wf}	=	mass (g) of water in the soil sample in the field
V_t	=	total volume (cm³) of the soil sample; for small rings: $V_t = 100$ cm³
V_w	=	volume (cm³) of water in the soil sample
V_{w0}	=	volume (cm³) of water in the soil sample when saturated ($pF = 0$)
V_{w1}	=	volume (cm³) of water in the soil sample at $pF = 1$
V_{w2}	=	volume (cm³) of water in the soil sample at $pF = 2$
V_{wf}	=	volume (cm³) of water in the soil sample in the field
θ	=	volumetric moisture content (–) of the soil sample = V_w divided by V_t
θ_0	=	volumetric moisture content (–) of the soil sample when saturated ($pF = 0$)
θ_1	=	volumetric moisture content (–) of the soil sample at $pF = 1$
θ_2	=	volumetric moisture content (–) of the soil sample at $pF = 2$
θ_f	=	volumetric moisture content (–) of the soil sample in the field
ρ	=	density of fresh water = 1000 kg m⁻³ = 1 g cm⁻³ ⇒ V_w (cm³) = M_w (g)
n	=	porosity (–) = θ_0
ρ_b	=	dry bulk density = M_s divided by V_t

whereas for $pF = 2.7$–4.2 a membrane pressure apparatus is used; in the latter apparatus, the loose samples collected in the field (step 1) are used instead of soil core samples; $pF = 6$ roughly equals the pF for a soil sample that is air-dry.

8. After equilibrium has set in, weigh the soil sample in its enveloping ring. For instance, for $\psi = -100$ cm ($pF = 2$), this yields a value for $M_s + M_{w2} + M_r$.

9. After having measured $M_s + M_w + M_r$ at many different pF values, put the soil sample, still in its enveloping ring, in a dry stove at a temperature of 105°C for a period of 24 hours.

10. After stove drying the sample in this way, weigh both the soil sample and the ring. This yields values for M_s and M_r.

11. Subtract the mass $M_s + M_r$ determined at step 10 from the mass determined at step 2, $M_s + M_{wf} + M_r$ to give the value of M_{wf}; also subtract the mass $M_s + M_r$ from the mass determined at step 4, $M_s + M_{w0} + M_r$, to give the value of M_{w0}; in the same way determine the values for M_{w1}, M_{w2}, and so on.

12. Because fresh water has a density of 1 g cm⁻³ (= 1 kg m⁻³), the moisture volume V_w in cm³ equals the mass M_w of the moisture in grams: $V_w = M_w$, and also V_{wf} (cm³) = M_{wf} (g), V_{w0} (cm³) = M_{w0} (g), V_{w1} (cm³) = M_{w1} (g); V_{w2} (cm³) = M_{w2} (g), and so on.

13. Divide the moisture volumes of step 12 by the volume of the soil core sample V_t (usually 100 cm³) to obtain the volumetric moisture contents θ_f, θ_0, θ_1, θ_2, etc.

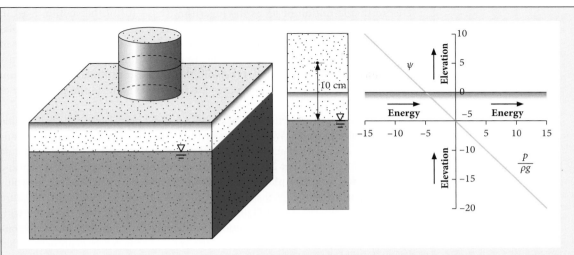

Figure B4.2. The sandbox method with a water table at 10 cm below the centre of the soil sample: during hydrostatic equilibrium conditions (no water flow), the matric potential ψ in the (centre of the) soil sample equals –10 cm ($pF = 1$)

14. Draw the moisture characteristic (main drying boundary curve; main drainage curve) by plotting the (θ, pF) positions $(\theta_0, 0)$, $(\theta_1, 1)$, $(\theta_2, 2)$, and so on.

15. One may obtain a rough idea of the pF of the soil sample during field sampling conditions from the value of θ_f: simply look it up in the moisture characteristic.

16. Finally, a number of other soil variables may be determined from the data collected above such as the **porosity** n, which simply equals θ_0, or the **dry bulk density** ρ_b of the soil sample, which is determined as M_s (determined at step 10) divided by V_t.

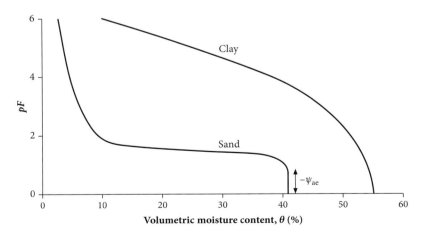

Figure 4.7. Soil moisture characteristic for a sand and clay soil; $-\psi_{ae}$ is the air-entry suction in pF units (log $-\psi$ units)

Due to the definition of *pF* as the logarithm (base 10) of the suction (−*ψ*) in centimetres (Equation 4.3), a suction of 0 cm cannot be presented in a soil moisture characteristic curve; this would give a *pF* value of −∞. Therefore the value of *pF* = 0 (−*ψ* = 10^0 = 1 cm) is taken as indicating saturation. The volumetric moisture content *θ* at *pF* = 0 then gives the volumetric moisture content at saturation *θ*$_s$ and thus – as practically all soil pores are filled with water – the porosity *n* of the soil.

Water held in a soil at suctions between *pF* = 0 and 2 percolates to the water table and is unavailable to plants. Water held at suctions larger than *pF* = 4.2 is also unavailable to plants, as explained above. Thus the portion of soil water available for plants lies between the *pF* values of 2.0 and 4.2. Because of this, one may simply determine the **available soil water for plants** (as a volume percentage) by subtracting the volumetric moisture content (0 < *θ* < 100%) at *pF* = 4.2 from the volumetric moisture content at *pF* = 2.0, as shown in Figure 4.8. Of course, plants subtract soil water with their roots; therefore, the above-mentioned difference in volumetric moisture content as a percentage can be interpreted as the amount of water, in centimetres, that would be held by the soil if the root zone was 100 cm deep. Thus, as an example, if the percentage between the *pF* values of 2.0 and 4.2 was equal to 20% (as shown in Figure 4.8), which is 0.2 as a fraction, then a root zone of depth 40 cm would hold 20% of 40 cm, or 0.2 × 40 cm = 8 cm of soil moisture.

Box 4.3 explains how the *pF* value at field capacity is linked to the depth to the water table. When the water table is near the surface – as, for instance, in the Dutch polders – it is better to use the volumetric moisture content at **pF = 1.7** as the field capacity. When the water table is located deeply, it is better to use the volumetric moisture content at **pF = 2.3** as the field capacity. When calculating the available soil water for plants, the volumetric moisture content at field capacity should then be determined at one of these *pF* values.

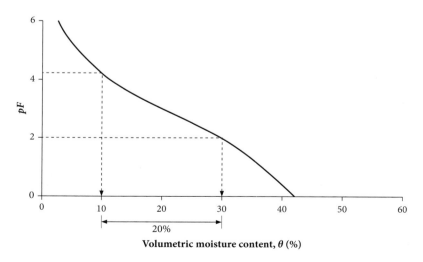

Figure 4.8. The available soil water for plants, determined from the soil moisture characteristic

BOX 4.3 *pF* at field capacity as a function of depth to the water table

As explained in the main text, the moisture content between the field capacity and the wilting point is available for plants. This is in fact a conservative estimate, because when a soil is wetter than the field capacity, this extra water (above the field capacity) is also available to the plants, but not for a long time, as the water will percolate into the groundwater.

The wilting point is usually taken at *pF* = 4.2 and the field capacity at *pF* = 2.0. The latter, however, depends on the depth of the water table as shown below.

Figure B4.3.1 shows a moisture characteristic curve of a sandy soil, and Table B4.3.1 gives the corresponding data for this curve, tabulated as *pF*, $-\psi$, and θ.

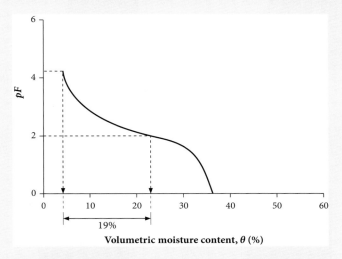

Figure B4.3.1. The soil moisture characteristic of a sandy soil: $\theta_{pF=2.0} - \theta_{pF=4.2} = 19\%$

Table B4.3.1 Data on *pF*, suction $-\psi$, and volumetric moisture content θ for a sandy soil

pF	$-\psi$ (cm)	θ (%)
0.0	0	36
1.0	10	34
1.5	32	32
1.7	50	29
2.0	100	23
2.3	200	16
2.7	501	11
3.0	1000	9
3.4	2512	6
4.2	15849	4

Figure B4.3.2a shows the volumetric moisture contents θ of the same sandy soil at different elevations above the water table under hydrostatic equilibrium conditions (no water flow). As there is no water flow, the matric potential ψ is linearly related to the elevation above the water table, as shown earlier in Figure 4.2: at an elevation of 10 cm above the water table, the matric potential equals –10 cm; at an elevation of 32 cm above the water table, the matric potential equals –32 cm; and so on. Figure B4.3.1 and Table B4.3.1 show that when the matric potential ψ = –10 cm, the volumetric moisture content θ equals 34%; when the matric potential ψ = –32 cm, the volumetric moisture content θ equals 32%; and so on. Thus, at an elevation of 10 cm above the water table, the volumetric moisture content θ equals 34%; at an elevation of 32 cm above the water table, the volumetric moisture content θ equals 32%; and so on. Using this approach, Figure B4.3.2a is constructed from the data of Table B4.3.1, with the elevation above the water table (cm) taken equal to $-\psi$ in Table B4.3.1.

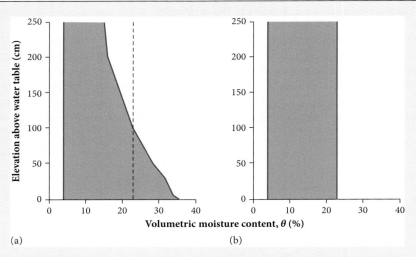

Figure B4.3.2. A physical model (a) and an approximation model (b) of the available soil water for plants with a water table at 2.5 m below the land surface – the approximation model uses $\theta_{pF=2.0} - \theta_{pF=4.2} = 23\% - 4\% = 19\% = 7.6$ cm water for a root depth of 40 cm

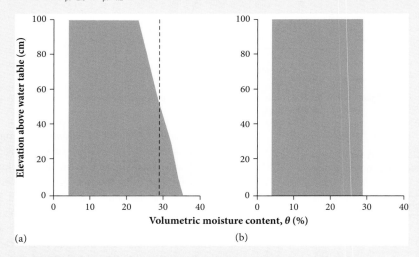

Figure B4.3.3. A physical model (a) and an approximation model (b) of the available soil water for plants with a water table at 1 m below the land surface – the approximation model uses $\theta_{pF=1.7} - \theta_{pF=4.2} = 29\% - 4\% = 25\% = 10$ cm water for a root depth of 40 cm

Both Figure B4.3.1 and Table B4.3.1 show us that the volumetric moisture content θ at $pF = 2.0$ equals 23%, and that the volumetric moisture content θ at $pF = 4.2$ equals 4%. The available soil water for plants equals $\theta_{pF=2.0} - \theta_{pF=4.2} = 19\%$, which is 19 cm of water if plant roots were to extend to 1 m below the land surface. Let us assume that we have trees with a root depth of 2.5 m; the available soil water for the trees would then be 2.5

times 19 cm = 47.5 cm of water. This amount of water is presented by the blue block of Figure B4.3.2b, showing water bound between $\theta_{pF=2.0} = 23\%$ and $\theta_{pF=4.2} = 4\%$ along the horizontal axis for a depth of 250 cm along the vertical axis.

If, however, we rethink our water availability for the trees, the blue region to the left of the $(\theta,-\psi)$ plotting positions of Figure B4.3.2a (the blue region stops

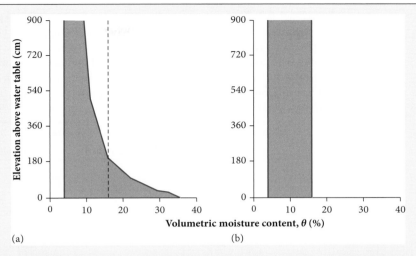

Figure B4.3.4. A physical model (a) and an approximation model (b) of the available soil water for plants with a water table at 9 m below the land surface – the approximation model uses $\theta_{pF=2.3} - \theta_{pF=4.2} = 16\% - 4\% = 12\% = 4.8$ cm water for a root depth of 40 cm

at the vertical line at $\theta_{pF=4.2} = 4\%$) gives the available soil water for trees in a direct physical manner. In Figure B4.3.2a, the vertical line at $\theta_{pF=2.0} = 23\%$ of Figure B4.3.2b is also given, but now as a broken line. Comparison of the blue block and region of the equally scaled Figures B4.3.2b and B4.3.2a shows that the areas covered by the block and region are the same.

From this, our conclusion should be that simply subtracting $\theta_{pF=4.2}$ from $\theta_{pF=2.0}$ and using this as an approximation for available soil water for trees is in good agreement with the physical model presented in Figure B4.3.2a.

However, if the depth to the water table were less, for instance 100 cm, or more, for instance 900 cm, our approximation model and the physical model just presented would not be in good agreement when still holding on to using the volumetric moisture content at $pF = 2.0$ as an approximation of the field capacity. Better results are then obtained when using the volumetric moisture content at $pF = 1.7$ as an approximation of the field capacity for shallow water tables and $pF = 2.3$ as an approximation of the field capacity for deeper water tables; Figure B4.3.3, for

a shallow water table at –100 cm, and Figure B4.3.4, for a deeper water table at –900 cm (both figures have been constructed following the same approach as for Figure B4.3.2), clearly demonstrate this.

Table B4.3.2 Calculated available soil water for the sandy soil of Figure B4.3.1 and Table B4.3.1, for plants with a root depth of 40 cm and with water tables at 1, 2.5, and 9 m below the land surface

Water table (m)	%	cm
–1.0	25	10.0
–2.5	19	7.6
–9.0	12	4.8

Finally, Table B4.3.2 summarizes the results of calculations of available soil water for the sandy soil of Figure B4.3.1 and Table B4.3.1 for plants with a root depth of 40 cm and with water tables at 1, 2.5, and 9 m below the land surface.

Exercise 4.4.1 A soil contains 2% moisture from 0 cm to –20 cm and 7% moisture from –20 cm to –40 cm. The soil moisture characteristic curve shows that the soil contains 17% moisture at $pF = 2$ (from 0 cm to –40 cm). How much water (cm) should minimally be added to the soil to establish a moisture content of 17% (field capacity) for both soil layers?

Exercise 4.4.2 Two soil core samples are saturated with water and placed upon a layer of fine sand in a sandbox. A matric potential ψ is applied to both soil core samples. After equilibrium has set in, the soil water content θ of the two core samples is determined. This procedure is repeated for increasingly more negative matric potentials (greater suctions), also using a sandbox filled with kaolinite clay and by using loose soil samples in a membrane pressure apparatus. The following results are obtained:

Matric potential, ψ (cm)	Volumetric moisture content, θ (%)	
	Soil sample a	Soil sample b
-10^0	42	57
-10^1	39	55
-10^2	12	51
-10^3	5	47
$-10^{4.2}$	2	35
-10^5	1.5	25
-10^6	1	9

a. Draw the soil moisture characteristic curve for both soil samples.

b. To what textural classes do the soil samples belong (an educated guess)?

c. Estimate the available soil water for plants (cm) for both soil types, assuming a moderately deep water table and plants that have a root depth of 40 cm.

4.5 Drying and wetting: hysteresis

The easiest way to determine the shape of a soil moisture characteristic is by starting with a saturated soil sample and then determining the **main drying boundary curve** (or **main drainage curve**) as explained in Box 4.2. In reality, soils may dry out or drain, but after a dry period a soil can of course also be wetted; for instance, by infiltrating and percolating rain. If we were to determine the soil moisture characteristic curve by starting at the dry end instead and then determining the shape of the curve by wetting (after waiting a sufficiently long time for equilibrium to set in for each established point on the curve), we would find a lower-lying soil moisture characteristic curve, as shown in Figure 4.9. We call this curve the **main wetting boundary curve** (or **main imbibition curve**). If we were to start at intermediate positions and dry or wet the soil sample, we would find intermediate (drying or wetting) **scanning curves**, some of which are shown in Figure 4.9. **Note that all points on and in between the curves are still equilibrium positions** (established after a few days or weeks). Figure 4.9 thus shows the equilibrium situation to be dependent upon the previous state and the process at hand, drying or wetting. The phenomenon of an equilibrium state being dependent on the history of the physical system is called **hysteresis**.

Probably the easiest way of explaining hysteresis in the soil moisture characteristic is by referring to the so-called 'ink bottle' effect. An ink bottle has a wide reservoir and a small opening. Figure 4.10 shows a connection of soil pores, or a wide pore with a small opening, shaped like an ink bottle. Water is held by different suctions in the large and small pore, as is also evident from Figure 4.6.

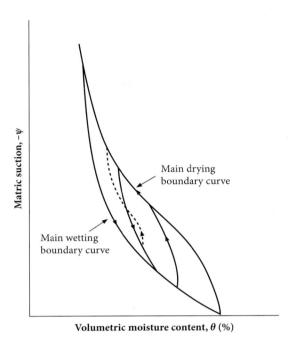

Figure 4.9. Hysteresis in the soil moisture characteristic: the main drying and wetting boundary curves, as well as some intermediate scanning curves, are shown

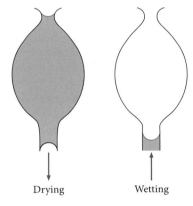

Figure 4.10. Explaining hysteresis of the soil moisture characteristic using the 'ink bottle' effect

When we determine the main drying boundary curve, thus starting at the saturated end, working towards increasingly drier situations, water first wants to drain from large pores in the soil. However, in the ink-bottle-shaped pore the water is held much more tightly (at a higher suction) in the adjacent small pore. This prevents the water from draining from the large (part of the) pore. Only when the soil has dried out further, to the point at which the small pore starts to drain, can the large pore also lose its water. Had we worked the other way around, from dry to wetter conditions, when determining the main wetting boundary curve, the first pore to fill with water would have been the small pore, and only later, under wetter conditions, would the large

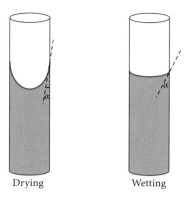

Drying   Wetting

Figure 4.11. Explaining hysteresis of
the soil moisture characteristic using the
'contact angle' effect

pore fill with water. The main wetting boundary curve therefore is a reflection of the
soil's pore size distribution, and contains less water at the same suction than the main
drying boundary curve. This provides an explanation for the occurrence of hysteresis
in a soil with ink-bottle-shaped pores.

There are other explanations for hysteresis of the soil moisture characteristic.
Another one is provided by the so-called **'contact angle' effect**. Figure 4.11 shows
that the contact angle in a pore tends to be smaller, and the air/water interface is
thus more concave, when a soil dries than when it wets. A smaller contact angle
or more concave air/water interface belongs to a smaller pore and/or a higher
suction. For the same moisture content, a smaller contact angle thus represents
a higher suction. As a consequence, the suction during drying is higher than during wetting, which is in accordance with Figure 4.9 when, for the same volumetric
moisture content, we compare values on the main drying boundary curve and the
main wetting boundary curve.

In the hydrological literature dealing with physical processes at the pore scale, the
expression **capillary pressure hysteresis** is often used; this expression is synonymous
with hysteresis of the soil moisture characteristic: the capillary pressure, written as p_c,
is equivalent to the suction $-\psi$. Also, the **wetting phase saturation** s_w is often used
instead of the volumetric moisture content θ. The wetting phase saturation s_w is that
part of the porosity n that is water-saturated; it links the volumetric moisture content
θ to the porosity n as follows:

$$\boxed{\theta = \text{volumetric moisture content (–)}}$$

$$s_w = \frac{\theta}{n}(\times 100\%) \qquad (4.8)$$

$$\boxed{n = \text{porosity (–)}}$$

An s_w value of 1 is thus equivalent to porosity; whereas, for example, an s_w value
of 0.5 is equivalent to the volumetric moisture when half of the volume of pores is
filled with water. The wetting phase saturation is also called the **relative moisture
content**.

Hysteresis of the soil moisture characteristic, or capillary pressure hysteresis, can be interpreted as being due to only representing two variables in presentations such as Figure 4.9 and missing out on (at least) one important additional variable that also significantly influences the relationship between suction and moisture content. In Figure 4.12, the **air/water interfacial area** per volume of soil (a_{wn}) is introduced in the light of research by Hassanizadeh and Gray (1990), Gray and Hassanizadeh (1991a,b), Held and Celia (2001), Hassanizadeh *et al.* (2002), and others. The air/water interfacial area a_{wn} differs with drying (drainage) or wetting (imbibition), as already explained by the 'ink bottle' and 'contact angle' effects. Rather than a set of boundary curves and intermediate scanning curves as in Figure 4.9, Figure 4.12 now shows the relationship as a three-dimensional surface determined by capillary pressure p_c, wetting phase saturation s_w, and air/water interfacial area a_{wn}. Chen *et al.* (2007) provide evidence that this three-dimensional surface is unique, and that **the observed hysteresis in soil moisture characteristics can be physically modelled by including the additional state variable of the air/water interfacial area per volume of soil**. The air/water interface also plays an important role in capturing pathogenic bacteria and viruses from soil water, as mentioned earlier, thereby protecting the underlying groundwater from pollution by these bacteria and viruses.

In a_{wn}, subscript 'w' stands for wetting phase fluid and subscript 'n' for non-wetting phase fluid; all liquids and gases are fluids (substances that flow).

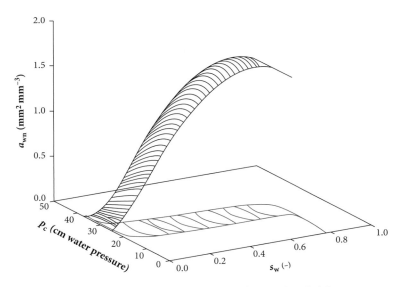

Figure 4.12. The unique relationship between the air/water interfacial area per volume of soil a_{wn}, the capillary pressure p_c, and the wetting phase saturation s_w (after Held and Celia 2001)

4.6 Unsaturated water flow

As earlier with groundwater (section 3.3), we may now extend hydrostatic equilibrium (no water flow) to water flow through a porous medium, the water flow experiencing friction. In the soil water zone, or unsaturated zone, this water flow is **unsaturated water flow**, water flow through a porous soil medium where pores are filled with water, a mixture of water and air, and/or air (two-phase flow). **As with groundwater flow, the direction of water flow is in the direction of the lower mechanical energy. Thus, for unsaturated water flow, water flow is in the direction of the more negative total potential, with this total potential h being made up of gravitational potential z and matric potential ψ.**

To quantify unsaturated water flow we need to introduce a proportionality factor to account for differences in soil material (e.g. soil texture), just as we did before for groundwater flow (section 3.7). This proportionality factor is the **unsaturated hydraulic conductivity**, which is the unsaturated zone equivalent of the saturated hydraulic conductivity K that we already know from Darcy's law (section 3.7). As some of the pores are partly or fully filled with air, the unsaturated hydraulic conductivity is lower than the saturated hydraulic conductivity. Also because of this, the volume flux density q through the soil matrix is usually lower than the volume flux density of saturated water flow through the same soil matrix unit; values for both volume flux densities and hydraulic conductivities in the unsaturated zone are therefore usually given in cm day^{-1} (or smaller units). The unsaturated hydraulic conductivity is a function of the matric potential ψ, and is therefore denoted by $K(\psi)$.

Figure 4.13 shows a declining unsaturated hydraulic conductivity $K(\psi)$ with a more negative matric potential ψ for two types of soil: a sand and a clay soil.

In Figure 4.13, the vertical axis with the unsaturated hydraulic conductivity is a logarithmic axis: the increments shown differ by a factor 10, or, regarded slightly differently: for the increments shown, the exponents (powers) of 10 differ linearly along the vertical axis.

In Figure 4.13, the matric potential ψ is given along a linear, horizontal axis. A matric potential of 0 cm indicates water saturation. Thus, the intersections of the

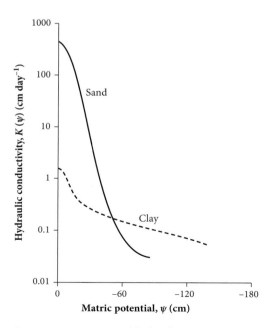

Figure 4.13. Unsaturated hydraulic conductivity $K(\psi)$ as a function of the matric potential ψ for a sand and clay soil (simplified from Bouma 1977)

curves with the vertical axis give the saturated hydraulic conductivities of sand and clay. We already know that the saturated hydraulic conductivity is larger for sand than clay (Table 3.2), and this is confirmed in Figure 4.13. However, when for the examples shown in Figure 4.13 the matric potential becomes more negative than −51 cm, the unsaturated hydraulic conductivity is larger for clay than sand.

In accordance with the above, the American physicist Edgar Buckingham (1867-1940) extended Darcy's law for groundwater flow to an equation for unsaturated water flow conditions. Using analogies to Ohm's law, Fourier's law, and Hagen–Poiseuille flow through a tube for unsaturated flow, but either unaware of Darcy's law or not recognizing the similarity between his capillary potential and the hydraulic head in saturated water flow (Rolston 2007), Buckingham in 1907 derived the following equation, usually referred to as the **Darcy–Buckingham equation**:

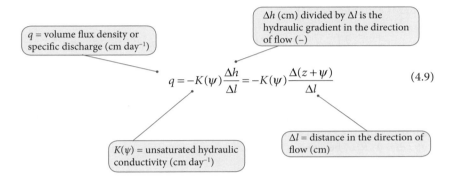

q = volume flux density or specific discharge (cm day^{-1})

Δh (cm) divided by Δl is the hydraulic gradient in the direction of flow (−)

$$q = -K(\psi)\frac{\Delta h}{\Delta l} = -K(\psi)\frac{\Delta(z+\psi)}{\Delta l} \qquad (4.9)$$

$K(\psi)$ = unsaturated hydraulic conductivity (cm day^{-1})

Δl = distance in the direction of flow (cm)

For horizontal unsaturated water flow, unsaturated water flow in the x direction, Equation 4.9 becomes

$$q_x = -K(\psi)\frac{\Delta h}{\Delta x} = -K(\psi)\frac{\Delta(z+\psi)}{\Delta x} = -K(\psi)\frac{\Delta z+\Delta \psi}{\Delta x} \qquad (4.10)$$

As $\Delta z = 0$ for horizontal unsaturated water flow, Equation 4.10 may be written as follows:

$$q_x = -K(\psi)\frac{\Delta \psi}{\Delta x} \qquad (4.11)$$

For vertical unsaturated water flow, unsaturated water flow in the z direction, Equation 4.9 becomes

$$q_z = -K(\psi)\frac{\Delta h}{\Delta z} = -K(\psi)\frac{\Delta(z+\psi)}{\Delta z} \qquad (4.12)$$

Using the notation for partial differential equations, the **Darcy–Buckingham equation** (Equation 4.9) can also be written for two- and three-dimensional problems as follows:

$$q = -K(\psi)\left(\frac{\partial h}{\partial x}+\frac{\partial h}{\partial y}+\frac{\partial h}{\partial z}\right) \qquad (4.13)$$

The **continuity equation** for unsaturated flow can be written as follows (see M8 near the end of this book):

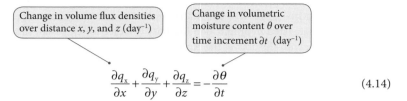

Change in volume flux densities over distance x, y, and z (day^{-1})

Change in volumetric moisture content θ over time increment ∂t (day^{-1})

$$\frac{\partial q_x}{\partial x} + \frac{\partial q_y}{\partial y} + \frac{\partial q_z}{\partial z} = -\frac{\partial \theta}{\partial t} \tag{4.14}$$

In 1931, the American soil physicist Lorenzo A. Richards (1904–1993) combined the Darcy–Buckingham equation (for q_x, q_y, and q_z) and the continuity equation (Equation 4.14) to provide the following non-linear partial differential equation, known as the **Richards equation** (see M8):

$$\frac{\partial}{\partial x}\left(K(\psi)\frac{\partial h}{\partial x}\right) + \frac{\partial}{\partial y}\left(K(\psi)\frac{\partial h}{\partial y}\right) + \frac{\partial}{\partial z}\left(K(\psi)\frac{\partial h}{\partial z}\right) = \frac{\partial \theta}{\partial t} \tag{4.15}$$

Equation 4.15, which can be written in other forms (see M9), is only valid for isotropic materials (as $K(\psi)$ is independent of direction). For steady water flow, $\partial\theta / \partial t = 0$ whilst under conditions of fully saturated water flow (groundwater flow) $K(\psi)$ can be replaced by the saturated hydraulic conductivity K. For steady groundwater flow ($\partial\theta / \partial t = 0$; $K(\psi)$ becomes K), the Richards equation (Equation 4.15) reduces to the **Laplace equation** (Equation 3.115):

$$\frac{\partial^2 h}{\partial x^2} + \frac{\partial^2 h}{\partial y^2} + \frac{\partial^2 h}{\partial z^2} = 0 \tag{4.16}$$

In hydrological literature, Equations 4.13 to 4.16 are usually given in shorthand form using the symbol ∇ (del); ∇ (pronounced as nabla) which represents the three-dimensional spatial gradient:

$$\nabla = \frac{\partial}{\partial x} + \frac{\partial}{\partial y} + \frac{\partial}{\partial z} \tag{4.17}$$

This shorthand notation makes it easier to write, comprehend, and/or remember many equations. Using the symbol ∇ (del), Equations 4.13–4.16 read as follows:

Darcy–Buckingham equation $\qquad q = -K(\psi)\nabla h$ $\qquad\qquad$ (4.18)

Continuity equation $\qquad\qquad \nabla q = -\dfrac{\partial \theta}{\partial t}$ $\qquad\qquad$ (4.19)

Richards equation $\qquad\qquad \nabla\big(K(\psi)\nabla h\big) = \dfrac{\partial \theta}{\partial t}$ $\qquad$ (4.20)

Laplace equation $\qquad\qquad \nabla^2 h = 0$ $\qquad\qquad$ (4.21)

For most practices and purposes, it suffices to consider the total potential h as made up of a gravitational potential z and a matric potential ψ when dealing with unsaturated water flow through the soil matrix. However, in some specific instances, other potentials may also be important, such as the **pneumatic potential**, when air is entrapped in the soil, or the **envelope potential**, caused by overburden pressure from overlying soil. Also, the osmotic potential may sometimes be an important potential to consider; for instance, in **irrigation**, the addition of water to dry soils. **Osmosis** is a natural process that causes water to move from locations with a low solute (ion) concentration to locations with a higher solute concentration in an attempt to rule out existing concentration differences. **Osmotic potential**, like the matric potential, is negative; in dry conditions, it acts to draw water into the soil and reduce the mechanical energy.

To quantify unsaturated water flow (two-phase flow), use of the Darcy–Buckingham equation (4.18) may suffice in many instances. However, this equation does not account for forces and energies at interfaces, such as the air/water interface. Therefore, when water flow is highly dynamic, Hassanizadeh and Gray (1990) propose the use of an **extended form of the Darcy–Buckingham equation** that links matric potential to the wetting phase saturation s_w and the air/water interfacial area per volume of soil a_{wn}; however, the use of this extended equation requires in-depth knowledge of the soil–water–air medium under study.

The coming sections 4.7–4.9 will further deal with water flow in the unsaturated zone, first upward flow and then downward flow, and will build upon the principles presented thus far. Also, soil water flow parallel to the land surface will be discussed.

4.7 Moving up: capillary rise and evaporation

Figure 4.14 shows the different potentials above the water table under hydrostatic equilibrium conditions (no water flow). The figure is the same as Figure 4.2, but the volumetric moisture content for a sand and clay soil has been added to the right. Note that at hydrostatic equilibrium (no water flow) the moisture content differs at different elevations above the water table (see also Box 4.3) and thus that **water does not flow from a wetter to a drier location by law of nature!** Of course, we already know that there is no water flow, as the total potential in Figure 4.14 is independent of the elevation above the water table.

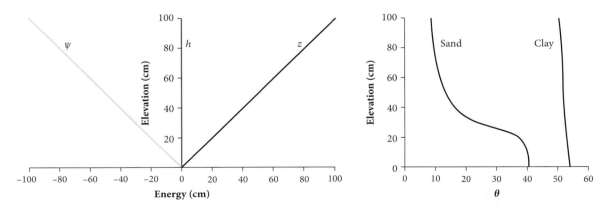

Figure 4.14. Total potential h, gravitational potential z, matric potential ψ, and volumetric moisture content θ in the unsaturated zone under hydrostatic equilibrium conditions (no water flow)

Also note that just above the water table there is a small zone where the volumetric moisture content is at a maximum, meaning that all pores are water-saturated. This is caused by pores sucking up water by capillary action (Figure 4.6) from the zero-pressure groundwater level below, a phenomenon known as **capillary rise**. Because of this, **just above the water table pores are saturated even though the matric potential ψ is less than zero**. The zone where water is sucked up from the water table by capillary forces is called the 'capillary zone' or **capillary fringe**. Because of the different pore sizes that exist in a soil, only the base of the capillary fringe is completely water-saturated.

The existence of the capillary fringe is related to the air-entry suction, a critical suction for air to enter a soil pore, as introduced before (see section 4.4). When, for a certain pore, the air-entry suction $-\psi_{ae}$ is larger than the existing matric suction $-\psi$, air will not enter and the sucked-up water will not drain from this pore. **At the top of the saturated part of the capillary fringe, the soil's matric suction is thus equal to the air-entry suction of the largest prevailing pore size just above the water table (as the largest pore has the lowest air-entry suction to be exceeded), and the air-entry suction for such a pore can simply be determined as the (equilibrium) suction $-\psi$ from Equation 4.6 (in section 4.3).**

From Equation 4.6, the height of the saturated part of the capillary fringe for well-sorted sand can be calculated to be of the order of centimetres. The saturated parts of capillary fringes are relatively high for fine-textured sediments with a predominant pore size – such as, for instance, loess soils. **Loess** is a well-sorted, mainly silt-textured, windblown deposit, which is common in and near formerly glaciated areas. Clay soils are more finely textured than loess soils, but usually harbour many different pore sizes. If there are many different pore sizes, the upper, unsaturated part of the capillary fringe may be several times thicker than the lower, saturated part of the capillary fringe. Also, soil pores are not straight and air may become uniformly entrapped in a soil by the process of capillary rise; therefore, the height of the saturated part of the capillary zone is usually less than the calculated value. In practice, the saturated part of the capillary fringe can extend to some 60 cm above the water table if the pore size is small and relatively uniform. Often, soils are 'effectively saturated' for some height above the water table, meaning that all but the largest pores are filled with water.

Theoretically, the air-entry suction is present in the main drying boundary curve (Box 4.2), as shown for sand in Figure 4.7, but in practice it is very difficult to determine the air-entry suction from a moisture characteristic curve, because of the many different pore sizes in soils and/or the heterogeneous character of soils.

In section 1.1 of Chapter 1, we have defined the unsaturated zone as the zone below the land surface, but above the water table, where soil pores may contain both air and water. This definition includes the saturated part of the capillary fringe as part of the unsaturated zone (vadose zone) and not as part of the saturated zone. Following this, we may redefine the **unsaturated zone (vadose zone)** as the zone between the land surface and the water table where pressure is less than atmospheric (matric potential $\psi < 0$), and where soil pores usually are not water-saturated, with the exception of the saturated part of the capillary fringe and temporarily saturated areas; for instance, from infiltrating water.

Some hydrologists exclude the saturated part of the capillary fringe from the unsaturated zone and define it as part of the saturated zone, while still others view the capillary fringe as a boundary condition that separates the water table from the unsaturated zone, not defining it as a part of either.

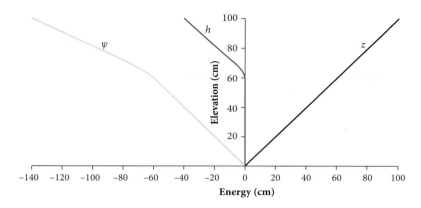

Figure 4.15. Total potential *h*, gravitational potential *z*, and matric potential ψ for upward water flow caused by evaporation

Figure 4.15 shows the effect that evaporation has on the matric potential ψ and the total potential *h*. Evaporation constitutes an upward water loss from the land surface to the atmosphere, with the upper soil becoming drier, and thus the matric potential at the land surface becoming more negative. Because of this, the total potential at the land surface is less than at some depth in the soil, causing water to flow upwards in the direction of the lower potential in an attempt to satisfy the atmospheric demand.

Both by capillary rise and evaporation, water is transported upwards in the direction of the soil surface. With upward water flow, salts contained in the water are also transported upwards. This may cause salts to accumulate in the upper soil and at the land surface, a phenomenon that is especially prominent in dry climates. If no counter-measures are taken, accumulated salts may be harmful to the growth of plants and crops. Counter-measures may comprise **combined irrigation and drainage** of the soil to flush out excess salts.

4.8 Moving down: infiltration and percolation

Knowledge of infiltration is important when studying **land degradation processes at the surface**, erosion due to the detachment of soil particles by rainfall and overland flow – for instance, when not all rain infiltrates – or **land degradation processes below the surface**, such as landslides triggered by high groundwater levels linked to both infiltration and percolation.

In section 3.5, the difference between the **infiltration rate** *f*, which is a volume flux density across a topographical soil surface, and the effective infiltration velocity v_e has already been discussed. In the following repeated example, assume that all rain infiltrates, all soil pores are interconnected and participate in the infiltration process, and that air can readily escape from the soil pores. The water or **wetting front** from 2 mm of rainfall *P* in an hour on a soil with an effective porosity n_e of 0.4 and a volumetric moisture content θ of 0.3 in the first instance will reach a depth of 20 mm, calculated as 2 mm divided by (0.4 – 0.3) (Equation 3.6). After this, part of the water may percolate under the influence of gravity or evaporate, as shown in Figure 4.15. The infiltration rate *f* in this example equals 2 mm hour^{-1}, whilst the effective infiltration velocity v_e equals 20 mm hour^{-1}. Another useful example: a soil that is compacted and therefore undergoes a reduction in porosity n from 0.35 to 0.29 loses, in its upper 100 mm, $(0.35 - 0.29) \times 100$ mm = 6 mm of **storage capacity**.

The Richards equation (Equation 4.20), whether for determining the infiltration rate, percolation rate, or evaporation rate – all of these being (defined as) volume flux densities – cannot be solved easily by hand. This is because both the unsaturated hydraulic conductivity $K(\psi)$ and the total potential h on the left-hand side of the equation, as well as the volumetric moisture content θ on the right-hand side of the equation are related to each other: this calls for some cumbersome trial-and-error problem solving.

Because of this, simpler analytical approximations (that can be physically and mathematically related to the Richards equation) have been developed in the past for estimating **ponded infiltration**, the changing infiltration rate with time from water ponded at the soil surface, and thus when the rainfall intensity (mm hour^{-1}) exceeds the infiltration rate (mm hour^{-1}). For **non-ponding infiltration**, thus when no ponding occurs and all rain infiltrates, the (non-ponding) infiltration rate (mm hour^{-1}) simply equals the rainfall intensity (mm hour^{-1}).

Measuring ponded infiltration

To study ponded infiltration, use can be made of an infiltrometer, preferably a double ring infiltrometer, and a Mariotte bottle, as shown in Figure 4.16. Water from a single ring infiltrometer does not only infiltrate vertically, but also slightly laterally; with a double ring infiltrometer, a water-filled area is created between the outer and inner ring to let the water infiltrate more or less vertically from the inner ring, which is the ring used for measuring the ponded infiltration rate.

Figure 4.16 shows the essence of an experimental set-up to determine the ponded infiltration rate with time. One should have several bottles filled with water at the ready, and at least one Mariotte bottle should be filled with water to a sufficient height, which is the maximum water-filling height at which the bottle's horizontal circumference does not change. The **Mariotte bottle** shown in Figure 4.16 is named after the French physicist Edme Mariotte (1620–1684): its design, which has many variations, was first reported by McCarthy (1934).

The ending of the soft plastic tube between the bottle's water outlet and the inside infiltrometer ring should lie just above the soil surface, well below the intended surface water level in the ring, as no air should enter this plastic tube once the experiment starts. (If air enters, the flow of water from the Mariotte bottle will halt, as should be clear from the following text.)

The experiment is started by filling the area within the rings with water to a level equal to the lower height of the bottle's air-inlet pipe, as shown in Figure 4.16. This should be done quickly but carefully, so as not to disturb the soil surface within the inner ring. For the inner ring, this can simply be done by slightly lifting the plastic cork (with air inlet) from the Mariotte bottle. As soon as the bottle is opened at the top, water will start to flow into the inner ring. After the desired water level in the inner ring is reached, quickly place the plastic cork back in to the bottle. The soft plastic tube between the bottle's water outlet and the infiltrometer ring will now also be filled with water.

When, during the experiment, water flows from the Mariotte bottle, air is taken in via the air inlet, and bubbles up to the air in the closed top of the bottle. The air inlet,

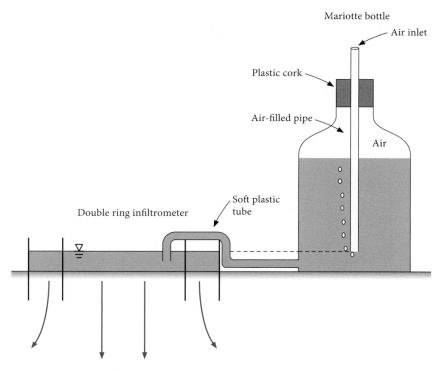

Figure 4.16. The double-ring infiltrometer and the Mariotte bottle

usually a hard plastic pipe, is dimensioned (the inside diameter, $ø$ = 14 mm) to let an air bubble pass easily. Both at the bottom (lower height) of the air inlet and at the free water surface in the inner ring, the water pressure equals the air pressure; thus, at the bottom of the air inlet, the water pressure equals zero.

When water infiltrates the soil, the water level in the inner infiltrometer ring falls slightly, causing water to flow from the Mariotte bottle into the inner ring until the water level in the inner ring again equals the lower height of the bottle's air inlet. In this way, a **constant head** water level is maintained within the inner ring.

By keeping track of the water level in the Mariotte bottle, one can determine the volume of water (in millilitres or cm^3) that flows from the bottle for a chosen time step; for instance, per minute. Dividing this water discharge ($cm^3\ min^{-1}$) by the horizontal area of the inner ring (cm^2) yields the infiltration rate f for each time step (in $cm\ min^{-1}$, which can be reworked to $mm\ hour^{-1}$).

The experiment ends, or should be continued as quickly as possible, with a new Mariotte bottle at the ready, when the water level in the Mariotte bottle reaches the level of the bottom of the air-inlet pipe.

As a final remark, the water between the outer and inner ring is kept at the same level as the water in the inner ring simply by pouring water from a hand-held bottle, or by using yet another Mariotte bottle with the lower end of its air-inlet pipe at the same height as the other Mariotte bottle.

Usually, circular-shaped rings are used in the above experiment, but of course any shape will do as long as the area of the shape is known – more specifically, its proportional relation to the inner, horizontal area of the Mariotte bottle.

Figure 4.17 shows an example of a **potential diagram**, a diagram showing the different potentials or heads with elevation (soil depth), for ponded infiltration. The land surface has been taken as the reference level (zero level) and the ponded infiltration is caused by a water layer at the surface with constant head h_0. Note that ponded infiltration causes water saturation from the soil surface downwards and thus positive water pressures in the upper part of the soil profile.

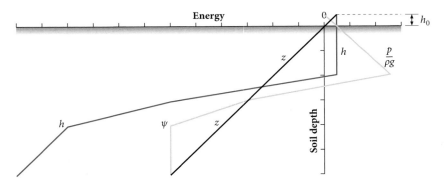

Figure 4.17. An example of a potential diagram for ponded infiltration

The Green and Ampt equation for ponded infiltration

The Australian scientists W.H. Green and G.A. Ampt (1911) developed a physical approximation (model) of ponded infiltration. They simplified the change in volumetric moisture content with soil depth at the wetting front (the broken curve in Figure 4.18) to a step-like approximation (the blue edges in Figure 4.18).

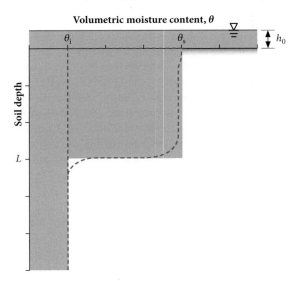

Figure 4.18. A momentary view of the change in volumetric moisture content θ with soil depth during ponded infiltration from a water layer with constant head h_0 and its Green and Ampt step-like approximation: θ_i = initial soil moisture content; θ_s = soil moisture content at saturation; L = depth of the wetting front

In Figure 4.18, the initial volumetric moisture content θ_i in the soil is taken as constant, and ponded infiltration evolves from a water layer with constant head h_0.

When we take the soil surface in Figure 4.18 as the reference or zero level ($z = 0$), then the wetting front in Figure 4.18 has reached a depth $z = L$ with $L < 0$. Because the soil above the wetting front is water-saturated and the soil below the front still has its initial moisture content, there exists a high suction gradient at the wetting front. The matric potential in the dry soil below the wetting front is represented by Green and Ampt as ψ_f ($\psi_f < 0$). Green and Ampt (1911) applied Bernoulli's law and Darcy's law as follows.

The gravitational potential z at the wetting front equals L ($L < 0$). The matric potential ψ equals ψ_f ($\psi_f < 0$). The total potential at the wetting front therefore equals $L + \psi_f$.

The elevation head or gravitational potential z at the soil surface equals 0. The pressure head $p / (\rho g)$ at the soil surface equals h_0. The total potential or hydraulic head h at the soil surface thus equals $0 + h_0 = h_0$.

The porous medium offering resistance to the infiltrating water flow stretches from the soil surface to the wetting front, and the difference in hydraulic head equals the total potential at the water-receiving end minus the hydraulic head at the water-despatching end, which is $L + \psi_f - h_0$. The distance covered is defined as the location to which water flows minus the location from where it is dispatched; thus $L - 0 = L$, and the hydraulic gradient i in Darcy's law for downward flow as portrayed in Figure 4.18 therefore equals $(L + \psi_f - h_0)$ divided by L. Multiplying the hydraulic gradient with the saturated hydraulic conductivity K yields the downward volume flux density q as follows:

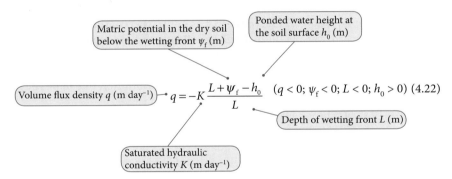

$$q = -K\frac{L + \psi_f - h_0}{L} \quad (q < 0;\ \psi_f < 0;\ L < 0;\ h_0 > 0) \quad (4.22)$$

In the hydrological literature, the infiltration rate f and the distance from the surface to the wetting front L are often taken to be positive; when we also replace the matric potential at the wetting front ψ_f by the wetting front soil suction head S_f ($S_f = -\psi_f$), this yields the **Green and Ampt equation** for ponded infiltration as follows:

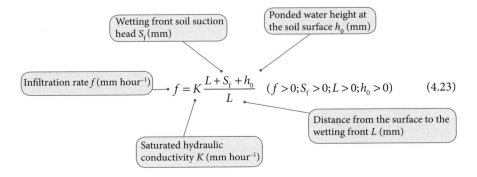

$$f = K\frac{L + S_f + h_0}{L} \quad (f > 0;\ S_f > 0;\ L > 0;\ h_0 > 0) \quad (4.23)$$

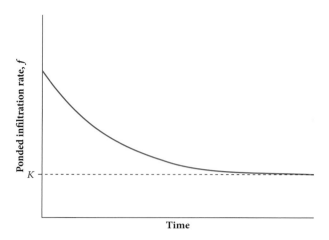

Figure 4.19. The decline of the infiltration rate during ponded infiltration

Values of the wetting front soil suction head S_f are of the order of 50 mm (10–254 mm) for sandy soils and 316 mm (64–1565 mm) for clay soils (Rawls *et al.* 1983 in Chow *et al.* 1988); the numbers in parentheses are one standard deviation around the given S_f value.

The Green and Ampt equation for ponded infiltration applies particularly to infiltration into uniform, initially dry, coarse-textured soils that exhibit a sharp wetting front. Although the Green and Ampt equation is derived from a model (a simplification of reality), it provides important insights into the infiltration process, which can be verified by infiltration experiments; for instance, by using a double-ring infiltrometer. **First of all, from Equation 4.23 it is clear that as infiltration continues, L becomes larger and larger still, causing the hydraulic gradient i, and consequently the infiltration rate f, to gradually decline during infiltration. Secondly, after quite some time has passed, L will be much larger than both S_f and h_0, and the hydraulic gradient will be reduced to a value of 1, causing the final infiltration rate to be equal to the saturated hydraulic conductivity K (gravity drainage).**

Figure 4.19 shows the gradual decline of the infiltration rate during ponded infiltration, as verified by infiltration experiments. This decline of the infiltration rate occurs during ponded infiltration or after ponding occurs during a real storm. When ponded infiltration lasts for long enough, the final infiltration rate will be equal to the saturated hydraulic conductivity K, as shown in Figure 4.19.

Infiltration capacity

Figure 4.20 shows the infiltration rate f during a storm with constant rainfall intensity i_r. At the beginning, when all rain water infiltrates, the infiltration rate f simply equals the rainfall intensity i_r. The infiltration rate is then smaller than the maximum rate at which rain falling on the soil surface can infiltrate; the maximum rate is presented by the broken curve in Figure 4.20.

After some time, however, due to packing of the soil surface by rain, swelling of the soil, and/or in-washing of fine materials into soil-surface openings (Beven 2004), the soil-surface porosity decreases, which may cause the upper few millimetres of a soil to become water-saturated and then water to start ponding, forming small pools and puddles at the soil surface. After ponding occurs (at t_p), the infiltration rate will decline as shown in Figure 4.20; the infiltration rate then equals the maximum rate at which rain falling on to the soil surface can infiltrate.

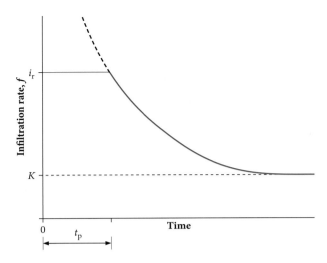

Figure 4.20. The infiltration rate during a storm with constant rainfall intensity: t_p = time to ponding

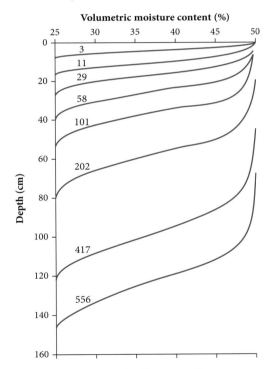

Figure 4.21. Computed soil water profiles at different times (hours) during ponded infiltration for a clay loam soil (after Philip 1964)

On sloping land surfaces, the difference between the rainfall intensity (mm hour^{-1}) and the declining infiltration rate after ponding (mm hour^{-1}) runs over the land surface as **infiltration-excess overland flow** (mm hour^{-1}), which is also called **Hortonian overland flow** after Robert E. Horton (see below).

In the hydrological literature, the maximum rate at which rain falling on the soil surface can infiltrate is referred to as the **infiltration capacity**. The decreasing curve drawn as a full line in Figure 4.20 thus shows the decline of the infiltration capacity after ponding occurs, and the broken curve in Figure 4.20 shows the decline of the infiltration capacity before ponding starts.

Figure 4.21 shows computed soil water profiles at different times during ponded infiltration, showing the slowing-down, downward movement of the wetting front with time.

After the rain stops, restoration of the infiltration capacity begins: the soil pores open due to wind action and/or differential temperatures near the soil surface, shrinkage of the soil, and/or due to earthworms and insects reinstating perforations in the soil (Beven 2004).

> **Exercise 4.8.1** The water level above flat terrain during an experiment with a double-ring infiltrometer is 20 mm. The saturated hydraulic conductivity K of the soil is 20 mm hour^{-1}. The wetting front soil suction head equals 60 mm. The initial moisture content θ_i is 20% and the effective porosity n_e equals 45%.
>
> **a.** Apply Bernoulli's law and Darcy's law to determine the infiltration rate (mm hour^{-1}) and the cumulative infiltration (mm) when the wetting front is 20, 40, 80, 160, 320, 640, and 1280 mm below the soil surface; the cumulative infiltration (mm) is the total volume of water (mm^3) added to the soil by infiltration per square unit of land surface (mm^2).
>
> **b.** Explain why the infiltration rate diminishes with time.

The Horton equation for ponded infiltration

The gradual decline of the ponded infiltration rate with time (Figure 4.19) has also been described by the American ecologist and soil scientist Robert E. Horton (1875–1945) as an exponential decrease that reaches a minimum constant rate f_c:

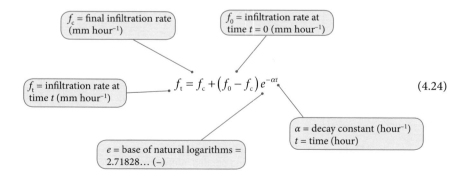

$$f_t = f_c + (f_0 - f_c)e^{-\alpha t} \tag{4.24}$$

- f_c = final infiltration rate (mm hour^{-1})
- f_0 = infiltration rate at time $t = 0$ (mm hour^{-1})
- f_t = infiltration rate at time t (mm hour^{-1})
- e = base of natural logarithms = 2.71828... (–)
- α = decay constant (hour^{-1})
- t = time (hour)

Equation 4.24 (Horton 1939) can be rewritten as follows:

$$f_t - f_c = (f_0 - f_c)e^{-\alpha t} \tag{4.25}$$

For $t + \Delta t$, we may write (substitute $t + \Delta t$ for t in Equation 4.25; Δt is constant):

$$f_{t+\Delta t} - f_c = (f_0 - f_c) e^{-\alpha(t+\Delta t)} = (f_0 - f_c) e^{-\alpha t} e^{-\alpha \Delta t} \tag{4.26}$$

From Equation 4.25, we already know that $(f_0 - f_c)e^{-\alpha t}$ in Equation 4.26 is equal to $f_t - f_c$; we can thus rewrite Equation 4.26 as follows:

$$f_{t+\Delta t} - f_c = (f_t - f_c) e^{-\alpha \Delta t} \tag{4.27}$$

or

$$\frac{f_{t+\Delta t} - f_c}{f_t - f_c} = e^{-\alpha \Delta t} \tag{4.28}$$

α is a decay constant; if t and Δt are in hours, then α has hour^{-1} as its unit. As α and Δt are positive, $-\alpha \Delta t$ is negative and $0 < e^{-\alpha \Delta t} < 1$.

If we look two time steps Δt ahead, Equation 4.25 becomes

$$f_{t+2\Delta t} - f_c = (f_0 - f_c) e^{-\alpha(t+2\Delta t)} = (f_0 - f_c) e^{-\alpha t} e^{-2\alpha \Delta t} = (f_0 - f_c) e^{-\alpha t} (e^{-\alpha \Delta t})^2 \tag{4.29}$$

Again, we know that $(f_0 - f_c) e^{-\alpha \Delta t}$ is equal to $f_t - f_c$, and thus rewriting in the same manner as before gives

$$\frac{f_{t+2\Delta t} - f_c}{f_t - f_c} = (e^{-\alpha \Delta t})^2 \tag{4.30}$$

Likewise, it can be shown that

$$\frac{f_{t+3\Delta t} - f_c}{f_t - f_c} = (e^{-\alpha \Delta t})^3 \tag{4.31}$$

and so on.

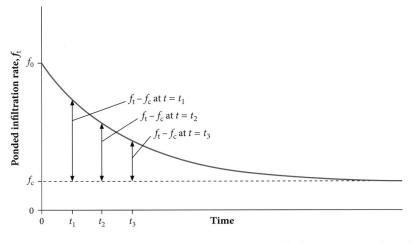

Figure 4.22. The decline of the ponded infiltration rate with time after Horton (1939): $t_3 - t_2 = t_2 - t_1 = \Delta t$

Thus, after every time step Δt, $f_t - f_c$ can be found by multiplying the earlier value for $f_t - f_c$ by a constant factor $e^{-\alpha \Delta t}$ (< 1). In the end, the ponded infiltration rate asymptotically reaches a minimum constant infiltration rate f_c, which is equivalent to the saturated hydraulic conductivity K mentioned earlier (in the Green and Ampt approximation). The above explanation in fact highlights a characteristic of a negative exponential curve $y = e^{-\alpha x}$, which is that after every step Δx a new value for y is obtained by multiplying the earlier value by a constant factor $e^{-\alpha \Delta x}$ (< 1).

If α is small, $\alpha \Delta t$ is only slightly negative, the factor $e^{-\alpha \Delta t}$ (< 1) is relatively large, and as a consequence the curve will only show a slight decline. If α is large, $-\alpha \Delta t$ has a more negative value, and as a consequence the factor $e^{-\alpha \Delta t}$ is small, causing the curve to decline steeply. Figure 4.23 shows this effect for two different values of α, with $\alpha_1 < \alpha_2$.

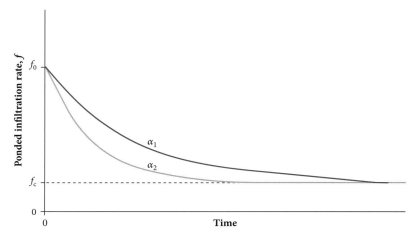

Figure 4.23. The decline of the ponded infiltration rate with time for two values of α $(\alpha_1 < \alpha_2)$

Exercise 4.8.2 Water ponded at the surface infiltrates into a soil. After 30 minutes, the infiltration rate f_t equals 45 mm hour^{-1}. After 60 minutes, the infiltration rate f_t equals 25 mm hour^{-1}. The final infiltration rate f_c equals 5 mm hour^{-1}. Start with the Horton equation for ponded infiltration given as $f_t - f_c = (f_0 - f_c) e^{-\alpha t}$ to determine the infiltration rate f_t after 90, 120, 75, and 65 minutes.

Different units for the sorptivity S are used in the hydrological literature; to convert, $\frac{1}{\sqrt{60}}$ mm $s^{-0.5} = 1$ mm min$^{-0.5} = 0.1$ cm min$^{-0.5} = 0.1 \times \sqrt{60}$ cm hour$^{-0.5}$.

The Philip equation for ponded infiltration

The Australian soil physicist John R. Philip (1927–1999) described the gradual decline of the ponded infiltration rate (Figure 4.19) mathematically as follows (Philip 1957):

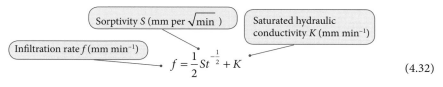

$$f = \frac{1}{2} S t^{-\frac{1}{2}} + K$$

(4.32)

Infiltration is governed by two forces, capillary action (Box 4.1) and gravity. Pores pull in water by capillary action, and this action is more prominent for smaller pores (Box 4.1) and drier soil (section 4.3). Small pores may even pull in water against the force of gravity, as is evident in the capillary fringe, discussed in section 4.7. Thus, if during an infiltration experiment or at the beginning of a rain event, we still have a relatively dry soil and ponded water at the surface, capillary action is very important. The **sorptivity** S in Equation 4.32 is a measure of the capacity of a porous medium to absorb or desorb liquid by capillary action (Philip 1957); note that the unit of sorptivity is length per square root of time. At the beginning of infiltration, when t in the equation is still small, the first term in the right-hand part of the equation, $\frac{1}{2}St^{-0.5}$, is important. As time passes, the wetting front moves down and t in the equation increases, causing this term $\frac{1}{2}St^{-0.5}$ ($\frac{1}{2}S$ divided by $\sqrt{t}$) to become less important. Because of this, the sorptivity S can only be reliably determined from the early stages of infiltration, and merely to emphasize the latter (as the choice of unit does not really matter) a 'per $\sqrt{min}$, unit (mm min$^{-0.5}$) is selected as the appropriate unit for S in this book. For very large times of ponded infiltration, $\frac{1}{2}St^{-0.5} = 0$ and the infiltration rate f will, as perceived before, equal the saturated hydraulic conductivity K.

When we mathematically integrate Equation 4.32 over time (section C2), we obtain the **cumulative infiltration** F (mm), the total volume of water (mm^3) added to the soil by infiltration per square unit of land surface (mm^2), as a function of the time:

$$\boxed{\text{Cumulative infiltration } F \text{ (mm)}}$$

$$F = St^{\frac{1}{2}} + Kt \tag{4.33}$$

Figure 4.24 shows an example of the ponded infiltration rate f (mm min^{-1}) versus time (minutes), and Figure 4.25 of the cumulative infiltration F (mm) versus time (minutes); both curves are the result of fitting a curve through data from infiltration experiments.

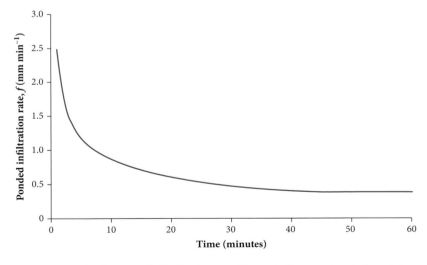

Figure 4.24. The fitted ponded infiltration rate f (mm min^{-1}) versus time (minutes) for a sandy loam soil

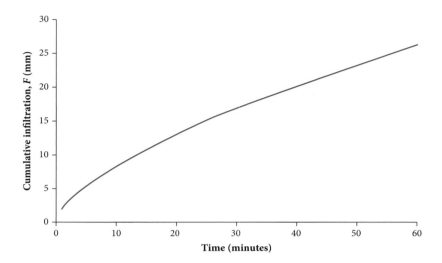

Figure 4.25. The fitted cumulative infiltration F (mm) versus time (minutes) for a loam soil

Both Equation 4.32 and Equation 4.33 can be used to estimate values for the sorptivity S and the saturated hydraulic conductivity K from infiltration experiments. One graphical way to go about finding these values for measured infiltration data is to first determine K, either as an asymptotic end value from Figure 4.24 or from Figure 4.25 as the slope of a straight part of the cumulative infiltration curve (ΔF divided by Δt) during the later stages of infiltration. After having estimated (an initial value for) K this way, simply look up the ponded infiltration rate f or the cumulative infiltration F at $t = 1$ minute (Bierkens, personal communication) from Figure 4.24 or Figure 4.25. Inserting $t = 1$ minute in Equations 4.32 and 4.33 yields the following:

Equation (4.32) for $t = 1$ min

$$f = \tfrac{1}{2}S + K \qquad (4.34)$$

Equation (4.33) for $t = 1$ min

$$F = S + K \qquad (4.35)$$

As we already know (an initial value for) K and f or F, we can simply calculate (an initial value for) S from one of the above equations.

After this, construct a graph for f or F by inserting the just calculated (initial) values for K and S in Equation 4.32 or Equation 4.33. Compare the calculated graph with the measured infiltration data and use a spreadsheet program to try to improve the correspondence between the calculated graph and the measured data by changing both K and S; by trial and error, select the values for K and S that provide the best correspondence between the measured data and calculated graph.

Exercise 4.8.3 Figure 4.24 shows fitted ponded infiltration rates f (mm min^{-1}) from an infiltration experiment on a sandy loam soil as a function of time. Use the Philip equation for ponded infiltration to do the following.

a. Estimate (initial values for) K (mm min^{-1}) and S (mm min$^{-0.5}$) using the method outlined in the text.

b. Algebraically find (initial values for) K (mm min^{-1}) and S (mm min$^{-0.5}$) using the ponded infiltration rate f (mm min^{-1}) at $t = 1$ and 60 minutes.

Exercise 4.8.4 Figure 4.25 shows the fitted cumulative infiltration F (mm) from an infiltration experiment on a loam soil as a function of time. Use the Philip equation for ponded infiltration to do the following.

a. Estimate (initial values for) K (mm min^{-1}) and S (mm min$^{-0.5}$) using the method outlined in the text.

b. Algebraically find (initial values for) K (mm min^{-1}) and S (mm min$^{-0.5}$) using the cumulative infiltration F (mm) at $t = 1$ and 60 minutes..

Another way to estimate the sorptivity S is from a horizontal infiltration experiment on a long soil core sample in the laboratory, as shown in Figure 4.26. Because the horizontal water flow in such an experiment is mainly controlled by capillary action, Equation 4.33 reduces to

$$F = S\sqrt{t} \qquad (4.36)$$

Equation (4.33) for horizontal infiltration

At time t, at the end of the experiment, the distance from the water inlet to the wetting front L is measured and the cumulative infiltration F (mm) may be determined from

θ_s = volumetric moisture content at saturation (–)

$$F = L\left(\theta_s - \theta_i\right) \qquad (4.37)$$

L = distance from the water inlet to the wetting front (mm)

θ_i = initial volumetric moisture content (–)

As t, the time the experiment ended, and F are known, S can simply be estimated from Equation 4.36.

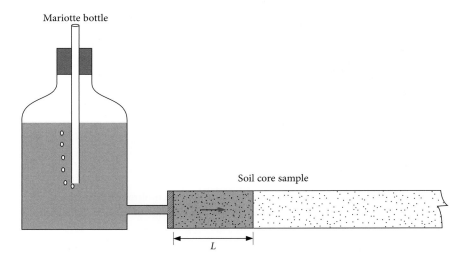

Mariotte bottle

Soil core sample

L

Figure 4.26. Estimation of the sorptivity S from a horizontal infiltration experiment on a long soil core sample in the laboratory

Exercise 4.8.5 The initial volumetric moisture content θ_i of the long soil core sample shown in Figure 4.26 equals 10%. The progression of the wetting front during a horizontal infiltration experiment is monitored: after 2 hours, the wetting front has progressed 20 cm. The test ends after 24 hours. The volumetric moisture content at saturation θ_s equals 35%.

a. Calculate the sorptivity S (mm min$^{-0.5}$).

b. Give the position of the wetting front after 6 and 24 hours.

In practice, sorptivity values are determined by both the wetness (volumetric moisture content) of the soil and the soil texture; sorptivity values measured in the field may therefore show a large variation in time and over short distances.

It should be emphasized here that the equations or models by Green and Ampt (1911), Horton (1939), and Philip (1957) have been introduced here primarily because they provide useful insights into the workings of the infiltration process, but that models (as always) are approximations of reality, and that detailed comments on the underlying assumptions, derivation, specific application, performance, and reliability of the models have (on purpose) been left out of this introductory text.

Rainfall simulator

Instead of using a double-ring infiltrometer and a Mariotte bottle, one of a number of other ways to measure the ponded infiltration rate, be it indirectly, is by using a rainfall simulator. Figure 4.27 shows a photograph of a small portable rainfall simulator. Rain is simulated by water dripping through small holes from a water container; the holes are positioned at regular spacing in a plastic plate at the bottom of the water container. By the same Mariotte bottle principle as explained earlier, simulated rain

Figure 4.27. A small portable rainfall simulator

is made to fall with constant intensity on a field plot. The rainfall intensity is determined by the height of the bottom of the hard plastic air inlet pipe within the Mariotte water container: moving the open air inlet pipe within the container slightly upwards increases the rainfall intensity; moving it down reduces the rainfall intensity.

One may determine the time to ponding t_p, and then the rate of runoff (overland flow) in cm^3 min^{-1} (millilitres per minute) from the wetted field plot by collecting the water that flows from it at regular time intervals – for instance, each minute – and measuring its volume (cm^3) in a measuring cylinder. Differences in water level in the Mariotte water container (with known volume) can be measured at the outside of the transparent water container at the same regular time intervals; this gives the simulated rainfall intensity in cm^3 min^{-1}. The ponded infiltration rate (cm^3 min^{-1}) can then simply be determined for all successive time intervals as the difference between the rainfall intensity (cm^3 min^{-1}) and the rate of runoff (overland flow; cm^3 min^{-1}). Dividing the calculated infiltration rate by the area of the wetted field plot (cm^2) gives the ponded infiltration rate in cm min^{-1} as a function of time, which can be reworked to values of mm $hour^{-1}$.

The calculated infiltration rates will show a gradual decline, and when the experiment lasts quite long enough the rate of runoff will become (reasonably) constant, indicating that the ponded infiltration rate f has reached its minimum constant infiltration rate f_c or saturated hydraulic conductivity K.

The use of a rainfall simulator may be regarded as an improvement compared with using a double-ring infiltrometer simply because the way in which rainfall is added to the soil is more natural. However, one of the disadvantages of a portable rainfall simulator, as shown in Figure 4.27, is that the altitude from which the raindrops fall is low, and the drops therefore do not reach their maximum fall velocity and full energetic impact on the soil surface, as may real rain when there is no vegetation to shield the soil surface.

Infiltration and erosion

On a slope that is receiving real rain, soil particles that are detached by the impact of falling raindrops, a process called **splash**, may cause a net downward movement of soil material along a hill slope. This net downward movement of soil material, **splash erosion**, is caused by the distance of splash along a slope being longer in a downward than an upward direction, as is evident from Figure 4.28. Splash and splash erosion may influence the characteristics of the soil surface and thus the infiltration process.

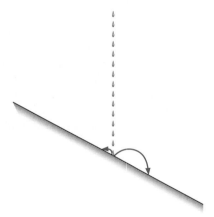

Figure 4.28. A falling raindrop and splash on a hill slope

As mentioned earlier, one may use a rainfall simulator to determine the time to ponding for different rainfall intensities: the time to ponding will be shorter when the rainfall intensity is higher. Figure 4.29 is an extension from Figure 4.20 to show the infiltration rate f and time to ponding t_p for storms with different (constant) rainfall intensities. Note that ponding will not occur when the rainfall intensity i_r remains below the value of the saturated hydraulic conductivity K of the upper soil. The curve shown in Figure 4.29, connecting times to ponding t_p for storms with different rainfall intensities i_r, is called an **infiltration envelope**. The infiltration envelope is an important curve and the time to ponding an important parameter in erosion studies: on a hill slope, ponding leads to overland flow, which in turn causes a further detachment and transport of soil particles down-slope (**erosion by overland flow**).

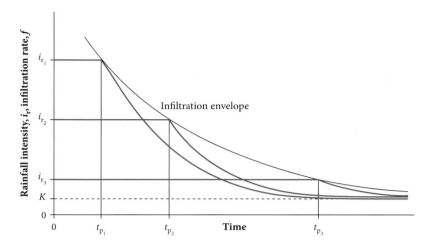

Figure 4.29. The infiltration envelope

Non-ponding infiltration

When the infiltration capacity exceeds the rainfall intensity, as for instance in well-vegetated areas, the infiltration rate i_r equals the rainfall intensity f and there will be no ponding of water at the soil surface. Figure 4.30 shows an example of a potential diagram for non-ponding infiltration, and thus without water saturation in the upper profile and with water pressures (matric potentials) in the profile remaining negative. The follow-up process of infiltration is percolation, and in order for water to percolate to the groundwater, the moisture content in the upper soil must be above field capacity, a moisture content at which some air still remains in the pore space. This is clearly different from ponded infiltration, which causes water saturation and positive water pressures near the soil surface, as is evident from Figure 4.17.

Figure 4.31 shows computed soil water profiles at different times during non-ponding infiltration. The profiles are similar in shape to Figure 4.21 for ponded infiltration, but (as already stated above) the soil does not become saturated. **Also here, for large infiltration times, the hydraulic gradient will decrease to a value**

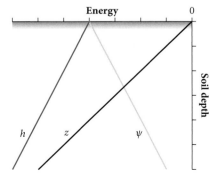

Figure 4.30. An example of a potential diagram for non-ponding infiltration

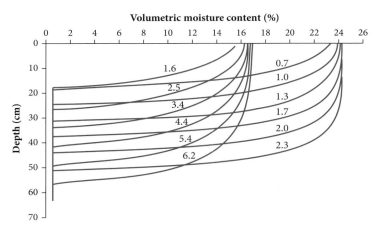

Figure 4.31. Computed soil water profiles at different times during non-ponding infiltration for two different, constant rainfall intensities (after Rubin 1966)

nearing 1, causing the infiltration rate to equal the unsaturated hydraulic conductivity $K(\psi)$ (the Darcy–Buckingham equation – Equation 4.18 – with a hydraulic gradient ∇h of 1), **or, put differently, causing the unsaturated hydraulic conductivity $K(\psi)$ to equal the rainfall intensity.**

The development of zero flux planes

When rainfall ceases, the zone wetted by infiltration continues to move downwards, whilst the upper soil already starts to dry by evaporation. Figure 4.32 shows an example of a potential diagram where soil moisture moves downwards by percolation below –20 cm, and upwards by evaporation above –20 cm depth. In our example, moisture thus moves both downwards and upwards from –20 cm, causing a no-flow situation at exactly –20 cm. If we take Figure 4.32 as representative for a surrounding flat soil,

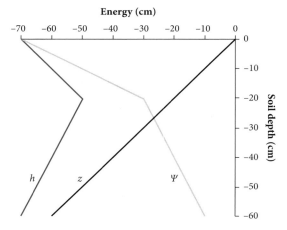

Figure 4.32. An example of a potential diagram with evaporation above –20 cm, a divergent zero flux plane at –20 cm, and percolation below –20 cm

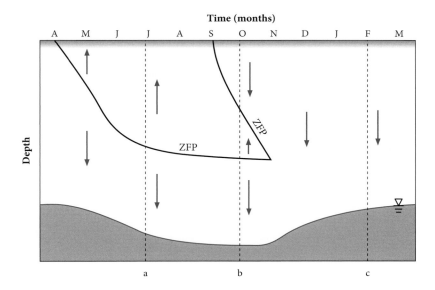

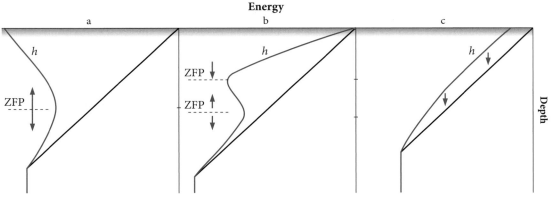

Figure 4.33. The development of zero flux planes in a temperate-climate soil profile throughout the year (Northern Hemisphere) (adapted from Wellings and Bell 1982)

the no-flow situation holds for a horizontal plane at −20 cm depth. Such a horizontal plane with no-flow or zero flux is called a **zero flux plane** – more specifically, a **divergent zero flux plane,** as moisture moves away from it.

In temperate climates a divergent zero flux plane typically starts to develop in spring, when evaporation starts to exceed rainfall. Figure 4.33 shows the divergent zero flux plane to move downwards as the soil dries out further in spring and summer. In autumn, when rainfall starts to exceed evaporation, the surface layers become wetter and a **convergent zero flux plane** develops. This convergent zero flux plane moves down rapidly to meet up with the earlier divergent zero flux plane. When they meet, both zero flux planes disappear, as the overall moisture flow in the wetted soil profile has become downward, which is the typical drainage situation for winter. This situation lasts until spring, when evaporation starts to exceed rainfall again.

Exercise 4.8.6 Figure E4.8.6 shows the matric potential ψ as a function of soil depth.

For $\psi < 0$ cm: $K(\psi) = 248.6 / (-\psi)^{2.11}$, with $K(\psi)$ in cm day^{-1} and ψ in cm.

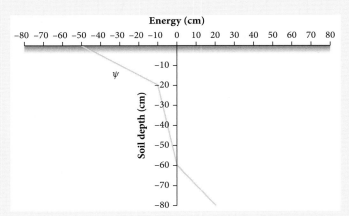

Figure E4.8.6

a. Draw the hydraulic head h with soil depth.

b. At what depth is the water table?

c. At what depth is a zero flux plane?

d. Determine the volume flux density in the groundwater.

e. Determine the hydraulic gradient in the evaporation zone.

f. Determine the unsaturated hydraulic conductivity in the evaporation zone using the K of the mean value of ψ, $K(\overline{\psi})$, in the evaporation zone as best estimate.

g. Determine the evaporation rate.

h. Why is the evaporation rate much smaller than the percolation rate?

Percolation and soil layering

Soils consist of a sequence of layers or **soil horizons**, layers caused by natural soil-forming processes. **Soil layering** has a pronounced effect on the movement of water through the soil profile, especially when soil horizons differ markedly in hydraulic conductivity and dominant pore sizes.

When a coarse-textured layer with high hydraulic conductivity overlies a fine-textured layer with lower hydraulic conductivity, the **percolation capacity**, the maximum rate at which water can percolate through a soil, is initially controlled by the coarse-textured layer. However, when the wetting front reaches the fine-textured layer, the percolation capacity will be controlled by the fine-textured layer and drops sharply. If percolation continues, a **perched water table** (section 3.4) may develop in the coarse soil above the impeding layer, which in sloping terrain may result in

throughflow, subsurface soil water flow parallel to the land surface in the direction of streams (watercourses).

However, percolation from a fine-textured layer overlying a coarse-textured layer also leads to **stagnation in the percolation process**, as is evident from Figure 4.34. This apparently surprising effect is caused by the fact that the smaller pores in the fine-textured upper layer hold on to the soil moisture with greater suction power. Only after continued wetting from above has reduced this suction power $-\psi$ to a value that matches with the capillary pore diameter $\varnothing$ of the coarse-textured layer below (a relation already provided by Equation 4.6) can water enter this lower layer. This reduced suction by wetting is visible in Figure 4.34 as a less negative matric potential ψ when water enters this lower layer. Further note that Figure 4.34 shows a discontinuity in moisture content between both layers, but that the matric potential (the pressure head) is continuous (as it physically should be!).

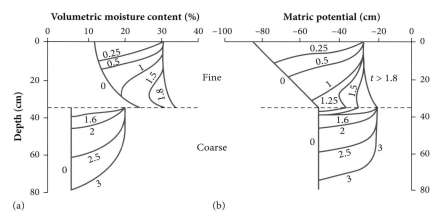

Figure 4.34. Profiles of the change with time (hours) of volumetric moisture content (a) and matric potential (b) during a constant rate of water application to a layered soil (after Vachaud *et al*. 1973)

Exercise 4.8.7 Figure E4.8.7 shows a schematic drawing of the soil profile of an agricultural field. The profile consists of a lower layer of 40 cm and an upper layer of 20 cm. A drainage pipe is located at the bottom of the lower layer. The water level in the field is maintained at 10 cm above the soil surface. The saturated hydraulic conductivity of the upper layer is $\frac{3}{8}$ times that of the lower layer. Select the bottom of the lower layer as the reference level.

a. Draw a potential diagram with the elevation head (gravitational potential), pressure head (matric potential), and hydraulic head (total potential) on millimetre paper.

b. Name three assumptions that need to be made in order to draw the potential diagram.

c. What is the volume flux density (mm min^{-1}) to the drainage system if the saturated hydraulic conductivity of the lower layer equals 1.2 mm min^{-1}?

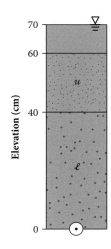

Figure E4.8.7

Percolation through a fine-textured layer that overlies a coarse-textured layer may give rise to the development of wet, finger-shaped areas in the coarse-textured layer; for instance, because, in reality, the capillary pore diameter ø of the coarse-textured layer below is not constant, but differs from place to place, or simply because the layers are not fully homogeneous –this type of flow is called **fingered flow** and is a type of preferential flow, which is the topic of the next section.

4.9 Preferential flow

Thus far we have considered **matrix flow**, water flow through the soil matrix, where the Richards equation applies. It is a quite well-established fact that nutrients, trace metals, manurial pathogens, pesticides, and other chemicals used in agriculture generally reach the water table much more rapidly and in higher concentrations than one would predict using the Richards equation or one of its analytical approximations. This is because a large part of the water generally flows through **vertical preferential pathways** such as cracks, root holes, or wormholes, effectively bypassing the soil matrix as it infiltrates and percolates (to recharge the groundwater). Preferential pathways usually only make up a small fraction of the soil, causing only a small fraction of the soil to participate in a large part of the flow, reducing the soil's potential to adsorb and degrade pollutants, and counteracting a prolonged stay of pathogenic bacteria and viruses in soil pores at solid/water and air/water interfaces (such a stay provides an effective means of inactivating these pathogenic bacteria and viruses, as mentioned in the Introduction to Chapter 4). In addition to vertical preferential pathways, soil water may bypass the matrix as **pipeflow**, subsurface flow through natural pipes parallel to the land surface and thus a rapid form of throughflow – for instance, when water flows through burrows made by moles or voles, or when it flows through artificial drains.

In general, the flow of soil water (and its solutes, substances dissolved in water) via preferential paths is called **preferential flow**. At least three major types of preferential flow may be distinguished: macropore flow (including pipeflow and flow through shrinkage cracks), fingered flow (due to soil heterogeneity, air entrapment and/or water repellency), and funnel flow.

Macropore flow

Two ranges of pore sizes may be distinguished to differentiate, albeit rather arbitrarily (Beven and Germann 1982), between matrix flow and macropore flow.

Micropores are pores with a diameter smaller than or equal to 30 µm and relate to porosity of the soil matrix (**textural porosity**), capillary binding of soil water, and matrix flow.

Macropores are pores with a diameter larger than 30 µm and relate to porosity around soil aggregates (**structural porosity**), drainage mainly through gravitational forces, and macropore flow. Macropores may be the result of biological activity (root channels, wormholes, mole and vole burrows), may be due to geological forces (subsurface erosion, soil shrinkage and cracking, fracturing), or may result from agro-technical practices (ploughing, drilling of boreholes or wells).

Soil aggregates are clusters of soil particles such as clods, crumbs, blocks, or prisms.

An important threshold for macropore flow initiation is the infiltration capacity of the soil matrix: as long as the matrix infiltration capacity is not reached, rainfall infiltrates into the matrix, but as soon as the infiltration capacity is reached, ponding at the land surface occurs and infiltration into macropores starts (Van Schaik *et al.* 2007).

Soil pipes (mole and vole burrows, artificial drains) are a special type of macropore, as they are continuous and prolonged in one direction, parallel to the land surface. During wet conditions, natural soil pipes (mole and vole burrows) may rapidly transport water parallel to the land surface in the direction of streams. During dry conditions, natural soil pipes located near the land surface (above the water table) contain little or no water.

Analogous to percolation from a fine-textured layer into a coarse-textured layer (Figure 4.34), where water can only enter the coarse-textured layer below if a critical threshold suction in the upper fine-textured layer, related to the capillary pore diameter ø of the coarse-textured layer below, is reached: water can only enter a soil pipe if a critical matrix threshold suction related to the natural soil pipe's diameter is reached. Thus, as a natural soil pipe's diameter ø (mole or vole burrow) is of the order of 1–2 cm, the matrix threshold suction $-\psi$ must be almost zero (Equation 4.6: ø = 1–2 cm gives $-\psi$ = 0.3–0.15 cm), meaning that the surrounding matrix must be nearly saturated. Similarly, artificial drainage pipes – used, for instance, to drain cultivated land – commonly are only effective when the soil matrix is nearly saturated or when the water table rises to the level of the drainage pipe.

The same mechanism as just discussed also holds for our starting example of macropore flow initiation with the soil matrix's infiltration capacity as an important threshold: ponding at the land surface causes near-saturation of the soil matrix slightly below the land surface, which is a prerequisite for macropores to start receiving water from the neighbouring soil matrix.

Figure 4.35 shows schematically two ways in which **pipeflow** can occur.

30 μm = 30 × 10^{-6} m = 0.003 cm relates to a matric suction $-\psi$ of 100 cm (Equation 4.6) or a *pF* value of 2 (Equation 4.3), which is approximately field capacity.

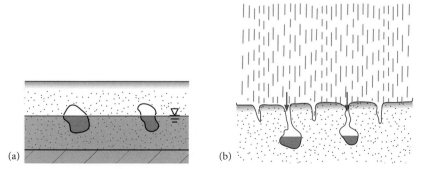

Figure 4.35. A schematic representation of two different schemes for the occurrence of pipeflow: (a) when a perched water table develops above an impeding layer under wet circumstances; and (b) when pipes receive rainfall directly from open shrinkage cracks under dry circumstances

Figure 4.35a shows the situation in which water is fed to a pipe from a perched groundwater body that has developed above an impeding layer. This is the situation discussed above, where water is fed to a pipe from a near-saturated soil matrix next to the pipe (albeit that part of the soil matrix here is evidently fully saturated).

Figure 4.35b, on the other hand, shows a situation in which pipeflow develops under dry circumstances. In soils prone to seasonal shrink–swell (clay soils), the occurrence of dry circumstances or thaw (after a freezing period) can cause **shrinkage cracks** to open; if the shrinkage cracks are deep enough and connect with the pipes, a very rapid, **short circuiting** of rainfall via shrinkage cracks and pipes to down-slope streams can evolve.

As an example, soils in forested areas of the Steinmergelkeuper marls in Gutland, Luxembourg are made up of a 30 cm-thick loam-textured layer overlying an impeding clay horizon. The topsoil above the impeding layer contains many mole burrows that are interconnected in a downslope direction for observed lengths of up to 120 m, and that support the short circuiting action just described during a large part of the year when shrinkage cracks are open and connected with these natural pipes (Hendriks 1990, 1993); water has been observed to gush out of pipe openings at downslope, near-stream positions in these forests during many storm events in the area, and subsurface flow through the pipes was observed as turbulent and erosive, further enhancing its short circuiting action. Similarly, subsurface flow through artificial drainage pipes in a clay soil in Oxfordshire, England was found to be more responsive to storms in summer than winter, also because in summer the soil was dry and cracked (Robinson and Beven 1983). Shrinking and swelling of clay is a seasonal phenomenon and, interestingly, very often shrinkage cracks, macropores formed by shrinkage, are observed to reappear at exactly the same location as before.

The relative importance of matrix flow and macropore flow generally depends on rainfall intensity and duration, infiltration capacity of the matrix, the shape and connectivity of pores, and the wetness of an area prior to rainfall (**antecedent moisture content**), but may also be linked to topographical position in the landscape and vegetation. As an example of the latter, in topsoils of forested Steinmergelkeuper marls in Gutland, Luxembourg, Cammeraat and Kooijman (2009) observed matrix flow to dominate in drier areas on the ridges, mainly occupied by beech trees, whereas pipeflow dominated in wetter topographical depressions on the hill slope, with mainly hornbeam.

Fingered flow

Fingered flow may be defined as unstable flow that results from a higher 'resistance' to water flow at certain locations in a soil. During infiltration and/or percolation, the wetting front may stagnate at certain locations; for instance, at the transition from a fine-textured layer to a coarse-textured layer, as discussed at the end of section 4.8 (Figure 4.34). In general, continued wetting due to the inflow of water in connection with a higher 'resistance' to water flow at certain locations causes the matric suction at these locations to decrease (the matric potential to become less negative) or, in simpler terms, causes the water pressure at these locations to increase. When a threshold water pressure value is exceeded, the water flow at that location is resumed. However, at locations where the threshold water pressure is not exceeded, water flow continues to stagnate. Because of all this, wet finger-shaped areas begin to form at locations where the water flow is resumed: with continued wetting, the fingers start to grow and develop into vertical flow paths, as shown in Figure 4.36. The width of fingers varies, but is often found to be 5–15 cm (max.20 cm). Figure 4.36 shows more fingers at the top of the soil profile than lower in the soil profile, which is a frequent observation for fingered flow in many soils. This is because a number of fingers do not carry enough water to keep growing; therefore the number of fingers usually decreases with increasing depth in the soil profile. On the other hand, if the water inflow into a finger becomes larger than the saturated hydraulic conductivity of the finger, the finger will start to widen, and if widening continues, fingered flow will disappear. A finger that has disappeared usually reoccurs at the same spot when circumstances are again favourable for the occurrence of fingered flow.

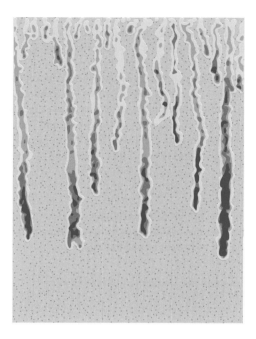

Figure 4.36. Fingered flow (modified from Cornell University 2002)

There are a number of circumstances in which fingered flow may come about. **Fingered flow due to soil heterogeneity** not only occurs at a transition from a fine-textured layer to a coarse-textured layer as discussed earlier, but may be due to any soil heterogeneity within a soil that causes the wetting front to become unstable and transform into fingered flow.

Yet another way for fingered flow to come about is **fingered flow due to air entrapment**. This occurs when a wetting front entraps and compresses air lower in the soil profile. When the air pressure ahead of the wetting front reaches a sufficiently high value, the **air-breaking value**, air penetrates into the water and escapes from the water-saturated layer at the surface. This leads to an immediate decrease in air pressure and an increase of water flow at the wetting front; during air eruptions the flow process is highly dynamic, causing wetting front instability and fingered flow at the wetting front (Wang *et al.* 1998).

If you decide to water your plants at home really well after having neglected them for quite some time, you will notice that the applied water does not infiltrate into the soil, but remains on the soil surface or fills small cracks in the soil and bypasses the soil matrix, which may have the unwanted effect that water runs into the saucer below your pot (or worse). This is an example of **water repellency**, which may be defined as the ability of a material to resist penetration by water. When a material repels water to a large extent, it is called **hydrophobic**; when it attracts water, it is called **hydrophilic**. The hydrophobic nature of materials is not static, but may change over time (remember your originally water-resistant raincoat?).

Soil water repellency is caused by long-chained organic compounds called **lipids**, mostly fatty acids and certain waxes, derived from living or decomposing plants or micro-organisms accumulating as coatings on soil particles, or as interstitial matter between soil particles: water repellency may also result from fires vaporizing and altering organic matter, some of which condenses into the soil profile (Doerr *et al.* 2000).

Fingered flow due to water repellency results from water repellency preventing or hindering downward water flow, directing the water flow into fingered flow paths. An increasing number of studies since the late 1980s have revealed that water repellency occurs in a variety of soils and under varying land uses and climatic conditions; the soil moisture content plays an important role, as water repellency generally is most pronounced when a soil is dry (as in our initial example).

Funnel flow

Figure 4.37 provides an example of **funnel flow**. For low unsaturated percolation rates, the tilting object in the middle of Figure 4.37 may either be a large stone, finer-textured material, or coarser-textured material.

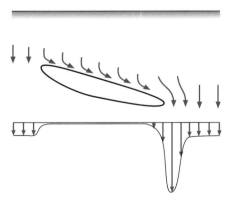

Figure 4.37. Funnel flow (modified from Cornell University 2002)

If the tilting object is a large stone or a finer-textured layer such as a clay lens, than it is obvious that water flow is funnelled to the right, where it concentrates as a vertical preferential pathway.

We know that unsaturated water flow does not easily enter a coarser layer, as greater pressure is required to push moisture into a large pore from small pores (as explained before). Thus, also when the object in the middle of Figure 4.37 constitutes coarser soil material, the effect for a low, unsaturated percolation rate will be that flow is funnelled to the right. However, when the unsaturated percolation rate is high, the inflow of water to the interface will become large enough to increase the water pressure to a level that enables flow to enter the coarser-textured layer, causing funnel flow to stop immediately.

Short rationale

Although most of the preferential flow paths occur in the top 1 m of soil profiles, some have been found to extend to depths of as much as 10 m. As stated at the beginning of this section, preferential flow may lead to the accelerated transport of water, nutrients, trace metals, manurial pathogens, pesticides, and so on towards the groundwater. This makes the study of preferential flow an important research topic.

Preferential flow is more the rule than the exception! Therefore it is important to come to a still better understanding of transport processes in the unsaturated zone; and for this, much more research is needed, on a continuing basis. Nowadays, efforts are being made by many researchers to incorporate preferential flow in infiltration and percolation models. Exercise 4.9 provides no more than a very first starting point for the current reader; some essential difficulties involved in building a valid modelling concept may, however, be evident from trying to come to grips with this exercise.

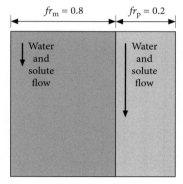

Exercise 4.9 Figure E4.9 provides a schematic presentation of a multi-domain (or multi-permeability) model with two domains: a matrix flow domain (m) and a preferential flow domain (p).

Some simplifying assumptions that we need to make to answer the questions below are that we have an initially completely dry soil, steady downward flow with a hydraulic gradient of 1, a step-like wetting front, and no interaction between the domains, other than that all excess water, resulting from the infiltration capacity of the matrix domain being exceeded, will immediately flow into the preferential flow domain. Table E4.9 provides the necessary data for both domains.

Figure E4.9. A schematic presentation of a multi-domain model with two domains: a matrix flow domain (m) and a preferential flow domain (p)

Table E4.9 Data for two domains

Matrix flow domain	Preferential flow domain
$fr_m = 0.8$	$fr_p = 0.2$
$n_{em} = 0.5$	$n_{ep} = 0.4$
$K_m = 3$ mm hour^{-1}	$K_p = 100$ mm hour^{-1}
fr = fraction coverage of the total soil by the flow domain (-)	
n_e = effective porosity as a fraction (-)	
K = saturated hydraulic conductivity (mm hour^{-1})	

a. Calculate the depth of the wetting front after 1 hour for both domains for a 1 hour rainfall depth of 3 mm.

b. Calculate the depth of the wetting front after 1 hour for both domains for a 1 hour rainfall depth of 6 mm.

c. Calculate the 1 hour precipitation depth that causes the volume of water (mm) in the preferential flow domain in 1 hour to reach the same value as K (mm hour^{-1}) of this domain; calculate the 1 hour infiltration depth.

Now define one domain with average characteristics of both domains.

d. Give the characteristics for the n_e and K of this one domain.

Answer the same questions (a, b, and c) as before, but now for this one domain with average characteristics. Thus:

e. Calculate the depth of the wetting front after 1 hour for this one domain for a 1 hour rainfall depth of 3 mm.

f. Calculate the depth of the wetting front after 1 hour for this one domain for a 1 hour rainfall depth of 6 mm.

g. Calculate the 1 hour infiltration depth that causes the volume of water (mm) in 1 hour to reach the same value as K (mm hour^{-1}) of this one domain.

h. What is your conclusion after comparing the results of e, f, and g (homogeneous infiltration) with those of a, b, and c (infiltration with preferential flow)?

Adapted from a Utrecht University Master's seminar on preferential flow by Loes van Schaik, January 2008

Exercise 4.9 provides a good example of the effect that preferential flow may have on the depth of infiltration under high rainfall intensities.

It is the author's hope that this chapter has clearly outlined the basics of soil water storage and flow as a logical follow-up on Chapter 3 on groundwater, and that it has provided useful information to all interested readers, and a head-start to those wanting to get more deeply entangled in physical hydrology and soil water flow.

→ Summary

- The total (water) potential (= soil water potential = hydraulic potential) h in the unsaturated zone (= vadose zone = zone of aeration) equals the sum of the gravitational potential z and the matric potential ψ: $h = z + \psi$. The total potential is fully equivalent to the term 'hydraulic head' for the saturated zone: the gravitational potential is the equivalent of the elevation head and the matric potential is the equivalent of the pressure head. Note that the matric potential is the pressure head with a negative sign.

- Suction $-\psi$ (cm) is the absolute value of the matric potential, and thus cm the matric potential without the negative sign. A low suction thus means a slightly negative matric potential, whereas a high suction indicates a strongly negative matric potential. In the soil water zone, the centimetre (cm) is selected as the unit of water potential, and thus for the total potential h, the gravitational potential z, and the matric potential ψ. To avoid large numbers at high suctions (strongly negative matric potentials), the pF has been introduced, which is the logarithm (base 10) of the suction $(-\psi)$ in cm: $pF = \log(-\psi)$. Suction is a unifying concept for all (types of) forces by which water is attached to solid soil particles, irrespective of the specific nature of the forces, and thus irrespective of water being held by capillary forces between the soil particles or by adsorption as a thin film of water around the soil particles.

- In this book, we have become acquainted with dealing with energy (J) terms per unit of weight (N), which yields a length unit ($J\,N^{-1} = N\,m\,N^{-1} = m$; $1\,cm = 10^{-2}\,m$). Another common unit is energy (J) per unit volume (m^3), which yields a unit of pressure ($J\,m^{-3} = N\,m\,m^{-3} = N\,m^{-2} = Pa$). Yet another unit may be energy per unit mass ($J\,kg^{-1}$).

- Because the total potential h in the unsaturated zone cannot be measured directly, the matric potential u is measured instead, and the total potential deduced from this. The matric potential ψ is measured with a so-called tensiometer, which consists of a permeable cup (plaster or ceramic) of height 5 cm or less beneath a pipe that is fully filled with water (Figure 4.5): at the top of the pipe, the water pressure is measured with a manometer (a pressure-measuring device). The water pressure at the level of the permeable cup (C) equals the sum of the water pressure at the level of the manometer (M) and the water pressure exerted by the column of water between C and M. This can be reworked as follows: $\psi_C = \psi_M + \Delta z$ (in units of cm; Δz = altitude of M minus the altitude of C).

- A prolonged stay of pathogenic bacteria and viruses in soil pores at solid/water and air/water interfaces, together with the forces acting on these interfaces, causes bacteria and viruses to become inactivated. The unsaturated zone thus protects groundwater (the saturated zone) from pathogenic bacteria and viruses, and provides a first line of defence against pollution of groundwater.

- The volumetric moisture content θ of a soil may be defined as the volume fraction ($0 < \theta < 1$) or volume percentage ($0 < \theta < 100\%$) of water-filled pores in a soil. Small pores have a larger suction power than large pores: when water is added to a dry soil (a soil with a low volumetric moisture content), the smaller pores will be the first to suck in the water (at high suctions), and only after the soil has become quite wet will the larger pores start to fill (at low suctions). When water is drained from a wet soil (a soil with a high volumetric moisture content), the larger pores will be the first to empty (at low suctions) and only when the soil has become quite dry will the smaller pores, where water is held at high suction (powers), start losing water.

- A soil moisture characteristic (soil moisture retention curve; pF curve) is the relation between the suction – usually on the vertical axis – and the volumetric moisture content θ, usually on the horizontal axis. All points on a soil moisture

characteristic curve describe equilibrium situations between suction and moisture content. Figure 4.7 shows soil moisture characteristics for a sand and clay soil.

- When a saturated soil starts to drain, a certain critical suction must be exceeded for air to enter the largest pores, causing water to be released from these pores. This critical suction is called the air-entry suction $-\psi_{ae}$ ($-\psi_{ae} > 0$ cm), and may be visible as the length of a vertical line on the right-hand side of the soil moisture characteristic curve. In practice, because a soil generally has many different pore sizes, it is very difficult to determine the air-entry suction of a soil from a moisture characteristic curve.

- The wilting point may be defined as the water content (volumetric moisture content) at which a plant starts to wilt and die when a soil dries out or desiccates; the plant can then simply no longer extract water from the soil, as the little soil moisture that is left in the soil is held there by too large a suction power. In practice, both the field capacity and the wilting point are linked to certain pF values: for field capacity, the moisture content at **pF = 2.0** is usually taken, whilst the moisture content at the wilting point is generally taken at **pF = 4.2**.

- The available soil water for plants as a volume percentage equals the volumetric moisture content (%) at field capacity (pF = 2.0) minus the volumetric moisture content (%) at the wilting point (pF = 4.2). When the water table is near the surface, it is better to use the volumetric moisture content at **pF = 1.7** as the field capacity; when the water table is located deeply, it is better to use the volumetric moisture content at **pF = 2.3** as the field capacity. The available soil water for plants as a volume percentage can be interpreted as the amount of water, in centimetres, that would be held by the soil if the root zone was 100 cm deep; for a root zone of, for instance, 40 cm depth, the outcome must be multiplied by 0.4.

- Figure 4.9 shows the effect of hysteresis on the soil moisture characteristic curve. Hysteresis is the phenomenon of an equilibrium state being dependent on the history of the physical system. Figure 4.9 shows the main drying boundary curve (or main drainage curve) and the main wetting boundary curve (or main imbibition curve), as well as some intermediate scanning curves. Again, note that all points on and in between the curves are equilibrium positions.

- Hysteresis in the soil moisture characteristic can be explained referring to the 'ink bottle' effect or, alternatively, to the 'contact angle' effect, but the best explanation is that hysteresis of the soil moisture characteristic is due to only representing two variables in presentations such as Figure 4.9 and missing out on an important additional variable: the air/water interfacial area per volume of soil (a_{wn}). Figure 4.12 shows the relationship between suction and moisture content as a unique, three-dimensional surface determined by capillary pressure p_c (= suction), wetting phase saturation s_w (which can be linked to the moisture content), and the air/water interfacial area a_{wn}.

- Water does not flow from a wetter to a drier location by law of nature! Water flow is in the direction of the lower mechanical energy; thus, for unsaturated water flow, the water flow is in the direction of the more negative total potential h. The unsaturated zone equivalent of Darcy's law, the Darcy–Buckingham equation, states that: $q = -K(\psi)\dfrac{\Delta h}{\Delta l}$ where $K(\psi)$ is the unsaturated hydraulic conductivity, which can be presented as a function of either the matric potential ψ or the volumetric moisture content θ. For highly dynamic water flow, Hassanizadeh and Gray (1990) propose the use of an extended form of the Darcy–Buckingham equation that links the matric potential to the wetting phase saturation s_w and the air/water interfacial area per volume of soil a_{wn}: the use of such an extended equation requires in-depth knowledge of the soil–water–air medium under study.

- Combining the Darcy–Buckingham equation and the continuity equation yields a non-linear partial differential equation, known as the Richards equation, which can be written in different forms (see section M9).

- Just above the water table is a small zone in which the volumetric moisture content is at a maximum, meaning that all pores are water-saturated. This is caused by pores sucking up water by capillary action from the zero-pressure groundwater level below, a phenomenon known as capillary rise. Because of this, just above the water table pores are saturated even though the matric potential ψ is less than zero. The zone where water is sucked up from the water table by capillary forces is called the 'capillary zone' or capillary fringe. The existence of the capillary fringe is related to the air-entry suction, a critical suction for air to enter a soil pore.

- At the top of the saturated part of the capillary fringe, the soil's matric suction is equal to the air-entry suction of the largest prevailing pore size just above the water table (as the largest pore has the lowest air-entry suction to be

exceeded); the air-entry suction for such a pore can simply be determined as the (equilibrium) suction $-\psi$ from Equation 4.6 (in section 4.3). In practice, the saturated part of the capillary fringe can extend to some 60 cm above the water table if the pore size is small and relatively uniform. Often, soils are 'effectively saturated' for some height above the water table, meaning that all but the largest pores are filled with water.

- Ponded infiltration occurs when the rainfall intensity (mm hour^{-1}) exceeds the infiltration rate (mm hour^{-1}). For non-ponding infiltration, and thus when no ponding occurs and all rain infiltrates, the (non-ponding) infiltration rate (mm hour^{-1}) equals the rainfall intensity (mm hour^{-1}).

- Figure 4.17 shows an example of a potential diagram for ponded infiltration, the changing infiltration rate (a volume flux density!) with time from water ponded at the soil surface. From the bottom of the ring, the soil water flow is directed downwards to the more negative total potential h.

- As ponded infiltration continues, the depth of the wetting front increases and the infiltration rate gradually declines: after quite some time has passed, the hydraulic gradient reduces to a value of 1, causing the final infiltration rate f (mm hour^{-1}) to equal the saturated hydraulic conductivity K (mm hour^{-1}) – the latter condition is called gravity drainage.

- As non-ponding infiltration continues, the hydraulic gradient will decrease to a value of 1, causing the infiltration rate to equal the unsaturated hydraulic conductivity $K(\psi)$; or, put differently, causing the unsaturated hydraulic conductivity $K(\psi)$ to equal the rainfall intensity.

- At the beginning of a rainfall event, when all rain water can still infiltrate, the infiltration rate f is not at maximum, but equals the rainfall intensity i_r. After some time, the soil-surface porosity decreases, which may cause the upper few millimetres of a soil to become water-saturated and then for water to start ponding, forming small pools and puddles at the soil surface. After ponding has occurred (at t_p), the infiltration rate will decline as shown in Figure 4.20; the infiltration rate then equals the maximum rate at which rain falling on the soil surface can infiltrate; this maximum infiltration rate is called the infiltration capacity. Similarly, the percolation capacity is the maximum rate at which water can percolate through a soil.

- On sloping land surfaces, the difference between the rainfall intensity (mm hour^{-1}) and the declining infiltration rate after ponding (mm hour^{-1}) runs over the land surface as infiltration-excess overland flow (mm hour^{-1}), which is also called Hortonian overland flow.

- To prevent the accumulation of salts in the upper soil and at the land surface, **irrigation must always be accompanied by drainage** to flush out excess salts.

- One application of a Mariotte bottle is to achieve a constant water head h_0 in an infiltrometer; another application is to obtain a constant water discharge Q from the bottle. The latter principle is also used to obtain a constant rainfall intensity from a rainfall simulator.

- The sorptivity S is a measure of the capacity of a porous medium to absorb or desorb liquid by capillary action, and is important at the beginning of ponded infiltration, when the soil is still relatively dry: the unit of sorptivity is length per square root of time; for instance, mm min$^{-0.5}$.

- The cumulative infiltration F (mm) is the total volume of water (mm^3) added to the soil by infiltration per square unit of land surface (mm^2). A curve showing the cumulative infiltration with time will become linear when the ponded infiltration equals the saturated hydraulic conductivity during the later stages of infiltration.

- An infiltration envelope is a curve that connects the times to ponding t_p for storms with different rainfall intensities i_r: the infiltration envelope is an important curve and the time to ponding an important parameter in erosion studies.

- A zero flux plane (ZFP) is a horizontal plane in the unsaturated zone with no flow or zero flux. When the soil water flow is directed away from the zero flux plane (both upwards and downwards), it is called a divergent zero flux plane. When the soil water flow is directed towards the zero flux plane (both upwards and downwards), it is called a convergent zero flux plane. Figure 4.33 shows the general movement of water and the development of zero flux planes in a temperate-climate soil profile throughout the year.

- Soil horizons are layers caused by natural soil-forming processes. Soil layering has a pronounced effect on the movement of water through the soil profile, especially when soil horizons differ markedly in hydraulic conductivity and dominant pore sizes. Both a coarse-textured layer overlying a fine-textured layer and a fine-textured layer overlying a coarse-textured layer (Figure 4.34) lead to **stagnation in the percolation process**.

- Throughflow is subsurface soil water flow parallel to the land surface in the direction of streams (watercourses).

- Preferential flow is the flow of soil water (and its solutes, substances dissolved in water) via preferential pathways such as cracks, root holes, or wormholes. Preferential flow effectively bypasses the soil matrix, and is **more the rule than the exception!** Preferential pathways usually only make up a small fraction of the soil, causing only a small fraction of the soil to participate in a large part of the flow.

- At least three major types of preferential flow may be distinguished: macropore flow (including pipeflow and flow through shrinkage cracks), fingered flow (due to soil heterogeneity, air entrapment, and/or water repellency), and funnel flow (section 4.9).

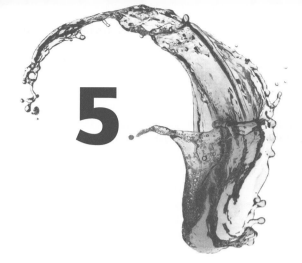

5. Surface water

Introduction

Surface water is water at the surface, whether stagnant in the form of **surface storage** (section 1.3) or flowing in brooks or rivers, or as overland flow on slopes. It is easier to imagine surface water flow than subsurface water flow. Yet, the basic principles of water flow are best explained for steady groundwater flow and as a follow-up for soil water flow, where the same principles apply. This provides the main reason why surface water is discussed after groundwater and soil water in this book. The notions that we have carefully built upon in Chapters 3 and 4, such as Bernoulli's law (conservation of energy), are now equally applicable to surface water. However, surface water flow may be rapid, with kinetic energy being important, and/or **turbulent**, varying errati-cally at any point, causing a more complicated type of flow than steady groundwater flow through sediments, which may simply be interpreted as **laminar**; that is, flowing in parallel layers with no disturbance between the layers (section 3.7). The notions 'turbulent' and 'laminar' are properties of the flow and not of the water. Osborne Reynolds, a British engineer and physicist (1842–1912), described these properties in 1883 by means of a dimensionless number that has come to carry his name, the **Reynolds number** (Re):

L = characteristic length (m)

v = flow velocity (m s^{-1}) ρ = water density (kg m^{-3})

$$Re = \frac{vL\rho}{\mu}$$

(5.1)

μ = dynamic viscosity (kg m^{-1} s^{-1})

Figure 5.1 shows a cross-section of a trapezoidal open channel. In open channels, the hydraulic radius of the channel is often taken as a measure for the **characteristic length** L (m) in Equation 5.1. The **hydraulic radius** R_h (m) of an open channel is defined as follows:

A = cross-sectional flow area (m^2) P_w = wetted perimeter (m)

$$R_h = \frac{A}{P_w}$$

(5.2)

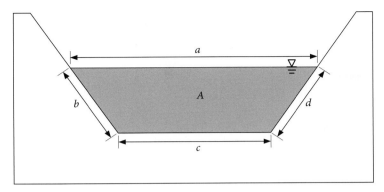

Figure 5.1. A cross-section of a trapezoidal open channel: A = cross-sectional flow area perpendicular to the water flow (m²); wetted perimeter P_w= part of the cross-section perpendicular to the water flow that is in contact with the water = $b + c + d$ (m)

As a rule of thumb, water flow in open channels is **laminar** for values of Re smaller than 500 and **turbulent** for values of Re larger than 2000 (CHO-TNO 1986); between these values, water flow cannot be classified as either one or the other. Note that the just mentioned Re values of 500 and 2000 are rather imprecise, as the transition from laminar to turbulent or vice versa is influenced by upstream conditions and the history of the water flow. Also, for different types of flow (for instance, open-channel flow or flow through a tube), different critical values for the Reynolds number (Re) are appropriate, explaining in part the different critical values for Re that are encountered in the hydrological literature and the literature on fluid dynamics.

Another useful definition for surface water is that of **steady flow**, when surface water velocities do not vary in time ($\partial v / \partial t = 0$), or when, in the case of turbulent flow, the statistical parameters (the mean value and the standard deviation of the water velocity) do not vary in time (Van Rijn 1994): if such conditions are not met, the flow is termed **unsteady**. Yet another useful definition is that of **uniform flow**, when surface water velocities are constant in the direction of flow ($\partial v / \partial x = 0$) (Van Rijn 1994), or, for an open channel with a constant cross-section, when the water height H remains constant throughout all cross-sections ($\partial H / \partial x = 0$): if such conditions are not met, the flow is **non-uniform**. A number of primary hydraulic equations for steady, uniform flow in an open channel, a channel with a free water surface, are presented in M10 near the end of this book. The water flow considered in the following section (5.1) is steady water flow.

It is important to study surface water. Too much surface water – flooding – can be a major problem in both mountainous and lowland areas, but also too little water, a water shortage, may be a major problem. Too much stagnant surface water may cause infestations of **water-related diseases** such as, for instance, **malaria**, a disease caused by parasites transmitted by female *Anopheles* mosquitoes, **dengue (hemorrhagic) fever** and **yellow fever**, viral diseases transmitted by *Aedes aegypti* mosquitoes, or **schistosomiasis** (also known as **bilharzia**), a disease caused by parasites carried by freshwater snails. Too little water, in itself a problem, may cause the little water that is available to be of bad quality. Also, too little surface water in a downstream country, as a result of water being captured upstream, may lead to political unrest and/ or conflicts between neighbouring countries. **All of the problems mentioned have unwanted effects on the economy and health, and therefore in-depth knowledge of** the overall water system – atmospheric water, groundwater, soil water, and surface water – is needed to manage our water resources in the best possible, sustainable, and peaceful manner.

5.1 Bernoulli revisited

We can use Bernoulli's law to determine the mechanical energy level of water: as water flow in the subsurface generally is slow, earlier on we could forget about the kinetic energy term for groundwater flow and matrix flow (sections 3.3 and 4.1). However, as surface water flow is usually rapid, we cannot disregard the kinetic energy term $\frac{1}{2}mv^2$ for this type of flow in Bernoulli's law:

$$\frac{1}{2}mv^2 + mgz + pV = \text{constant} \tag{5.3}$$

After dividing by the volume V (m³), the water density ρ (kg m⁻³), and the acceleration due to gravity g (N kg⁻¹ or m s⁻²), and thus dividing in total by the weight (N), earlier on (section 3.3) we obtained:

$$\frac{v^2}{2g} + z + \frac{p}{\rho g} = \text{constant} \tag{5.4}$$

For surface water, the first term in Equation 5.4, the kinetic energy per unit weight (J N⁻¹ or m), may thus not be neglected: this first term, $v^2/2g$, is called the **velocity head** (m).

The Pitot tube

Figure 5.2 shows a Pitot tube, an invention by the Italian-born, French hydraulic engineer Henri Pitot (1695–1771): the tube is held with its lower opening opposing the water flow, which in Figure 5.2 is from left to right. Held in this way, a Pitot tube converts the kinetic energy of the surface water flow into a pressure head that is equal to the velocity head h_v (see Box 5.1):

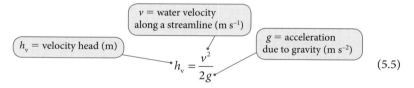

$$h_v = \frac{v^2}{2g} \tag{5.5}$$

By measuring h_v in the Pitot tube, we can determine the water velocity along the streamline in Figure 5.2 as follows:

$$v = \sqrt{2gh_v} \tag{5.6}$$

Thus, for instance, when the velocity head h_v=2.5 cm (g = 9.8 m s⁻¹), then the water velocity v along the streamline equals 0.7 m s⁻¹. For precise measurements using a Pitot tube, v must be larger than 0.2 m s⁻¹ (Van Rijn 1994) or h_v > 2 mm.

Figure 5.2 shows a striking resemblance to Figure 3.4, which shows the energy terms at the location of a piezometer screen in groundwater. A Pitot tube measures the **total head** h_{total} of surface water at the location of the Pitot tube's water inlet,

BOX 5.1 The Pitot tube

Over very short flow distances, frictional energy losses may be disregarded. We may write Bernoulli's law for the short channel section or **reach** of streamline 1–2 in Figure 5.2 as follows:

$$\frac{v_1^2}{2g}+\frac{p_1}{\rho g}+z_1 = \frac{v_2^2}{2g}+\frac{p_2}{\rho g}+z_2 \qquad (B5.1.1)$$

As $z_1 = z_2$, $v_2 = 0$, and $\frac{p_2}{\rho g}=\frac{p_1}{\rho g}+h_v$, Equation B5.1.1 reduces to

$$\frac{v_1^2}{2g}=h_v \qquad (B5.1.2)$$

v_1 is the water velocity v along the streamline and thus

$$h_v = \frac{v^2}{2g} \qquad (B5.1.3)$$

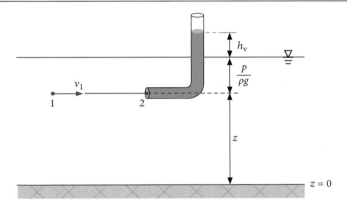

Figure 5.2. A Pitot tube

thereby incorporating the velocity head h_v, which is caused by the stream's kinetic energy or energy of flow. The total hydraulic head h_{total} at the inlet point of the Pitot tube is thus as follows:

$$h_{total} = z + \frac{p}{\rho g} + \frac{v^2}{2g} \qquad (5.7)$$

Total head (m)   Elevation head (m)   Pressure head (m)   Velocity head (m)

To measure the water velocity, usually a combined Pitot tube is used; Figure 5.3 shows such a tube. The combined Pitot tube has an opening towards the flow that measures the total head h_{total}, and vertical openings on both sides of the horizontal part of the tube that measure the hydraulic head h. Subtracting the hydraulic head h from the total head h_{total} yields the velocity head h_v, and from this the water velocity can be determined as shown above.

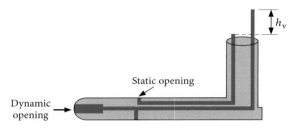

Figure 5.3. A combined Pitot tube (adapted after Van Rijn 1994)

An artificial lake with a lower opening in the dam

Figure 5.4 shows a water-filled reservoir – for instance, an artificial lake – with water exiting through a lower opening in the dam. The water velocity v through the dam opening is related to the square root of the water height H above the centreline of the dam opening:

v = water velocity through the dam opening (m s^{-1})

H = water height above the centreline of the dam opening (m)

$$v = \sqrt{2gH}$$

(5.8)

This equation is derived from an experiment by Evangelista Torricelli (1608–1647), an Italian physicist and mathematician: the derivation of Equation 5.8 is given in Box 5.2. Equation 5.8 is the same as that for a solid particle dropped a distance H in a vacuum (Box 5.3).

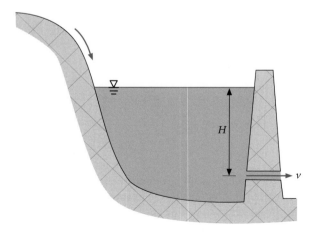

Figure 5.4. A water-filled reservoir with a lower opening in the dam

BOX 5.2 Torricelli's experiment

Figure B5.2 shows an experiment carried out by Torricelli in 1643. The water surface level is kept at a constant level. By disregarding frictional energy losses, we may write Bernoulli's law for streamline 1–2 as follows:

$$\frac{v_1^2}{2g} + \frac{p_1}{\rho g} + z_1 = \frac{v_2^2}{2g} + \frac{p_2}{\rho g} + z_2 \qquad \text{(B5.2.1)}$$

v_1, the water velocity at the water surface (location 1), is much smaller than v_2, the water velocity at the reservoir opening (location 2); this is caused by the water surface being much larger than the cross-sectional area of the outlet opening at the bottom. Because of this, we can assume that $v_1 = 0$.

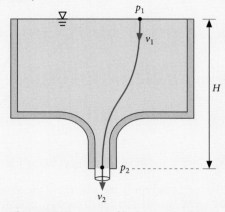

Figure B5.2. A water-filled reservoir with an opening at the bottom

The pressure at the water surface (location 1) equals the existing air pressure, as does the pressure at the water-outlet opening (location 2): the small difference in air pressure between these locations due to their difference in elevation can be neglected. Because of this, the water pressures p_1 and p_2 at locations 1 and 2 can be set at zero: $p_1 = p_2 = 0$. Further, as $z_1 = H$ and $z_2 = 0$, Equation B5.2.1 reduces to

$$H = \frac{v_2^2}{2g} \qquad \text{(B5.2.2)}$$

or

v = water velocity at the outlet opening (m s^{-1})

H = water height (m)

$$v = \sqrt{2gH} \qquad \text{(B5.2.3)}$$

BOX 5.3 A falling object in a vacuum

Figure B5.3 shows the increase in velocity (m s^{-1}) with time (s) for a falling object in a vacuum; or, in other words, the increase in velocity of an object if we disregard frictional energy losses.

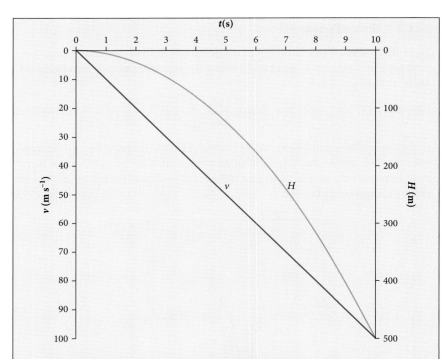

Figure B5.3. t (s), v (m s^{-1}), and H (m) for an object falling in a vacuum

At $t = 0$ seconds, thus at the beginning, the velocity v of the object is 0 m s^{-1}. For ease of computation, we will set the acceleration due to gravity g at 10 m s^{-2}. This means that every second, the falling object gains 10 m s^{-1} in velocity. Thus, at $t = 1$ s, $v = 10$ m s^{-1}; at $t = 2$ s, $v = 20$ m s^{-1}; and so on, as shown in Figure B5.3. In equation form:

$$v = gt \quad \text{or} \quad t = \frac{v}{g} \tag{B5.3.1}$$

How far does an object fall in, let us say, the first 10 seconds of its fall? To answer this question, we simply determine the average velocity over these 10 seconds in m s^{-1} and multiply this by the number of seconds involved, which is 10 seconds. As the starting velocity at $t = 0$ seconds equals 0 m s^{-1}, the average velocity over the first 10 seconds is simply half the velocity of the object at $t = 10$ seconds. The falling distance H after t seconds is thus as follows:

H = falling distance after t seconds (m) v = velocity after t seconds (m s^{-1})

$$H = \frac{1}{2}vt \tag{B5.3.2}$$

Combining Equations B5.3.1 and B5.3.2 gives

$$H = \frac{v^2}{2g} \quad \text{or} \quad v = \sqrt{2gH} \tag{B5.3.3}$$

If we multiply the water velocity v through the dam opening by the area of this opening perpendicular to the water flow, and if we incorporate this area, the factor $\sqrt{2g}$, and the frictional energy losses of water flow through the dam opening in a 'constant' C, we obtain the following **Q–H relation for an artificial lake with a lower opening in the dam:**

Q = discharge through the dam opening (m³ s⁻¹) C = 'constant' (m²·⁵ s⁻¹)

$$Q = C\sqrt{H} \tag{5.9}$$

The term 'constant' for C is placed in quotes because the frictional energy losses in the dam opening are of course higher for larger values of v and H: in practice, one can account for this.

The storage S (m³) in the reservoir above the centreline of the dam opening is approximated by

S = reservoir storage (m³) above the centreline of the dam opening A = horizontal area of the reservoir (m²)

$$S = A\,H \tag{5.10}$$

We may relate the reservoir storage S (m³) and the discharge from the reservoir Q (m³ s⁻¹) by combining Equations 5.9 and 5.10 to give the following **S–Q relation for an artificial lake with a lower opening in the dam:**

$$S = \frac{Q^2}{\alpha} \tag{5.11}$$

α = 'constant' (m³ s⁻²)

or, in a general form (which we will encounter again later), as

For an artificial lake with a lower opening in the dam: $n = 2$

$$S = \frac{Q^n}{\alpha} \tag{5.12}$$

Note that the horizontal area A of the reservoir is not quite constant with depth (A decreases when H diminishes) and that $\alpha = C^2/A$ for a small artificial lake should thus be adjusted accordingly. However, for large artificial lakes the horizontal dimensions dominate, and then differences in A with depth are minimal.

Artificial lakes may serve many purposes, such as the provision of water for irrigation, regulation of river discharges, mitigation of high river discharges through the buffering capacity of the lake, provision of hydroelectricity, availability of water for aeroplanes used in extinguishing fires, touristic activities, and so on.

Ripples in the water

Whenever you throw a stone into a natural stream, study the movement of the ripples at the water surface that this causes! You may observe part of the ripples moving upstream from the point of contact with the water: the propagation velocity of the surface wave v_w caused by the stone's impact on the water then is larger than the flow velocity v of the water (if not, the ripple could not move upstream) – we call this type of water flow **subcritical**

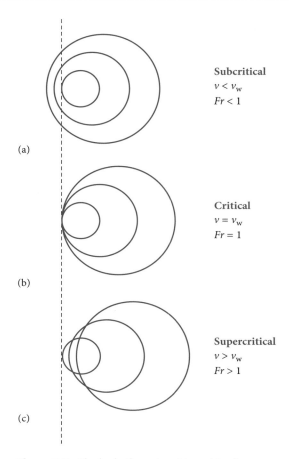

Figure 5.5. Ripples in the water at t_1 and t_2 after an initial disturbance at t_0

flow (Figure 5.5a). You may, however, also observe the ripples to be totally swept down-stream; the flow velocity of the water v then is larger than the propagation velocity of the surface wave v_w – we call this type of water flow **supercritical flow** (Figure 5.5c). When the water flow velocity v and the propagation velocity of the surface wave v_w are equal, the water flow is called **critical flow**: in theory, we would then have to observe that the upstream part of the ripple remained stationary at its point of initiation (Figure 5.5b).

The following relationship can be established for the **propagation velocity of a surface wave** (sometimes called the **surface wave celerity**) v_w (Box 5.4):

v_w = propagation velocity of a surface wave (m s⁻¹)

$$v_w = \sqrt{gH} \qquad (5.13)$$

Note that for the propagation velocity of a surface wave there is no factor 2 below the square root sign, which you may have become accustomed to in Equations 5.6, 5.8, B5.2.3, and B5.3.3. Equation 5.13 can also be used to determine the propagation velocity of a **tsunami** (Japanese for 'harbour wave'), which is a huge sea or ocean wave that is approaching the coast, and is usually caused by an underwater earthquake or volcanic eruption, or by a coastal landslide (Box 5.5).

BOX 5.4 The propagation velocity of a surface wave

The surface wave in Figure B5.4a has a propagation velocity v_w. An observer moving with the surface wave observes the water as having a velocity v_w in the opposite direction, as shown in Figure B5.4b.

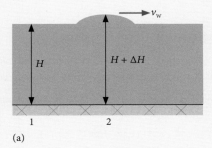

(a)

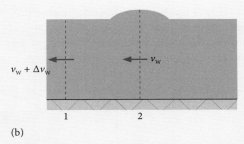

(b)

Figure B5.4. The propagation of a surface wave

From continuity, it follows that

$$Q'\,(\mathrm{m^2 s^{-1}}) = \text{constant} \implies Q'_1 = Q'_2 \tag{B5.4.1}$$

For cross-section 1:

$$Q'_1 = (v_w + \Delta v_w)H \tag{B5.4.2}$$

For cross-section 2:

$$Q'_2 = v_w(H + \Delta H) \tag{B5.4.3}$$

Inserting Equations B5.4.2 and B5.4.3 in Equation B5.4.1 yields

$$(v_w + \Delta v_w)H = v_w(H + \Delta H) \tag{B5.4.4}$$

which can be reworked to

$$\Delta v_w = \frac{v_w\,\Delta H}{H} \tag{B5.4.5}$$

Over short distances of flow, we can assume that frictional energy losses are negligible and apply Bernoulli's law as follows:

$$\frac{(v_w + \Delta v_w)^2}{2g} + H = \frac{v_w^2}{2g} + H + \Delta H \tag{B5.4.6}$$

Rearranging, taking $(\Delta v_w)^2$ as negligible, yields

$$v_w \Delta v_w = g \Delta H \qquad \text{(B5.4.7)}$$

Inserting Equation B5.4.5 in Equation B5.4.7 gives

$$v_w = \sqrt{gH} \qquad \text{(B5.4.8)}$$

BOX 5.5 The propagation velocity of a tsunami

Figure B5.5 presents a cross-section of a tsunami in the ocean. By combining the continuity equation and Bernoulli's law (the energy equation), we have deduced in Box 5.4 that

v_w = propagation velocity of a tsunami (m s^{-1}) H = ocean depth (m)

$$v_w = \sqrt{gH} \qquad \text{(B5.5)}$$

This relation also holds for a tsunami wave. Actually, Equation B5.5 holds for a surface wave, when the water depth H is less than 5% of the wavelength L_w. The wave length of a tsunami is hundreds of kilometres, and its amplitude a_w in deep water is less than a metre, which is why tsunamis generally pass unnoticed at sea.

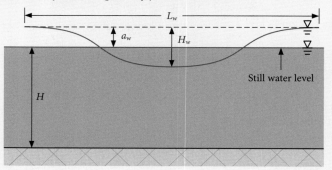

Figure B5.5. A cross-section of a tsunami in the ocean (a_w = wave amplitude; H = water depth; H_w = wave height; L_w = wave length)

If we assume the water depth H of an ocean to be 4.5 km = 4500 m, it is obvious from the foregoing information that a tsunami is classified as a surface wave, a ripple on the vast ocean surface, and that we may use Equation B5.5 to determine its propagation velocity. A simple calculation using these data then teaches us that the propagation velocity v_w of a tsunami equals $\sqrt{9.8 \times 4500} = 210$ m s$^{-1} \approx 750$ km hour^{-1}, the speed of a jet airliner!

A tsunami wave involves the movement of water all the way to the ocean floor and it therefore differs totally from a wind-driven ocean wave, which only has an effect at or near the ocean surface. The front of an approaching tsunami wave usually consists of a trough, which is why a greater than normal and unusual retreat of the sea or ocean along the coast provides an important warning sign of an approaching tsunami. When a tsunami approaches the coast, the water depth H diminishes and its propagation velocity v_w decreases. However, because of increased friction, its wave height H_w at the coastline may grow to heights of tens of metres, but also, because of the enormous mass of water that has been set in motion (again, a tsunami wave involves the movement of water all the way to the ocean floor), tsunamis have a devastating effect on coastal areas.

The ratio of the water velocity v and the surface wave propagation velocity v_w is called the **Froude number** (*Fr*), after the British engineer and naval architect William Froude (1810–1879):

$$Fr = \frac{v}{v_w} = \frac{v}{\sqrt{gH}}$$

(5.14)

In summary, a surface wave cannot propagate against the water flow direction when $v > v_w$ or when, using Equation 5.14, $Fr > 1$, and the water flow is

- **subcritical** for $v < v_w$ or $Fr < 1$,
- **critical** for $v = v_w$ or $Fr = 1$, and
- **supercritical** for $v > v_w$ or $Fr > 1$.

Figure 5.6 shows a photograph of the bottom of a kitchen sink when the water tap is turned on, showing the transition from supercritical to subcritical water flow in everyday life: the transition is accentuated by the water rim (**hydraulic jump**), where the flow is critical.

Figure 5.6. A photograph of the bottom of a kitchen sink when the water tap is turned on, showing the transition from supercritical, radial to subcritical, turbulent water flow: in between, at the water rim (hydraulic jump), the water flow is critical

The following section will show us that we can make good use of critical water flow, because only then are the water height H and the total energy level h_{total} of the water flow related in a unique manner.

Specific energy

Figure 5.7 shows the theoretical distribution of the flow velocity with the water height in an open channel. In practice, however, this distribution can be disturbed by a rough bed, weed growth, obstructions, turbulence of the flow, and so on. As there usually is no such thing as a constant water velocity in a channel, we use the average flow velocity, which equals the discharge Q ($m^3 \ s^{-1}$) divided by the area of the channel cross-section perpendicular to the water flow. With this is mind, we may rewrite Equation 5.7 (Bernoulli's law) for a channel cross-section perpendicular to the water flow as follows:

$$Q = \text{discharge } (m^3 \ s^{-1})$$

$$h_{total} = \frac{Q^2}{2gA^2} + \frac{p}{\rho g} + z \qquad (5.15)$$

$$A = \text{cross-sectional flow area } (m^2)$$

Using the channel bed as the reference level and introducing the **specific energy** h_e as the energy per unit weight ($J \ N^{-1} = m$) of the flowing water relative to the stream bottom, we may rewrite Equation 5.15 as follows:

h_e = specific energy (m) = energy per unit weight of the flowing water relative to the stream bottom

H = water height of the stream (m)

$$h_e = \frac{Q^2}{2gA^2} + H \qquad (5.16)$$

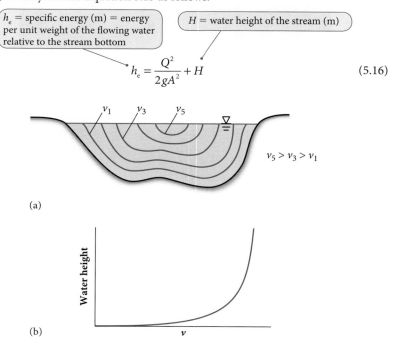

(a)

(b)

Figure 5.7. The theoretical distribution of the flow velocity with the water height in an open channel: (a) in cross section; (b) in longitudinal section

Defining the **specific discharge** q_w (m² s⁻¹) as the discharge Q (m³ s⁻¹) per unit channel width w (m) of the stream, we may replace Q with q_w, and A with H in Equation 5.16 to give

$$q_w = \text{specific discharge (m}^2\text{ s}^{-1}) = \text{discharge per unit channel width } w$$

$$h_e = \frac{q_w^2}{2gH^2} + H \qquad (5.17)$$

Equation 5.17 is a cubic equation in H and has three roots. However, as negative heights of H have no physical meaning, (h_e,H) positions below the horizontal $H = 0$ axis are disregarded and are not shown in Figure 5.8.

Figure 5.8 shows a **specific energy diagram**, a graph of the water height H (m) versus the specific energy h_e (m), for a specific discharge q_w of 0.5 m² s⁻¹ as established from Equation 5.17. Figure 5.8 shows that there are two **alternative water heights**, two possible (positive) water heights H for each specific energy value h_e, with the exception of the minimum specific energy value h_e, the left-most (h_e,H) position on the graph, that relates to only one water height, H. Figure 5.2 has already shown us that in order for water to flow, the total head h_{total} must be larger than the water height H or, likewise: $h_e > H$, which explains why the graph in Figure 5.8 is located beneath the 1:1 line (and, logically, above the horizontal $H = 0$ line).

Figure 5.9 is the same as Figure 5.8, but now with the water height H shown on the horizontal axis and the specific energy h_e on the vertical axis. The rate of change of h_e with H, dh_e / dH, is equal to the slope of a tangent line touching the curve of Figure 5.9. At a local minimum, the tangent line touching a curve is horizontal and the slope of the tangent line then equals zero (see C2.7). In mathematical notation:

$$\frac{dh_e}{dH} = 0 \qquad (5.18)$$

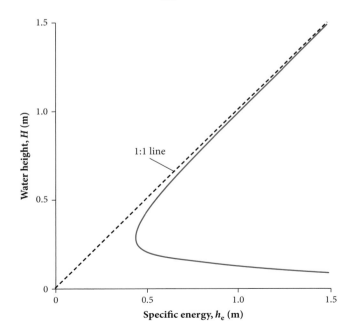

Figure 5.8. The specific energy diagram for a specific discharge q_w of 0.5 m² s⁻¹

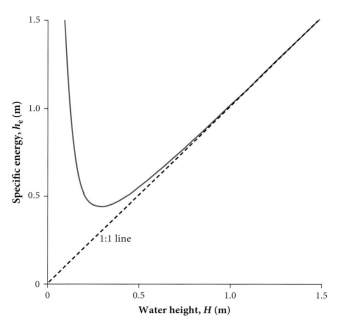

Figure 5.9. Specific energy h_e (m) versus the water height H (m) for a specific discharge q_w of 0.5 m^2 s^{-1}

Mathematical differentiation of Equation 5.17 (Bernoulli's law) yields

$$\frac{dh_e}{dH} = \left(\frac{q_w^2}{2g}\right)\left(-2H^{-3}\right) + 1 \tag{5.19}$$

Substituting Equation 5.19 in Equation 5.18 yields

$$q_w = H\sqrt{gH} \tag{5.20}$$

As $v = q_w / H$, Equation 5.20 can also be written as follows:

$$v = \sqrt{gH} \tag{5.21}$$

Thus, when the specific energy h_e is at a minimum, the flow velocity v equals $\sqrt{gH}$, which is the propagation velocity v_w of a surface wave, as we already know from Equation 5.13 and Box 5.4. This means that only at this **minimum specific energy** $h_{e,min}$, which is the only position on the graph of Figure 5.8 that relates to a single value of the water height H and not to alternative water heights H, we have **critical flow** ($v = v_w$ or $Fr = 1$), a transition flow between subcritical and supercritical flow or vice versa. We can now deduce which parts of the graph of Figure 5.8 describe subcritical and supercritical flow.

In the upper part of the specific energy curve of Figure 5.8 (near the 1:1 line), the water heights H are relatively large and differences between h_e and H are small: thus the velocity head h_v and water flow velocities v are relatively small, and the energy of the water is predominantly made up of pressure energy related to the water height H.

In the lower part of the specific energy curve (near the horizontal $H = 0$ axis), the water heights H are low and differences between h_e and H are large: thus the velocity head h_v and water flow velocities v are relatively large, and the energy of the water is predominantly made up of kinetic energy related to the flow velocity v.

Thus, for the upper part of the graph of Figure 5.8, we must have **subcritical flow** ($v < v_w$ or $Fr < 1$), and for the lower part of the graph of Figure 5.8 **supercritical flow** ($v > v_w$ or $Fr > 1$).

The above Equations 5.20 and 5.21 only hold for critical flow; and thus when the specific discharge q_w equals the **critical specific discharge** q_c (m² s⁻¹), when the water height H equals the **critical water height** H_c (m), and when the water velocity v equals the **critical flow velocity** v_c (m s⁻¹). Therefore, the equations should be written accordingly as follows:

$$\boxed{q_c = \text{critical specific discharge (m}^2\text{ s}^{-1})} \qquad \boxed{H_c = \text{critical water height (m)}}$$

$$q_c = H_c^{\frac{3}{2}}\sqrt{g} \tag{5.22}$$

and

$$\boxed{v_c = \text{critical flow velocity (m s}^{-1})}$$

$$v_c = \sqrt{gH_c} \tag{5.23}$$

Substitution of Equation 5.22 in Equation 5.17 (Bernoulli's law) for $h_e = h_{e,\text{min}}$, $q_w = q_c$, and $H = H_c$ yields:

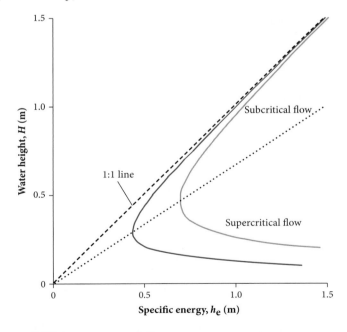

Figure 5.10. Specific energy diagrams for two specific discharges q_w of 0.5 (left-hand graph) and 1.0 m² s⁻¹ (right-hand graph): the points of minimum specific energy $h_{e,\text{min}}$, where critical flow occurs for different specific discharges q_w (Equation 5.24), are given by the lower broken line

$$h_{e,\text{min}} = \text{minimum specific energy (m)}$$

$$h_{e,\text{min}} = \frac{3}{2}H_c \quad \text{or} \quad H_c = \frac{2}{3}h_{e,\text{min}} \tag{5.24}$$

Thus, when critical flow occurs, the water height H (m) equals two-thirds of the specific energy h_e (m).

Figure 5.10 is an extension of Figure 5.8 to include a specific energy diagram for a specific discharge q_w of 1.0 m. The shapes of specific energy diagrams for different specific discharges are similar, and for higher specific discharges the graphs lie more to the right (when the axes are chosen as in Figures 5.8 and 5.10). The (h_e, H) positions where critical flow occurs for different specific discharges q_w are given by Equation 5.24 and are shown as the lower striped line in Figure 5.10.

Critical flow at the dip

When water flows from a gently sloping reach, for which we can assume or test the flow to be subcritical (throw a stone in the water!), over a natural step such as in Figure 5.11, the water flow accelerates and becomes supercritical. Therefore, the

Figure 5.11. A rapid in the Souloise river, Dévoluy, France: at the dip, the water flow is critical, as it changes from subcritical upstream to supercritical in the rapid

water flow at the dip is critical, as it changes from subcritical upstream of the dip to supercritical in the rapid. As the water flow is critical at the dip, it is possible to make a rough estimate of the river discharge Q (m³ s⁻¹): estimate the critical water height H_c (m) at the dip and calculate the critical flow velocity v_c (m s⁻¹) using Equation 5.23; then estimate the cross-sectional flow area A (m²) perpendicular to the water flow at the dip ($A = w_c H_c$) and multiply v_c by this area A to obtain the river discharge Q (m³ s⁻¹). Thus, the **Q–H relation for critical flow at the dip** reads as follows:

$$w_c = \text{channel width (m)}$$

$$Q = w_c H_c \sqrt{g H_c} \qquad\qquad (5.25)$$

Exercise 5.1.1 Determine the discharge of the Souloise river if we estimate the water at the dip in Figure 5.11 to be 70 cm deep and the channel to be 10 m wide.

Exercise 5.1.2 The **Venturi effect** is a drop in pressure head when an incompressible fluid such as water flows through a length of tube with a smaller diameter, as shown in Figure E5.1.2. The effect is named after the Italian physicist Giovanni Battista Venturi (1746–1822).
    Disregard frictional losses and calculate the discharge Q (m³ s⁻¹) through the tube as a function of the drop in pressure head

$$\frac{p_1}{\rho g} - \frac{p_2}{\rho g}$$

and the cross-sectional areas A_1 and A_2 (m²) in the tube.

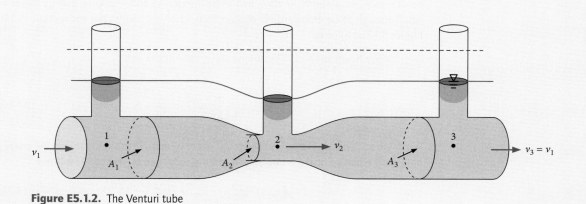

Figure E5.1.2. The Venturi tube

Water flow over a natural step in the river bed

Estimating the discharge by visual inspection as just presented is a rather crude method, as it is difficult to accurately estimate the critical water height H_c from a falling water surface as shown in Figure 5.11. For a more accurate estimate of the discharge, we would rather like to find a relation between the discharge Q (m³ s⁻¹) and

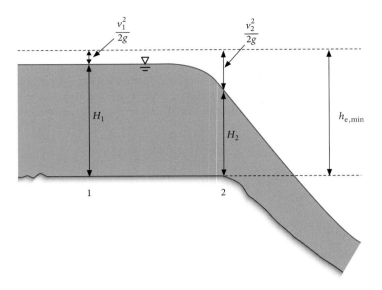

Figure 5.12. A side view of water flowing over a natural step in a river bed

the water height H_1 (m) slightly upstream of the rapid, where the water flow is still relatively slow and deep, and where the water height can be measured more accurately. Using Bernoulli's law (the energy equation) and the earlier established relation that the critical water height H_c (m) equals two-thirds of the minimum specific energy $h_{e,\,min}$ (m) (Equation 5.24), we can deduce such a relation.

Figure 5.12 shows a side view of water flowing over a natural step in the **river bed**, the channel bottom of a river. Over a very short distance of flow upstream of the natural step, we assume the river bed to be horizontal and frictional energy losses as small enough to be disregarded. We can then write Bernoulli's law for section 1–2 in Figure 5.12 as follows:

H_1 = water height upstream of the natural step in the river bed (m)

$$\frac{v_1^2}{2g} + H_1 = \frac{v_2^2}{2g} + H_2 \tag{5.26}$$

For Figure 5.12, the relation between the water height H (m) and the specific energy $h_{e,\,min}$ (m) when critical flow occurs (Equation 5.24) can be written as follows:

$$H_2 = H_c = \frac{2}{3} h_{e,\,min} = \frac{2}{3}\left(H_1 + \frac{v_1^2}{2g} \right) \tag{5.27}$$

Because the water flow upstream of the natural step is relatively slow and deep, we may assume the flow there to be subcritical and v_1 to be negligibly small ($v_1 = 0$). Equations 5.26 and 5.27 then reduce to

$$H_1 = \frac{v_2^2}{2g} + H_2 \tag{5.28}$$

and

$$H_2 = \frac{2}{3} H_1 \tag{5.29}$$

Substituting Equation 5.29 in Equation 5.28 yields the following relation between the water velocity v_2 and the water height H_1 measured upstream:

$$H_1 = \frac{3 v_2^2}{2g} \tag{5.30}$$

or

$$v_2 = \sqrt{\frac{2}{3} g H_1} \tag{5.31}$$

Furthermore,

w_c = width of the water stream at the dip where critical flow occurs (m)

$$Q = v_2 w_c H_2 \tag{5.32}$$

Substituting Equation 5.31 in Equation 5.32 and applying Equation 5.29 yields

$$Q = w_c H_2 \sqrt{\frac{2}{3} g H_1} = w_c \frac{2}{3} H_1 \sqrt{\frac{2}{3} g H_1} \tag{5.33}$$

which can be rearranged to give the following **Q–H relation for water flow over a natural step in the river bed**:

H_1 = water height (m) upstream of the natural step in the river bed (with the top of the natural step in the river bed taken as the reference level or $z = 0$ level)

$$Q = \left(\frac{8}{27} g \right)^{\frac{1}{2}} w_c H_1^{\frac{3}{2}} \tag{5.34}$$

Using Equation 5.34, we can estimate the discharge Q by measuring the water height H_1 upstream of the natural step in the river bed; note that H_1 is the upstream water height above the top of the natural step (taken as the reference level or $z = 0$ level) as evident from Figure 5.12.

In the following sections, we will see that Equation 5.34 is equally well applicable to flow over a dam and to flow over a rectangular weir – a man-made, rectangular-shaped, fixed construction; also, Equation 5.34 may be modified to estimate the discharge from water height readings upstream of a number of other man-made, fixed constructions in the river bed, such as V-notch weirs and flumes.

An artificial lake with overflow

Figure 5.13 shows a water-filled reservoir – for instance, an artificial lake – with water flowing over the top of the dam. The water velocity v over the dam is related to the

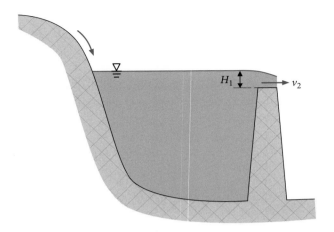

Figure 5.13. A water-filled reservoir with overflow

upstream water height H_1 (m) with the top of the dam as the reference level, in the same way as presented in Equation 5.31 for a natural step in the river bed. Thus:

$$v_2 = \sqrt{\frac{2}{3}gH_1} \tag{5.35}$$

Compare this equation to Equation 5.8 for an artificial lake with a lower opening in the dam.

As the upstream water height H_1 with the top of the dam as the reference level is equivalent to the upstream water height H_1 above the elevation of a natural step in the river bed, as just presented, we may combine Equation 5.35 with Equation 5.32, repeated here as follows,

w_c = width of the water stream at the top of the dam dip where critical flow occurs (m)

$$Q = v_2 w_c H_2 \tag{5.36}$$

and Equation 5.29 to obtain the **Q–H relation for an artificial lake with overflow** as

H_1 = upstream water height (m) with the top of the dam taken as the reference level or $z = 0$ level

$$Q = CH_1^{\frac{3}{2}} \tag{5.37}$$

C in Equation 5.37, of course, equals $w_c \sqrt{\frac{8}{27}g}$. Compare Equation 5.37 to Equation 5.9 for an artificial lake with a lower opening in the dam.

For the reservoir storage S (m³) above the level of the top of the dam, we may write

S = reservoir storage above the level of the top of the dam (m³) A = horizontal area of the reservoir (m²)

$$S = AH_1 \tag{5.38}$$

We may relate the reservoir storage S (m³) above the level of the top of the dam and the discharge from the reservoir Q (m³ s⁻¹) by combining Equations 5.37 and 5.38 to give the following S–Q relation for an artificial lake with overflow:

$$S = \frac{Q^{\frac{2}{3}}}{\alpha} \tag{5.39}$$

or, in a generalized form,

For an artificial lake with overflow: $n = \frac{2}{3}$

$$S = \frac{Q^{n}}{\alpha} \tag{5.40}$$

Equation 5.40 is similar to Equation 5.12 for an artificial lake with a lower opening in the dam, albeit that different values for n and α should be inserted to relate the storage S (m³) to the discharge Q (m³ s⁻¹) that flows from this storage. Note that S equals the reservoir storage (m³) above the level of the top of the dam in Equations 5.39 and 5.40, whilst S is the reservoir storage (m³) above the centreline of the dam opening in Equations 5.11 and 5.12; also, the units of C and α in Equations 5.37, 5.39, and 5.40 are different from those in Equations 5.9, 5.11, and 5.12. Importantly, both the values of C and α may be adjusted to account for the frictional energy losses that occur when water flows over the top of the dam.

Water flow over a weir

Figure 5.14 shows a **weir**, a man-made, fixed construction to force water in a brook or river to overflow and the water flow to become critical. More specifically, Figure 5.14 shows both the upstream and side view of a **rectangular weir**, a rectangular shape cut into a plate of aluminium (or other material), which is placed perpendicular to the water stream, forcing the water to flow through the rectangular opening. The edge or surface over which the water flows to become critical is called the **crest**. When, as in Figure 5.14, the length of the crest in the direction of the water flow is less than

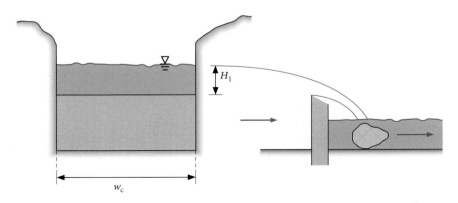

Figure 5.14. Upstream and side views of a rectangular weir

2 mm, the weir is classified as a **sharp-crested weir**; when the weir has a horizontal or nearly horizontal crest that is sufficiently long in the direction of flow to support the overflowing sheet of water, it is called a **broad-crested weir**.

Equation 5.34 is equally applicable as the **Q–H relation for a rectangular weir** and may be written for different weir widths w_c as follows:

H_1 = upstream water height (m) above the level of the crest of the weir (thus with the elevation of the crest of the weir taken as the reference level or $z = 0$ level)

$$Q = C w_c H_1^{\frac{3}{2}} \tag{5.41}$$

H_1 should now be taken as representing the water height (m) above the level of the crest of the weir, and thus with the elevation of the crest of the weir taken as the reference level or $z = 0$ level. The value of C may be adjusted (diminished) to account for frictional energy losses due to contractions in the area of flow and at the edges of the weir.

Figure 5.15 shows both an upstream and a side view of a **V-notch weir**, a V-shape cut into a plate of aluminium (or other material), which is placed perpendicular to the water stream, forcing the water to flow through its notch-opening. The cross-sectional flow area A in the weir perpendicular to the water flow is related to the angle θ of the V-notch as follows:

A = cross-sectional flow area in the weir perpendicular to the water flow (m²)

$$A = H_2^2 \tan\left(\frac{\theta}{2}\right) \tag{5.42}$$

H_2 = critical water height above the level of the crest of the weir (m)

Using this cross-sectional flow area A and by following the same reasoning as for water flow over a natural step in the river bed, and thus replacing H_2 by H_1 (Equation 5.29),

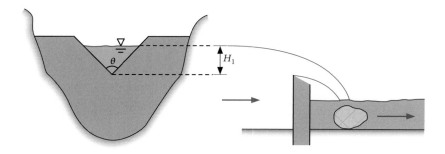

Figure 5.15. Upstream and side views of a V-notch weir

the **Q–H relation for a V-notch weir** with different angles θ of the V-notch can be deduced as follows:

> H_1 = upstream water height (m) above the level of the crest of the weir (thus with the elevation of the crest of the weir taken as the reference level or $z = 0$ level)

$$Q = C \tan\left(\frac{\theta}{2}\right) H_1^{\frac{5}{2}}$$

(5.43)

The value of C may (similarly to the case of a rectangular weir) be adjusted (diminished) to account for frictional energy losses due to contractions in the area of flow and at the edges of the weir.

As a practical tip, always place a large boulder downstream of a (V-notch) weir to prevent the force of the falling water from undermining the built construction.

> **Exercise 5.1.3** Verify that the above Q–H relation for a V-notch weir (Equation 5.43) is correct.

Water flow through a flume

Figure 5.16 shows yet another man-made, fixed construction that forces water flow in a brook or river to become critical: a flume. A **flume** differs from a weir in that critical flow is not attained by water flowing over a crest under the influence of gravity, but by water flowing through a constricted structure. The constricted section (**throat**) of the flume is formed by raising the channel bottom or by contracting the sides of

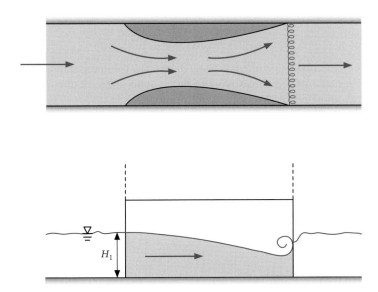

Figure 5.16. Top and side views of a flume

the channel, or both. Figure 5.16 shows that the throat creates a drop in water level or pressure head analogous to the Venturi effect (see Exercise 5.1.2): this causes first supercritical flow and then a hydraulic jump when the water flow decelerates to sub-critical flow.

The **hydraulic jump** (see also Figure 5.6) is caused by supercritical flow (caused by a natural step in the river bed, a weir or flume, etc.) reaching slower-moving, down-stream water, which causes the water to be swept upwards and to fall back on itself like a crashing wave, but falling in the upstream direction. In other words: at the hydraulic jump, kinetic energy is explosively converted into turbulence and potential energy.

An advantage of a flume over a weir is that a flume largely cleans itself. The **bed load** of a stream consists of particles (sand, stones) that are moved along the stream bed by rolling, pushing, and/or saltation (advancement by leaps): the **suspended load** of a stream consists of particles (clay, silt) that are carried in suspension (floating in the water column) above the stream bed. Bed load in particular, but also part of the suspended load, may accumulate upstream of a weir, below its crest. As an example, sandstone areas are characterized by high bed-load transports, causing a large accu-mulation of material behind a weir in a short time. As this accumulation has an effect on the Q–H relation, the accumulated material has to be shovelled out periodically (which is hard work!): hardly any such action is needed to maintain a flume in opera-tion. However, a disadvantage of a flume over a weir may be its higher complexity of construction.

Several types of flume exist, the rectangular standing wave flume (Figure 5.16) probably being the simplest type, as water height measurements H_1 upstream of the throat may be directly related to the discharge by a Q–H relation identical to Equation 5.37. The **Q–H relation for a rectangular standing wave flume** is thus as follows:

$$H_1 = \text{water height measurement upstream of the throat of the flume (m)}$$

$$Q = CH_1^{\frac{3}{2}} \tag{5.44}$$

In practice, extensive experimental research has been done, yielding a vast set of empirical equations for different types of flume under different circumstances, and the **general empirical Q–H relation for a flume** would be

$$Q = CH_1^{n} \tag{5.45}$$

The factor C may incorporate the channel width, different flume widths, and/or other key variables. The exponent n in Equation 5.45 for different throat widths of the well-known **Parshall flume**, a flume developed and calibrated in 1921 by Dr Ralph L. Parshall (1881–1959) at the then-named Colorado Agricultural College's hydrology laboratory, circulates around a value of 1.5; however, for other types of flume other values may be appropriate.

Concluding remark

As a concluding remark for this section, where we have extensively applied Bernoulli's law, note that H in the theoretical Q–H relations just derived always carries an exponent of $\frac{1}{2}$ or a multiple of $\frac{1}{2}$ ($\frac{3}{2}$ or $\frac{5}{2}$), which, of course, has to do with the velocity head playing an important role in surface water flow. This observation clearly distinguishes surface water flow equations from groundwater and soil water flow equations, where the exponent of ∇h (see section 4.6) in Darcy's law and the Darcy–Buckingham, Laplace, and Richards equations is always a full integer, 1 or 2, and never a $\frac{1}{2}$ or a multiple of $\frac{1}{2}$. For an extensive review of weirs and flumes, the reader is referred to Bos (1989).

5.2 Measuring stage, water velocity, and discharge

There are many methods to determine the water level, water velocity, and discharge of a brook or river. Here, we will discuss the workings of a limited number of methods.

Stage

It is important to have knowledge of the reaction of the water level of a brook or river on precipitation and of the susceptibility of certain areas to flooding. Therefore, measuring the water level – or the **stage**, as it is often called – is an important activity, first to obtain a record of the water stages and, secondly, as we can link the discharges Q of a brook or river at specific locations to stages H slightly upstream, as we have seen in the previous section. This Q–H relation is also called (in reversed terms) the **stage–discharge relation**.

There exist a variety of ways to measure stage, the low-budget way being to install a concrete iron in the channel bed near the channel side and to simply determine the water height from measuring along the concrete iron.

A slightly more sophisticated method to measure the stage would be the use of a **staff gauge**, a graduated scale placed in a brook or river as shown in Figure 5.17.

For continuous measurement of the stage, a **water level recorder** can be used: Figure 5.18 shows that such a recorder consists of a float and counterweight connected by a cable or metal tape that moves over a pulley system. In paper chart water level recorders, a paper chart is attached to a slowly rotating drum and when the stage changes, the pulley system moves a pen up or down over the paper chart, pencilling the change of stage with time on the chart. The paper chart is changed daily, weekly, or monthly, depending on the period of revolution of the rotating drum. Nowadays, a digital backup is usually made of the paper registration, or the paper registration is replaced altogether by a **shaft encoder and data logger** that register the movement of the pulley (shaft rotation) and consequently the water level (see Figure 5.18).

Nowadays, increasing use is made of **pressure sensors** (pressure transducers): a pressure sensor is lowered into a pipe that has openings to enable water to flow in; the sensor is lowered in the pipe to the bottom of the stream to measure the total pressure (water pressure + air pressure); and another pressure sensor is positioned above the water to measure the air pressure. The difference between the total pressure and the air pressure gives the water pressure, which is linearly related to the stage H; a similar set-up was described in section 4.2 for determining the hydraulic head h of groundwater in a piezometer.

Figure 5.17. A staff gauge

Vertical float recorder

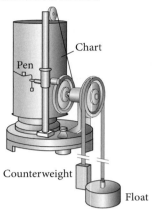

Horizontal float recorder

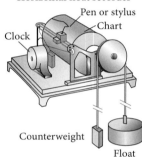

Shaft encoder

Figure 5.18. Water level recorders: paper chart recorders (from Gregory and Walling 1973) and an example of a shaft encoder

In the developed countries, paper chart recorders are becoming largely obsolete. For instance, in the UK the Environment Agency (EA) uses pressure transducers and shaft encoders that record the water level in digital code on data loggers, and these data are transmitted electronically via landlines or, increasingly, via GSM- or GPRS-based telemetry networks.

GSM = Global System for Mobile Communications; GPRS = General Packet Radio Services.

A good way to continuously measure the water stage is in a **stilling well installation**, a stage recording station connected to a brook or river by water intake pipes as shown in Figure 5.19. Such a set-up has the advantage that oscillations due to the turbulence of the water flow are subdued. Figure 5.19 shows a stilling well installation equipped with a shaft encoder, but of course pressure transducers can equally well be used in such a set-up.

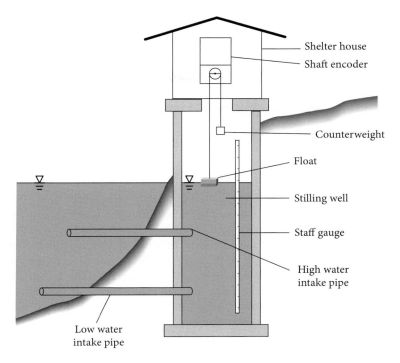

Figure 5.19. A stilling well installation

Water velocity

In section 5.1 the Pitot tube (Figure 5.2) and the combined Pitot tube (Figure 5.3) have been introduced for measuring the water velocity along a streamline.

A low-budget way to measure the water velocity at the surface of a stream is by using a **float**, a buoyant object on the water: the water flow velocity is simply determined by timing the movement of the float over a set distance. A number of objects can be used as float – for instance, an apple (not your computer!) or an orange. Care should be taken to ensure that the object used for measuring the water velocity should not be too light or too buoyant: for the precision of the measurements, it is best when the floating object sits at some depth whilst floating. It is good practice to repeat the measurements several times and to average the trustworthy results to obtain a best estimate. Interestingly, the Italian polymath Leonardo da Vinci (1452–1519) measured the velocity distribution across stream sections using a float, measuring the distance with an odometer and time by rhythmic chanting (Chow *et al.* 1988). Note that when using a float, one measures the water velocity at the surface of a stream. As a rule of thumb, the average water velocity for a flow distribution as shown in Figure 5.7b equals about 85% of the water velocity at the surface.

A well-known instrument for determining the water velocity at different locations and depths in a stream is the **Ott-type current meter**, named after Albert Ott, who developed this type of current meter in 1875. One may simply determine the water velocity using an electronic device that keeps track of the number of revolutions of the current meter's propeller during a set time interval: the flow velocity can then simply be determined from a calibration equation provided by the manufacturer

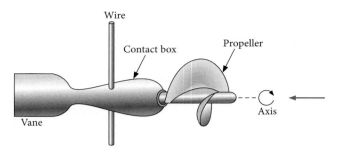

Figure 5.20. An Ott-type current meter

that links the water velocity to the number of revolutions. The average water velocity for a flow distribution as shown in Figure 5.7b can be found by averaging the velocity measurements at different depths. If only one measurement in the vertical is made, a measurement at 40% of the water height (measured from the channel bottom upwards) provides the best estimate of the average water velocity. If two measurements in the vertical are made, standard practice is to determine the water velocity at 20 and 80% of the water height and to average these. If three measurements in the vertical are made, these are best made at 20, 40, and 80% of the water height (measured upwards). One should then determine the average water velocity as follows:

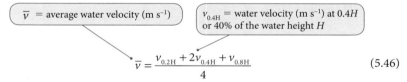

$$\bar{v} = \frac{v_{0.2H} + 2v_{0.4H} + v_{0.8H}}{4} \tag{5.46}$$

Another well-known instrument for determining the water velocity at different locations and depths in a stream, which is becoming more common, is the **electromagnetic current meter**. Such a current meter measures the water velocity by generating a magnetic field in the water around a sensor, with the water acting as electrical conductor. The water velocity is proportional to the voltage measured by the sensor's electrodes, and the average water velocity for a flow distribution as shown in Figure 5.7b may be determined in exactly the same way as just explained for an Ott-type current meter.

Other instruments for determining water velocities that are worth mentioning are the **Acoustic Doppler Current Profilers** (ADCPs) shown in Figure 5.21. An ADCP is a **sonar** (originally an acronym for **s**ound **n**avigation **a**nd **r**anging) that can be used to produce a record of water velocities for a range of depths.

Discharge

Due to slowing down of the water flow at the channel sides, the average flow velocity of a river is about 60–70% of the water velocity at the surface and in the middle of a river as, for instance, measured using a float. Multiplying the average flow velocity of a brook or river (m s^{-1}) by the cross-sectional flow area perpendicular to the water flow (m^2) of course yields an estimate of the discharge (m^3 s^{-1}).

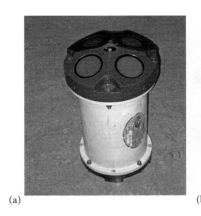

(a)   (b)

Figure 5.21. Acoustic Doppler Current Profilers (ADCPs)
An ADCP uses sound waves to determine the water velocity, making use of the so-called
Doppler effect, or Doppler shift, named after Austrian physicist Christian A. Doppler (1803–
1853): a number of discrete beams are oriented in oblique directions relative to the central
axis of the ADCP and to one another – in upward- or downward-listening mode (a), an ADCP
provides information about horizontal velocity components; and in side-listening mode (b), it
provides information about vertical velocity components

Volumetric gauging

After selecting and/or modifying a cross-section of a brook in such a way that the
brook discharge can be collected in a large bucket, low discharges (less than 10 litre s^{-1})
can be measured by **volumetric gauging**, which is simply measuring the volume of
water collected in the bucket after a set period of time. It is recommended to repeat
the measurements some three to five times and to average the trustworthy results
to obtain a best estimate; for measuring low discharges, volumetric gauging offers
a precise method.

The velocity–area method

Using an Ott-type or electromagnetic current meter, the discharge of a brook or river
can be determined using the **velocity–area method**. The essence of this technique
is to sum the discharges through segments of the cross-sectional flow area to obtain
the discharge for the whole cross-sectional flow area. Figure 5.22 shows an example
with velocity measurements at the two water heights $0.2H$ and $0.8H$ at equally spaced
verticals of the cross-sectional flow area perpendicular to the water flow.

As an example, the segment discharge Q_{23} (m^3 s^{-1}) for a segment in between the
verticals 2 and 3 is as follows:

$\overline{v}_2 = $ average water velocity in vertical 2 (m s^{-1}) $H_2 = $ water height of vertical 2 (m)

$$Q_{23} = \left(\frac{\overline{v}_2 + \overline{v}_3}{2}\right)\left(\frac{H_2 + H_3}{2}\right)(w_3 - w_2) \qquad (5.47)$$

$w_3 = $ width between left bank and vertical 3 (m)

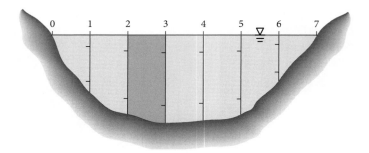

Figure 5.22. The velocity–area method of discharge measurement: an upstream view of the cross-sectional flow area perpendicular to the water flow

Consequently, the (total) discharge Q (m³ s⁻¹) can be written as follows:

$$n = \text{number of segments from bank to bank}$$

$$Q = \sum_{i=1}^{n} \left(\frac{\overline{v}_{i-1} + \overline{v}_i}{2} \right) \left(\frac{H_{i-1} + H_i}{2} \right) (w_i - w_{i-1}) \qquad (5.48)$$

The graphical method

Figure 5.23 shows how the discharge Q can also be determined using a **graphical method**. For all verticals, draw a graph of the velocity v (m s⁻¹) with the water height H (m) on millimetre paper as shown in Figure 5.23a; following our above example, such a graph would have to be constructed from velocity measurements at the two water heights $0.2H$ and $0.8H$. Then, simply by counting the millimetre squares (or programming a spreadsheet program to do so), determine the area (m² s⁻¹) to the left of the curve. Such an area is, in fact, the specific discharge q_w (m² s⁻¹) for the vertical under study: in the same way, determine the specific discharges of the other verticals under study.

> **Exercise 5.2.1** The cross-sectional flow area of a brook perpendicular to the water flow is 80 cm wide. Table E5.2.1 gives the data from water velocity measurements in three verticals of the cross-section at 20, 40, and 60 cm distance from the bank; for each vertical, the water velocity is measured at 20, 40, and 80% of the water height (measured upwards).
>
> **Table E5.2.1** The water height H (cm) and water velocity v (cm s⁻¹) at 20, 40, and 60 cm from the bank of a brook
>
> | | 20 cm | 40 cm | 60 cm |
> | ------------ | ----- | ----- | ----- |
> | H | 15 | 23 | 17 |
> | v at $0.8H$ | 18 | 26 | 20 |
> | v at $0.4H$ | 13 | 20 | 15 |
> | v at $0.2H$ | 8 | 14 | 10 |
>
> Determine the discharge Q of the brook in litre s⁻¹ using the velocity–area method.

Next, construct a top view of the cross-section as shown in Figure 5.23b, where the values of the specific discharges q_w (m² s⁻¹) just determined are represented by arrow lengths; the arrows begin at their exact location in the cross-section and are drawn in the direction of flow. Connect the arrow points to draw a graph of the specific discharge q_w (m² s⁻¹) with the width w (m) of the stream, and determine the area to the left of the curve as shown in Figure 5.23b. This area equals the discharge Q in m³ s⁻¹.

Exercise 5.2.2 Determine the discharge Q of the brook of Exercise 5.2.1 in litre s⁻¹, but now using the graphical method.

Salt dilution gauging

In mountain streams, the operation of an Ott-type current meter may be difficult due to the turbulence of the water flow, high velocities, or rocks and shallow sections in the channel: under such circumstances, the velocity–area or graphical method cannot be used. However, a well-suited technique for estimating the discharge in turbulent mountain streams is provided by **salt dilution gauging**, where the discharge is determined from the degree of dilution by the flowing water of an added solution of sodium chloride (table salt; NaCl). There are basically two methods of salt dilution gauging: constant rate injection and slug injection.

Using **constant rate injection**, a salt solution is added to the stream with a constant rate or discharge; this method works well for determining the discharge in small mountain streams with flows less than 100 litre s⁻¹ (0.1 m³ s⁻¹).

At high discharges, it becomes difficult to accurately determine the discharge. Also, one should ensure that the people performing the necessary measurements are kept safe, as the power of flowing water is never to be underestimated. A relatively safe method of salt dilution gauging then is provided by **slug injection** (gulp injection), where a slug or gulp of salt solution is emptied instantaneously into the stream: this method is well-suited for discharges up to about 10 m³ s⁻¹.

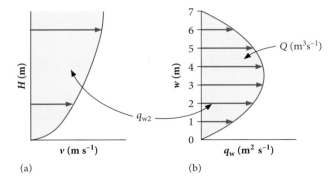

Figure 5.23. The graphical method of discharge measurement: (a) a side view for vertical 2; (b) the top view of the cross-section perpendicular to the water flow

Constant rate injection

Figure 5.24 shows a schematized top view of a stream during constant rate injection of a salt solution to estimate the discharge of the stream. A salt solution is added from a Mariotte bottle (Figure 4.16), with a constant rate or discharge Q_i (litre s^{-1}). The salt solution in the bottle has a concentration C_i (mg litre^{-1}) that must be much higher than the upstream salt concentration C_u (mg litre^{-1}) of the stream. Prior to the constant rate injection, a fluorescent coloured dye (for instance, Uranine or Rhodamine) may be added to the stream at the location of the Mariotte bottle, to visually determine the downstream location at which complete mixing of the injected salt solution with the flowing water will occur during the experiment. After equilibrium sets in during the injection experiment, a water sample may be taken at this downstream location to determine the salt concentration C_d (mg litre^{-1}) of the water.

The **salt load** (mg s^{-1}) of the river upstream of the injection point equals the product of the river discharge Q (litre s^{-1}) and the upstream salt concentration C_u (mg litre^{-1}). Likewise, the injected salt load (mg s^{-1}) from the Mariotte bottle equals the product of the constant injection rate Q_i (mg litre^{-1}) and the Mariotte bottle's salt concentration C_i (mg litre^{-1}). At the downstream point of complete mixing, the salt load (mg s^{-1}) equals the product of the total discharge $Q + Q_i$ (litre s^{-1}) and the sampled downstream salt concentration C_d (mg litre^{-1}).

A simple mass balance or **chemical mixing model** tells us that adding together the salt load (mg s^{-1}) of the river upstream of the injection point and the salt load (mg s^{-1}) from the injection point must yield the salt load (mg s^{-1}) at the downstream measuring location (if there is no downward or upward seepage along the measuring reach). We can thus write the mass balance or chemical mixing model for constant rate injection of a salt solution into a stream as

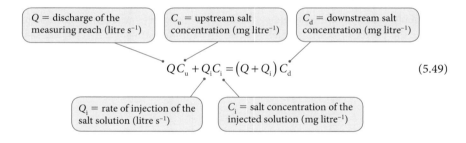

Q = discharge of the measuring reach (litre s^{-1})

C_u = upstream salt concentration (mg litre^{-1})

C_d = downstream salt concentration (mg litre^{-1})

$$QC_u + Q_i C_i = (Q + Q_i) C_d \qquad (5.49)$$

Q_i = rate of injection of the salt solution (litre s^{-1})

C_i = salt concentration of the injected solution (mg litre^{-1})

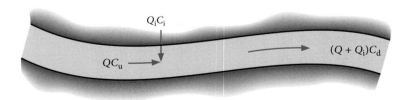

$Q_i C_i$

QC_u

$(Q + Q_i)C_d$

Figure 5.24. A schematized top view of a stream during constant rate injection of a salt solution to estimate the discharge of the stream

which can be rewritten as

$$Q = Q_i \frac{C_i - C_d}{C_d - C_u} \tag{5.50}$$

Because $C_i \gg C_d$ and $C_d \gg C_u$, Equation 5.50 can be simplified to

$$Q = Q_i \frac{C_i}{C_d} \tag{5.51}$$

Note that C_i / C_d is the **salt dilution factor** from the salt solution in the Mariotte bottle to the mixed stream water at the downstream location. Theoretically, the discharge Q follows from Equation 5.51 after measuring the values of Q_i, C_i, and C_d, with Q_i simply determined from the falling water level in the Mariotte bottle.

In practice, the concentrations C_i and C_d are not determined, as this would involve demanding sampling and laboratory analyses, but instead the electrical conductivity or *EC* (section 3.13; Box 3.4) is measured at the downstream measurement location with an *EC* measuring device that is easy and quick to operate. The **electrical conductivity** or *EC* (μS cm^{-1}) is a measure of the water's ability to conduct electricity, and therefore a measure of the water's ion concentration or **total dissolved solids** (*TDS*). When equilibrium sets in during the injection experiment, the measured *EC* at this downstream location becomes constant. To explain how to determine the discharge Q from this constant *EC* value (and the value for Q_i), we first need to introduce the concept of the **relative concentration** C_r, which is a fictitious concentration measure that replaces the real concentration for ease of use but, importantly, that still respects the salt dilution factor as introduced above.

For ease of use, the salt concentration in the Mariotte bottle is set to a relative concentration value of 1. Equation 5.51 can then be reworked as follows:

C_{ri} = relative salt concentration in the Mariotte bottle, set to 1

$$Q = Q_i \frac{C_i}{C_d} = Q_i \frac{C_{ri}}{C_{rd}} = Q_i \frac{1}{C_{rd}} \tag{5.52}$$

C_{rd} = relative salt concentration at the downstream location

C_{rd} in Equation 5.52 is the relative concentration at the downstream measurement location at which we measure the equilibrium, constant *EC* value. Because the *EC* and the salt concentration C, and thus also the *EC* and the relative salt concentration C_r, are linearly related, the C_{rd} value at the downstream location can simply be determined from an *EC*–C_r **calibration equation**: Box 5.6 explains how this linear relationship is established for salt dilution gauging (constant rate injection as well as slug injection). Translating the equilibrium, constant *EC* value to its C_{rd} value by means of the calibration equation and inserting the measured value for Q_i in Equation 5.52 yields the discharge Q (litre s^{-1}) of the measuring reach.

> **BOX 5.6 Determining the _EC-C_$_r$ calibration equation for salt dilution gauging**
>
> Use two buckets, one large one (> 20 litres) and a smaller one; both buckets should be clean. Add 2–3 kg of fine table salt to the large bucket filled with 20 litres of water taken from the brook (2–3 kg and 20 litres are indications). Stir the solution well and thoroughly mix the salt and water. The relative salt concentration C_r in the large bucket is set at 1.
>
> With a pipette, take 5 ml of the salt solution out of the large bucket and dilute this with 250 ml of water taken from the brook. Empty this newly made salt solution into the small bucket. Do not measure the _EC_ in the small bucket yet, as the salt concentration is too high, damaging the quick-to-operate _EC_ measuring device. Again, add 250 ml of water from the brook to the small bucket. Now measure the _EC_ with the _EC_-measuring device. The relative concentration C_r in the small bucket equals $5\ \text{ml} / (2 \times 250\ \text{ml}) = 10^{-2}$. Proceed in this way, taking care after each addition of water from the brook that the solution is well mixed: thus, add another 250 ml of water from the brook and after stirring measure the _EC_ at $C_r = 5\ \text{ml} / (3 \times 250\ \text{ml}) = 0.\overline{6} \times 10^{-2}$, and so on, until 3000 ml of water from the brook has been added; the C_r value then equals $5\ \text{ml} / 3000\ \text{ml} = 0.1\overline{6} \times 10^{-2}$.
>
> Now take 250 ml of the solution out of the small bucket, empty and clean the small bucket with water from the brook, empty and clean the small bucket again, add the 250 ml of solution to the bucket, and again add 250 ml water from the brook. Measure the _EC_ of this new solution, now at a C_r value of $0.001\overline{6} / 2 = 0.8\overline{3} \times 10^{-3}$. Again add 250 ml water from the brook; measure the _EC_ of this new solution, now at a C_r value of $0.001\overline{6} / 3 = 0.\overline{5} \times 10^{-3}$, and so on.
>
> Plot all of the points in an _EC-C_$_r$ (spreadsheet) scattergram (with C_r on the vertical axis) and determine the linear regression equation, which is the **_EC-C_$_r$ calibration equation**.
>
> Visually inspect how well the regression line fits the data points and/or determine the squared correlation coefficient (R^2); the line should fit well and/or R^2 should be near 1.

Exercise 5.2.3 The discharge of a stream is determined by the constant rate injection of a salt solution from a Mariotte bottle with a constant discharge of 0.1 litre s^{-1}. The constant, equilibrium _EC_ at the downstream location with complete mixing equals 765 μS cm^{-1}. The _EC-C_$_r$ calibration equation is determined (Box 5.6) as: $C_r = 6.486 \times 10^{-6}\,EC - 2.871 \times 10^{-3}$ (_EC_ in μS cm^{-1}; $R^2 = 0.998$).

Determine the discharge Q of the stream in litre s^{-1}.

Slug injection

With a **slug injection** (gulp injection), a slug or gulp of salt solution is emptied instantaneously into the stream: again, as with the above-discussed constant rate injection, the concentration of the added salt solution must be much higher than the base concentration C_b (mg litre^{-1}) of the stream. The slug causes a saltwater wave to pass downstream through the channel. Figure 5.25 shows the passing of the saltwater wave at the downstream measurement point with complete mixing. The blue area A under the C_d curve, but above the base level concentration C_b, can be written mathematically as follows:

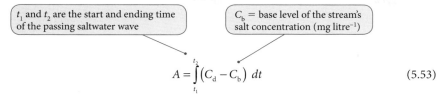

$$A = \int_{t_1}^{t_2} \left(C_d - C_b \right) dt \qquad (5.53)$$

The blue area A in Figure 5.25 has units of mg litre^{-1} (vertical axis) multiplied by seconds (horizontal axis), thus mg litre^{-1} s. Theoretically, if we know the mass of salt M (mg) added to the stream by the slug injection, we can determine the discharge Q (litre s^{-1}) of the measuring reach as follows:

M = mass of slug-injected salt (mg) C_i = concentration of salt solution (mg litre^{-1}) V = volume of slug-injected salt solution (litre)

$$Q = \frac{M}{A} = \frac{C_i V}{A} \qquad (5.54)$$

In practice, again, the EC (μS cm^{-1}) is measured at the downstream measurement point with complete mixing (and not C_d minus C_b), as measuring the EC is much less demanding. Again, we have to make use of the relative concentration C_r, set the salt concentration for the slug-injected salt concentration to 1, and use the EC–C_r calibration equation to determine the discharge Q (litre s^{-1}) of the measuring reach. To do so, Equations 5.53 and 5.54 need to be rewritten as follows:

C_{rd} = relative salt concentration at the downstream location C_{rb} = relative base level salt concentration of the stream

$$A' = \int_{t_1}^{t_2} \left(C_{rd} - C_{rb} \right) dt \qquad (5.55)$$

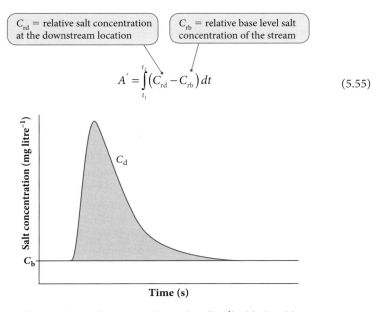

Figure 5.25. Salt concentration C_d (mg litre^{-1}) with time (s) at the downstream measuring location after a slug injection of a salt solution

C_{ri} = relative salt concentration of the slug-injected salt concentration, set to 1

$$Q = \frac{C_{ri}V}{A^{'}} = \frac{1V}{A^{'}} = \frac{V}{\displaystyle\int_{t_1}^{t_2}\left(C_{rd} - C_{rb}\right)dt} \tag{5.56}$$

By means of the EC–C_r calibration equation, the EC values with time are reworked to values for C_{rb} and C_{rd}, yielding the area $A^{'}$ (as determined in a spreadsheet program) and thus the discharge Q (litre s^{-1}) of the measuring reach.

Slug injection is also known as the **ionic wave method**, referring to the passing saltwater wave, and as the **integration method**, referring to the mathematical integration of $C_{(r)d}$ – $C_{(r)b}$ against time as the main part of the method.

> **Exercise 5.2.4** 18.5 litres of a salt solution is emptied instantaneously into a stream. The EC is monitored at the downstream measurement point with complete mixing, and the EC–C_r calibration equation is used to translate the EC values to C_r values. Using a spreadsheet program, the area $A^{'} = \displaystyle\int_{t_1}^{t_2}\left(C_{rd} - C_{rb}\right)dt$ is determined from the C_r values as 0.405 'mg s litre^{-1}'.
>
> Determine the discharge Q of the stream in litre s^{-1}.

EC-routing

An EC-**measuring device** is a small, easy and quick to operate, stick-like apparatus for measuring the electrical conductivity or EC (μS cm^{-1}). Measuring the EC at a large number of locations in the longitudinal profile of one or more streams and pencilling in the EC data obtained in this way on a map or in a sketch of the longitudinal profile of a stream is called EC-**routing**. Such routing can, for instance, provide useful information on the upward seepage of groundwater with a strongly differing chemical composition into a stream.

Figure 5.26 shows the top view of two streams numbered as 1 and 2 that meet and continue as one stream, numbered as 3; when we know the discharge of one of these branches or tributaries 1, 2, or 3, measuring the EC in all tributaries may enable us to figure out the discharges of the other two tributaries. To do so, the EC values of the confluencing tributaries should differ significantly – for instance, one tributary drains a forest and the other tributary drains arable land with a high nitrate load. To obtain trustworthy results, the EC of the downstream tributary should be measured at a location at which the water from both upper branches is completely mixed. The method involves setting up two mass balance equations, one for water (continuity equation) and one for the total dissolved solids presented as the EC (chemical mixing model):

$$Q_1 + Q_2 = Q_3 \tag{5.57}$$

$$Q_1\,EC_1 + Q_2\,EC_2 = Q_3\,EC_3 \tag{5.58}$$

As the EC values (EC_1, EC_2, and EC_3) are measured as part of the EC-routing, knowing the value of one of the discharges (Q_1, Q_2, or Q_3) means that we have two Equations 5.57 and 5.58 with two unknowns, which can be solved mathematically as shown in Box 5.7.

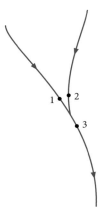

Figure 5.26. The top view of two streams, numbered 1 and 2, meeting and continuing as one stream, numbered 3

BOX 5.7 Solving two equations with two unknowns

Use Equations 5.57 and 5.58; the unknown variables in this example are Q_2 and Q_3.
The known variables in this example are $Q_1 = a$, $EC_1 = b$, $EC_2 = c$, and $EC_3 = d$.
From Equation 5.57, it follows that $Q_3 = Q_2 + a$.
Substitute the above expression for Q_3 in Equation 5.58 and also insert the known variables.
This yields one equation with one unknown variable Q_2:

$$ab + cQ_2 = d(a + Q_2) \Rightarrow ab + cQ_2 = ad + dQ_2 \Rightarrow ab - ad = dQ_2 - cQ_2 \Rightarrow$$

$$ab - ad = Q_2(d - c) \Rightarrow Q_2 = \frac{ab - ad}{d - c} \Rightarrow Q_2 = \frac{a(b-d)}{d-c}$$

From the above expression $Q_3 = Q_2 + a$, it follows that

$$Q_3 = \frac{a(b-d)}{d-c} + a \Rightarrow Q_3 = \frac{a(b-d)}{d-c} + \frac{a(d-c)}{d-c} \Rightarrow$$

$$Q_3 = \frac{ab - ad + ad - ac}{d-c} \Rightarrow Q_3 = \frac{ab - ac}{d-c} \Rightarrow Q_3 = \frac{a(b-c)}{d-c}$$

Thus,

$$Q_1 = a \Rightarrow Q_2 = \frac{a(b-d)}{d-c} \quad \text{and} \quad Q_3 = \frac{a(b-c)}{d-c}$$

Exercise 5.2.5 In a stream, the electrical conductivity (EC) is measured in two upstream branches at locations 1 and 2 and downstream at location 3, as shown in Figure 5.25. The EC is linearly related to the salt concentration (mg litre^{-1}) of the stream. At location 3, the discharge Q_3 is measured.

$EC_1 = 1200\ \mu\text{S cm}^{-1}$; $EC_2 = 500\ \mu\text{S cm}^{-1}$; $EC_3 = 900\ \mu\text{S cm}^{-1}$; $Q_3 = 17.5$ litre s^{-1}.

Determine the discharge of the stream at locations 1 and 2.

Instead of using the EC, one may also use the concentration C (mg litre^{-1}) of a **conservative ion** – that is, a non-reactive ion – as it passes through a catchment; for instance, chloride (Cl^-). Equation 5.58 then changes to

$$Q_1 C_1 + Q_2 C_2 = Q_3 C_3 \tag{5.59}$$

Similar to the above description for electrical conductivity, Equations 5.57 and 5.59 may be used to extend discharge data, now after sampling the concentration of a conservative ion along the tributaries of a river.

Exercise 5.2.6 Blawan is located in the Ijen caldera, Java, Indonesia. Figure E5.2.6.1 shows that the freshwater stream Kalisat (stream 1) and the very acid stream Kalipahit (bitter stream; stream 2) join in the vicinity of Blawan. The Kalipahit (stream 2) has its origin in an area situated below an old, man-made, leaking dam of the Ijen crater lake. The mixture of fresh and acid water continues its route downstream under the name Kaliputih (white stream; stream 3). Further downstream, the Kaligedang (stream 4) joins the Kaliputih (stream 3). The resulting stream (stream 5) leaves the Ijen caldera via a waterfall; downstream, the acid water is used for the irrigation of rice fields (Bogaard and Hendriks 2001; Löhr *et al.* 2004; Löhr 2005).

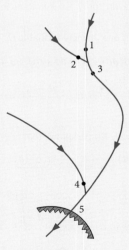

Figure E5.2.6.1. A top view of the drainage network near Blawan, Ijen caldera, Java, Indonesia

Velocity measurements are made with an Ott-type current meter (Figure E5.2.6.2 below) and the discharge Q_1 of the Kalisat (stream 1) is determined, using the velocity–area method, to amount to 2.9 m³ s⁻¹.

Table E5.2.6 gives the sampled chloride (Cl⁻) concentrations of the streams in mg litre⁻¹.

Table E5.2.6. Sampled chloride (Cl⁻) concentrations of the streams near Blawan, Ijen caldera, Java, Indonesia; concentrations are in mg litre⁻¹

Kalisat (stream 1)	110
Kalipahit (stream 2)	995
Kaliputih (stream 3)	222
Kaligedang (stream 4)	86
waterfall (stream 5)	191

Determine the discharge at the waterfall (stream 5).

The stage–discharge relation

Man-made, fixed constructions in the river bed, such as weirs and flumes (section 5.1), need to comply with a number of conditions, such as a straight upstream reach of some length leading to the construction. Also, the measurement of the stage must be carried out at a distance of two to three times the maximum water level in the crest, to avoid the effects of drawdown from water flowing over the crest. Often, in reality, the prescribed conditions are not fully met and because of this Q–H relations in practice, **Q–H rating curves,** may differ from theoretically derived equations. Therefore, it is good practice to calibrate the stage–discharge relation in the field by measuring the discharge Q at different stages H using one or more of the methods outlined above.

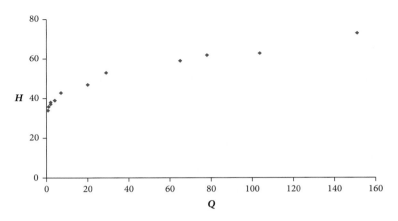

Figure 5.27. *Q-H* measurements for a stream

Figure 5.27 shows *Q–H* measurements for a stream. The stage *H* (cm) is measured with a staff gauge upstream of a weir, and the discharge *Q* (litre s^{-1}) is measured using one of the methods just discussed. The stage measurements in Figure 5.27 stand uncorrected; that is, the maximum water level H_0 along the staff gauge at which the discharge *Q* equals zero has not yet been determined. For water flow through a weir, the H_0 level is determined by the altitude of the weir crest, and it is possible to determine the H_0 level from land surveying. However, the **maximum stage with zero discharge** H_0 may also be estimated from visual inspection of plotted *Q–H* data, and in Figure 5.27 a value of H_0 of 30 cm appears to be a good estimate. Note that the validity of the estimate can best be checked after fitting a trend line through the $(Q, H - H_0)$ data, as shown in Figure 5.28 below.

When we subtract $H_0 = 30$ cm from the stages presented in Figure 5.27, we can construct the *Q–H* relation for the measured data points with $H - H_0$ (cm) as the stage or water level above the weir crest. In Figure 5.28, a **power function** trend line

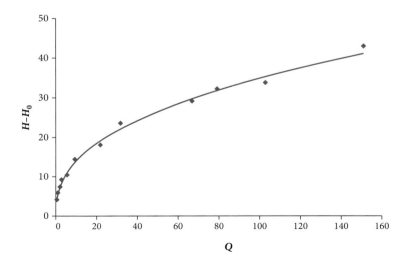

Figure 5.28. The *Q-H* relation for a stream

has been fitted through the $(Q, H - H_0)$ data points (using a spreadsheet program). From the curvature of the trend line in Figure 5.28, it follows that $H - H_0$ equals zero when the discharge reaches zero and that $H_0 = 30$ cm is thus a good estimate; also, a power function as a trend line produces a curve that fits the measured Q–H data well. Mathematically, the power function shown in Figure 5.28 can be written as

$$Q = a\left(H - H_0\right)^b \tag{5.60}$$

The values of a and b in Equation 5.60 can best be found after rewriting Equation 5.60 in a logarithmic form:

$$\log Q = \log a + b \log\left(H - H_0\right) \tag{5.61}$$

As shown in Figure 5.29, the $(\log(H - H_0),\ \log Q)$ data will plot as a straight line ($y = \text{intercept} + \text{slope} \times x$) in a graph with $\log Q$ on the vertical (y-) axis and $\log(H - H_0)$ on the horizontal (x-) axis; we can then simply find $\log a$ as the intercept (at $\log(H - H_0) = 0$) and b as the slope of the fitted line.

In the United Kingdom, **Acoustic Doppler Current Profilers** (ADCPs) are being used on a widespread basis by the Environment Agency (EA) to provide spot gaugings at remote sites and to collect discharge data to check the stage–discharge relation

$$\log Q = \log(a(H - H_0)^b)$$
$$= \log a + \log(H - H_0)^b$$
$$= \log a + b \log(H - H_0)$$

$$b = \frac{\Delta \log Q}{\Delta \log(H - H_0)}$$

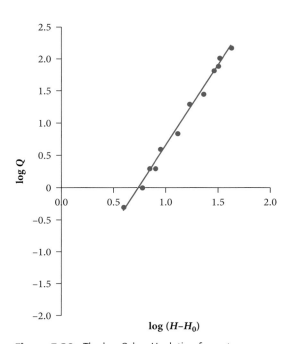

Figure 5.29. The log Q-log H relation for a stream

Exercise 5.2.7 Table E5.2.7 presents stage–discharge measurements for a stream in 2007 and 2008. The stage H (cm) is measured with a staff gauge slightly upstream of a weir, and the discharge Q (litre s^{-1}) is measured using one (or more) of the methods described in section 5.2. In the winter of 2007/2008, the staff gauge is repositioned. Figure E5.2.7 shows data points and trend lines for the Q–H measurements in 2007 and 2008.

Table E5.2.7 Stage-discharge measurements in a stream in 2007 and 2008

Date	H (cm)	Q (litre s^{-1})
15 January 2007	43	29.0
14 March 2007	33	7.0
23 May 2007	29	4.0
05 June 2007	26	1.0
23 September 2007	63	151.0
22 October 2007	52	78.0
12 April 2008	59	65.0
30 June 2008	34	0.5
05 July 2008	37	2.0
12 August 2008	38	2.0
02 September 2008	63	103.5
08 October 2008	47	20.0

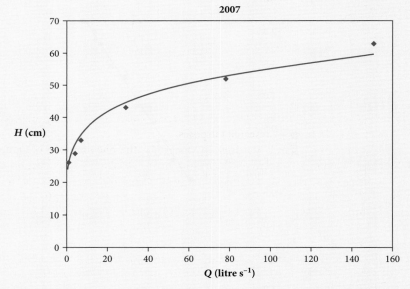

Figure E5.2.7. Data points and trend lines for the Q–H measurements in 2007 and 2008

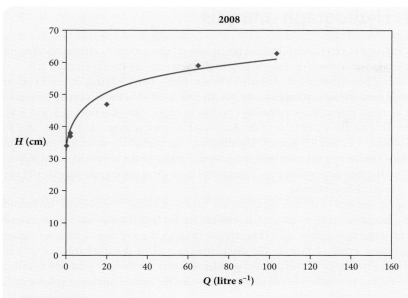

Figure E5.2.7 (*Continued*)

a. Estimate H_0 in 2007 and 2008.

b. Plot the Q–H relation logarithmically.

c. Determine the coefficients a and b in the equation $Q = a\left(H - H_0\right)^b$ with the discharge Q in litre s^{-1} and the stage H in centimetres.

d. Determine the coefficients a and b in the equation $Q = a\left(H - H_0\right)^b$ with the discharge Q in m^3 s^{-1} and the stage H in metres.

at permanent gauging stations across England and Wales, as well as by universities and other organizations for research purposes.

The Q–H rating curve is sensitive to **hysteresis**, the phenomenon of an equilibrium state being dependent on the history of the physical system (section 4.5): hysteresis of the Q–H rating curve stems from the reality that a change in discharge Q can be linked to both a change in water velocity and a change in the wetted cross-sectional flow area perpendicular to the water flow. The presence of vegetation in the water can also be important in this respect; for instance, when vegetation stands upright at the start of a storm, causing resistance to the water flow, but is flattened towards the end of the storm, with the opposite effect.

Thus, if enough data is available, it may be recommended to construe separate rating curves for rising and falling water levels – in other words, the rising and falling limbs of a **hydrograph**, which is a graph of the changes in discharge (or water level) as a function of time, and which is the topic of the next section.

5.3 Hydrograph analysis

The discharge of a stream results from precipitation falling on to the stream's catchment or drainage basin, the geographical area that drains into the stream (section 1.3).

Channel precipitation falls directly on to a stream – but, for instance, precipitation that falls near the stream may also be quickly transported to the stream by processes such as rapid throughflow (pipeflow) and/or overland flow (sections 4.8 and 4.9), and/or because the water table at the foot of a hill slope is temporarily raised due to precipitation (section 5.5 below). The discharge (or part of the discharge; $m^3\ s^{-1}$) of a stream caused by processes delivering water rapidly to the stream is called **quickflow**; for the discharge volume (m^3) that stems from these rapid processes, we will use the term **quickflow volume**.

During periods with no precipitation, many streams – especially in humid areas – continue to carry water; the streams are fed by processes such as sustained ground water flow and/or slow throughflow through the soil matrix from an earlier precipitation event or events. The discharge (or part of the discharge; $m^3\ s^{-1}$) of a stream caused by processes delivering water slowly to the stream could best be called **slowflow**, but is usually referred to as **baseflow**; for the related discharge volume (m^3) we will use the term **baseflow volume**.

Streams that hold water throughout the year are called **perennial**, whereas a channel that only holds water during and immediately after a rain or snowmelt event is called **ephemeral**; an ephemeral stream thus has no baseflow, only quickflow. In between these two extremes lie **intermittent** streams, streams that only carry water during the wet part of the year. The definitions just given are not conclusive, as river regimes may change with time, but they suffice here to present a general setting.

Figure 5.30 shows a **hydrograph**, a graph of the changes in discharge with time, for a precipitation event, together with a number of hydrological terms used to describe hydrographs; Table 5.1 provides descriptions of the terms.

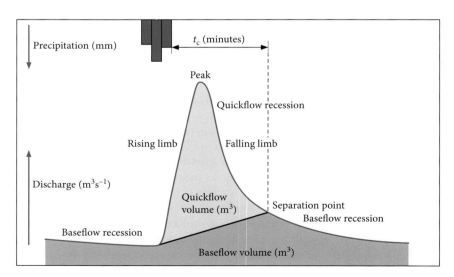

Figure 5.30. A hydrograph of a precipitation event (horizontal axis: time; $t_c =$ time of concentration)

Table 5.1 Some hydrological terms used for describing hydrographs

baseflow	The discharge (or part of the discharge) caused by processes delivering water slowly to the stream ($m^3 \, s^{-1}$) (also called **delayed flow** or **delayed runoff**)
baseflow recession	The part of the hydrograph with gradually declining baseflow
baseflow volume	The discharge volume that stems from processes that slowly deliver water to the stream (m^3)
falling limb	The part of the hydrograph with decreasing discharge (after the hydrograph has reached its peak)
inflection point	The separation point determined by assuming that the recession curve can be modelled as the discharge from linear reservoirs representing quickflow and baseflow (see text)
peak	The maximum discharge during a precipitation event ($m^3 \, s^{-1}$)
quickflow	The discharge (or part of the discharge) caused by processes delivering water rapidly to the stream ($m^3 \, s^{-1}$) (also called **direct flow** or **direct runoff**)
quickflow recession	The part of the hydrograph with gradually declining quickflow
quickflow volume	The discharge volume that stems from processes that rapidly deliver water to the stream (m^3)
recession curve	A curve showing a gradual decrease in discharge (after the hydrograph has reached its peak)
rising limb	The part of the hydrograph with increasing discharge (before the hydrograph has reached its peak)
separation point	The point on the recession curve separating the quickflow recession from the baseflow recession
time of concentration	The time required for surface water (or other water contributing to the quickflow of a stream) to flow from the farthest point in the watershed to the outlet (minutes)

Hydrograph recession analysis

Figure 5.31 shows a hydrograph of heavy rain on a **wadi**, a usually dry river bed (baseflow = 0 $m^3 \, s^{-1}$) in a (semi-)arid climate with the exception of periods during and shortly after heavy rainfall; a wadi is thus a type of ephemeral stream and its discharge is made up solely of quickflow.

For an ephemeral stream, the recession curve can conceptually be interpreted as outflow from a quickflow reservoir. The recession curve can mathematically be described as an exponential decrease starting at the hydrograph peak, similar to ponded infiltration with time according to Horton (section 4.8):

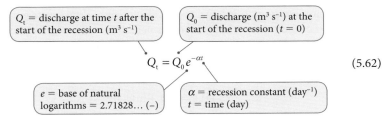

Q_t = discharge at time t after the start of the recession ($m^3 \, s^{-1}$)

Q_0 = discharge ($m^3 \, s^{-1}$) at the start of the recession ($t = 0$)

$$Q_t = Q_0 e^{-\alpha t} \qquad (5.62)$$

e = base of natural logarithms = 2.71828... (–)

α = recession constant (day^{-1})
t = time (day)

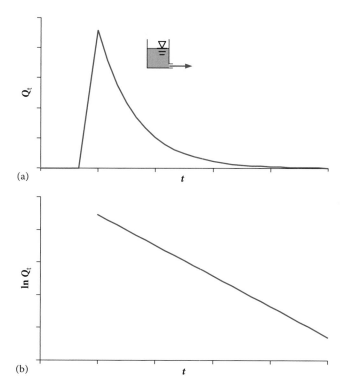

Figure 5.31. A hydrograph of a precipitation event for an ephemeral stream: (a) discharge Q_t versus time t; (b) ln Q_t versus t

By analogy with section 4.8, Equation 5.62 can be rewritten as (Δt is small and constant):

$$Q_{t+\Delta t} = Q_0 e^{-\alpha(t+\Delta t)} = Q_0 e^{-\alpha t} e^{-\alpha \Delta t} = Q_t e^{-\alpha \Delta t} \tag{5.63}$$

$$Q_{t+2\Delta t} = Q_0 e^{-\alpha(t+2\Delta t)} = Q_0 e^{-\alpha t} e^{-2\alpha \Delta t} = Q_0 e^{-\alpha t} \left(e^{-\alpha \Delta t}\right) = Q_t \left(e^{-\alpha \Delta t}\right)^2 \tag{5.64}$$

$$Q_{t+3\Delta t} = Q_0 e^{-\alpha(t+3\Delta t)} = Q_0 e^{-\alpha t} e^{-3\alpha \Delta t} = Q_0 e^{-\alpha t} \left(e^{-\alpha \Delta t}\right) = Q_t \left(e^{-\alpha \Delta t}\right)^3 \tag{5.65}$$

Remarks made for the decay constant α in the Horton equation for ponded infiltration (section 4.8: Equations 4.24–4.31) similarly hold for the recession constant α in the above equations (α and Δt are positive: $0 < e^{-\alpha \Delta t} < 1$; also see Exercise 4.8.2). Thus, if the **recession constant** α is large, the recession curve shows a steep decline and the depletion of the reservoir is fast; if α is small, the recession curve shows a slight decline and the depletion of the reservoir is slow; α ranges from 0.001 day^{-1} for deep groundwater (De Zeeuw 1973) to over 3 day^{-1} for overland flow (Bonell *et al.* 1984).

Equation 5.62 can be rewritten in semi-logarithmic form as

$$\ln Q_t = -\alpha t + \ln Q_0 \tag{5.66}$$

$\ln Q_t = \ln(Q_0 e^{-\alpha t})$
$\quad = \ln Q_0 + \ln e^{-\alpha t}$
$\quad = -\alpha t + \ln Q_0$

$\Rightarrow \ln Q_t = -\alpha t + \ln Q_0 \Rightarrow$
$y = \text{slope} \times x + \text{intercept with}$
$y = \ln Q_t, \text{slope} = -\alpha, x = t,$
and the intercept $= \ln Q_0$

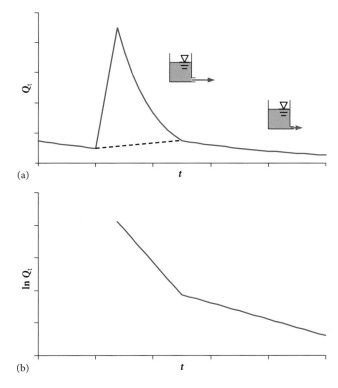

Figure 5.32. A hydrograph of a precipitation event for a perennial stream: (a) discharge Q_t versus time t – the drawn baseflow separation line is based on the constant-slope inflection-point method (see text); (b) ln Q_t versus t

Equation 5.66 plots as a straight line in a graph with ln Q_t on the vertical axis and t on the horizontal axis; the value of α can simply be determined as the absolute or positive value of the slope ($\Delta \ln Q_t / \Delta t$) of a straight line as presented in Figure 5.31.

Figure 5.32 shows the hydrograph of a perennial stream; for instance, a stream in a humid climate setting – the discharge of the perennial stream is made up of both quickflow and baseflow.

The recession curve of the hydrograph shown in Figure 5.32 can conceptually be interpreted as outflow from both a quickflow reservoir and a baseflow reservoir, with the larger outflow per unit of time from the quickflow reservoir. At the separation point the quickflow reservoir has emptied, and from then on there is only outflow from the baseflow reservoir. As the outflow from the quickflow reservoir is dominant, the first part of the recession curve after the hydrograph peak can simply be interpreted as outflow from the quickflow reservoir alone.

The recession curve can therefore mathematically be described as first an exponentially decreasing outflow from the quickflow reservoir and then, after the separation point, an exponentially decreasing outflow from the baseflow reservoir. For both parts of the recession, the above Equations 5.62–5.66 can be used, albeit with a different value of the recession constant α for the quickflow and baseflow reservoirs. Note that the quickflow reservoir in Figure 5.32 has been drawn with a larger opening at the bottom than the baseflow reservoir; the size of the opening and the friction are incorporated in the value of the **recession constant** α, α being largest for the quick-flow reservoir with its fast depletion.

Water flow from the quickflow and baseflow reservoirs is conceptually similar to water flow from an artificial lake with a lower opening in the dam (section 5.1). However, at least for the baseflow reservoir, n in Equation 5.12 will equal 1 (and not 2), as we will come to learn in section 5.4.

Hydrograph separation

There are many methods for hydrograph separation into quickflow and baseflow: all methods have in common that they are unavoidably arbitrary. Importantly, one should always stick to the method one sets off with when analysing hydrographs, or else the basis for comparison is lost.

For all methods, the area under the hydrograph curve, but above the **baseflow separation line** (see Figures 5.30 and 5.32) or curve, gives the quickflow volume (m³) during the precipitation event, whereas the area under the separation line or curve gives the baseflow volume (m³). There are many methods of hydrograph separation: Figure 5.33 shows three graphical methods.

The **constant discharge method** assumes that baseflow during a precipitation event is constant and equal to the discharge level at the start of the precipitation event. A considerable disadvantage of this method is that the discharge may not return to this starting level because of the start of a new precipitation event.

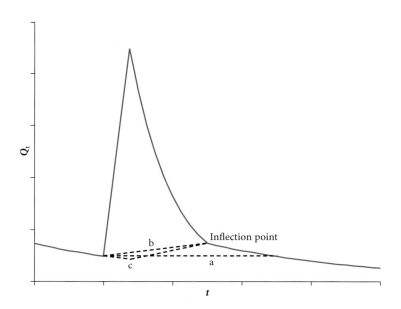

Figure 5.33. Three graphical methods of hydrograph separation: (a) the constant discharge method; (b) the constant-slope inflection-point method; and (c) the concave-curve inflection-point method – methods (b) and (c) use the inflection point as a separation point

The **constant-slope inflection-point method** assumes an immediate response in baseflow to precipitation and connects the start of the rising limb with the inflection point on the recession curve. **Figure 5.32a shows a baseflow separation line obtained using the constant-slope inflection-point method: the position of the inflection point is found as the point in time at which the slope of the semi-logarithmic plot of ln Q_t versus t (Figure 5.32b) changes.** The method has been extensively used by the author of this book when comparing hydrographs of brooks in small (1–2 km²), forested and grassland drainage basins in the Keuper marls of Luxembourg (Hendriks 1990): because more than one inflection point is usually found on a recession curve, it was decided to use the inflection point with the lowest discharge as the selected inflection point to distinguish between baseflow and quickflow.

The **concave-slope inflection-point method** assumes a continued decrease in baseflow during the rising limb of the hydrograph and projects the declining antecedent baseflow recession prior to the precipitation event to directly under the hydrograph peak: the minimum obtained this way is then connected to the (lowest) inflection point on the recession curve.

Length units in surface water hydrology

Similar to precipitation (depth) and evaporation, the discharge volume of a river may be presented as a length unit; for instance, millimetres. Discharge volumes in cubic metres (m³) can simply be converted to metres by dividing by the total drainage basin area in square metres (m²) (see also section 1.4); of course, multiplying this result by 1000 mm m⁻¹ yields an answer in millimetres. One should divide by the total drainage area irrespective of whether all or only part of the drainage basin actually contributes to the discharge; this is because the same total drainage basin area is involved when translating precipitation or evaporation from volume units to length units or vice versa. **Importantly, a discharge presented, for instance, in mm hour⁻¹ is not a velocity (which it may seem at a first glance), but a discharge in m³ hour⁻¹ adjusted for the drainage basin area by dividing by the total drainage basin area (m²) and multiplying by 1000 mm m⁻¹.** Sometimes, discharges expressed in units of

$$\frac{\text{litre}}{\text{s}\,\text{km}^2}$$

may be found in hydrology texts, meaning litres both per second and per the unit drainage basin area, in km²:

$$1\,\frac{\text{litre}}{\text{s}\,\text{km}^2} = 3.6 \times 10^{-3}\,\text{mm hour}^{-1}$$

Runoff coefficient

The **runoff coefficient** of a stream can simply be determined by dividing the discharge volume in millimetres by the received precipitation volume (precipitation depth) in millimetres and then multiplying by 100 to convert from a fraction to a percentage. The term '**runoff**' in 'runoff coefficient' should be taken as the discharge (volume) of a stream: another use of the term 'runoff' is as a synonym for overland flow.

One should distinguish between the runoff coefficient on an annual basis and the runoff coefficient on a storm basis, 'storm' being a general term used in hydrology as a synonym for precipitation event. In the hydrological literature, one may read passages such as the following:

> ... globally an average percentage of 36 per cent of the total precipitation falling on the land reaches the oceans as runoff. Of this amount, quickflow accounts for about 11 per cent and delayed flow accounts for the remaining 25 per cent of precipitation. (Ward and Robinson 2000; delayed flow is the same as baseflow)

or

> In general, it would seem that in the Keuper region of Luxembourg ~30% of annual rainfall under forest eventually becomes runoff and that between 20% and 30% of this is quickflow. (Bonell *et al.* 1984)

In line with this, the **runoff coefficient on an annual basis** is defined as the *total discharge volume* (mm or m³) as a percentage of the precipitation volume (mm or m³) on an annual basis, whereas the **runoff coefficient on a storm basis** is defined as only the *quickflow volume* (mm or m³) as a percentage of the precipitation volume (mm or m³) on a storm basis. Thus, referring to the above examples, the global runoff coefficient equals 36% on an annual basis and 11% on a storm basis, and the runoff coefficient in the Keuper region of Luxembourg equals 30% on an annual basis and between 20 and 30% on a storm basis.

Usually in plots of storm hydrographs, discharge and precipitation are presented using different units (for instance, m³ s⁻¹ versus mm hour⁻¹) and thus along differently scaled axes (see, for instance, Figure 5.30). The first impression of such figures may be that a large percentage of precipitation is discharged as quickflow during storms, which may be misleading! Often, runoff coefficients on a storm basis are surprisingly low, as interception storage and evaporation, as well as storages at the surface and in the subsurface are usually quite high. However, in situations when a small clayey drainage basin is pre-wetted by earlier precipitation, or when rain falls on a frozen soil, or when snow melts, storms with a high runoff coefficient may occur. To get the perception right, Figure 5.34 shows a hydrograph of the quickflow Q_q for a storm with a runoff coefficient of 30%, with (just for once) precipitation and quickflow presented (in mm hour⁻¹) along equally scaled axes.

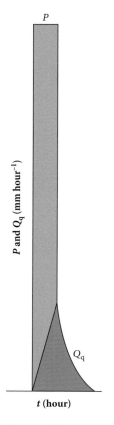

Figure 5.34.
A hydrograph of the quickflow discharge Q_q for a storm with a runoff coefficient of 30%; precipitation P and quickflow discharge Q_q are shown equally scaled

Exercise 5.3.1 The Tollbaach drainage basin near Haller, Luxembourg, has an area of 2 km². Figure E5.3.1 shows a hydrograph with a half-hour rainfall event. The discharge on the vertical axis is in litres per second. Time on the horizontal axis is in hours. In the upper part of Figure E5.3.1, the rainfall intensity is given in mm hour⁻¹. Approximately 1% of the drainage basin area is occupied by the stream channel.

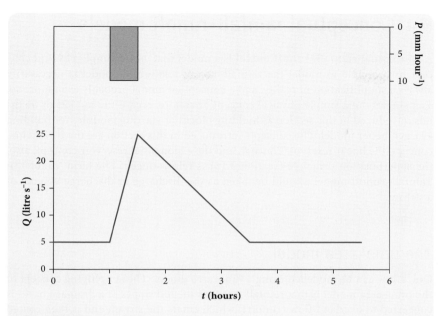

Figure E5.3.1

a. Determine the rainfall depth in millimetres.

b. Determine the rainfall volume in m³.

c. Determine the baseflow discharge of the Tollbaach in litre s⁻¹.

d. Determine the quickflow volume of the Tollbaach due to the rainfall event in m³.

e. Determine the quickflow volume of the Tollbaach due to the rainfall event in millimetres.

f. Determine the runoff coefficient for this rainfall event.

g. What process(es) contribute(s) to the quickflow of this rainfall event?

Exercise 5.3.2 In the vicinity of Consdorf, Luxembourg, a drainage basin with an area of 10 km² receives 25 mm of rainfall. The losses due to interception and evaporation during the rainfall event amount to 5 mm. On tracks near the stream, overland flow occurs: 2% of the rain falls directly into the stream as channel precipitation. A very large part of the rainfall infiltrates into the sandy soil and does not contribute quickly to the discharge of the stream: 4×10^4 m³ of water leaves the drainage basin at the outlet; $\frac{3}{4}$ of this water is baseflow.

a. What percentage of rainfall becomes quickflow?

b. How much overland flow (mm) is produced?

c. How much overland flow (m³) is produced?

5.4 Conceptual rainfall–runoff models

A **conceptual rainfall–runoff model** is a model that uses a simple physical concept or principle to model the rainfall–runoff relation. A model is necessarily always a simplification of reality, and a conceptual model probably even more so, as it singles out a simple physical concept; therefore, regard the models or methods introduced in this section as building blocks or starting modules for building a larger, better model. The concepts introduced in this section are the travel-time concept, the linear reservoir concept, and the exponential reservoir concept; also, the superposition principle (section 3.15) is reintroduced. The term 'runoff' in 'rainfall–runoff model' should be taken as the discharge or discharge volume of a stream.

The time–area model

Quickflow can be modelled using a time–area model. The simplifying concept of the time–area model is that rainfall input on the hill slopes of a drainage basin is converted to overland flow (runoff), which enters the stream and is then routed through the channel network to the drainage basin outlet. Figure 5.35 shows a drainage basin divided into five zones on the basis of **isochrones**, lines of equal travel time for water to reach the drainage basin outlet. For the sake of ease, we will set the isochrones in the coming discussion at 1 hour intervals: the quickflow produced in time zone 1, nearest to the drainage basin outlet, then reaches the outlet in 1 hour, and the quickflow produced in time zone 2 takes 2 hours to reach the drainage basin outlet, and so on. The time needed for the entire drainage basin to contribute to the quickflow is called the **time of concentration** (see also Figure 5.30 and Table 5.1): thus, in our example, the time of concentration equals 5 hours.

In 1851, Thomas Mulvaney, an Irish land drainage expert, developed a method, known as the **rational method**, to predict the quickflow part of the discharge Q_q from a drainage basin as a function of the rainfall intensity I:

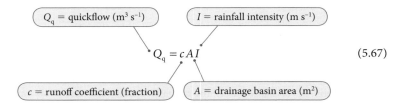

$$Q_q = c A I \qquad (5.67)$$

Q_q = quickflow (m³ s⁻¹)

I = rainfall intensity (m s⁻¹)

c = runoff coefficient (fraction)

A = drainage basin area (m²)

The runoff coefficient c is expressed as a fraction, and thus has a range between 0 and 1, and represents losses and storages in the drainage basin area; c may be as low as 0.01 in very permeable drainage basins and as high as 0.9 in impermeable urban catchments (Kirkby *et al.* 1987). Of course, the method is extremely simple, as the rainfall intensity I is assumed constant and homogeneously distributed over the drainage basin, and no reference whatsoever is made to actual processes.

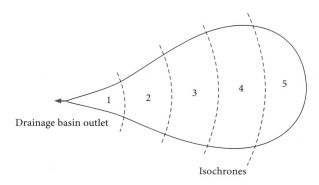

Figure 5.35. The top view of a drainage basin divided into five time zones

Applying the rational method to time zone 1 gives

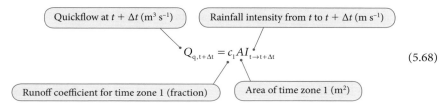

$$Q_{q,t+\Delta t} = c_1 A I_{t \to t+\Delta t} \qquad (5.68)$$

Equation 5.68 shows that the quickflow proportion of the discharge Q_q that is produced in a time zone at $t + \Delta t$ is proportionally related to the rainfall intensity I in that time zone during the preceding time step Δt, and thus in the interval from t to $t + \Delta t$ – in our example, the preceding hour.

Let us assume that we have arrived at a point in time when the entire drainage basin contributes to the quickflow arriving at the drainage basin outlet. The quickflow from time zone 2 takes 2 hours to arrive at the drainage basin outlet and thus stems from rain that has fallen from $t - \Delta t$ to t; the quickflow from time zone 3 takes 3 hours to arrive at the drainage basin outlet and thus stems from rain that has fallen from $t - 2\Delta t$ to $t - \Delta t$, and so on. Thus, in general, the quickflow from time zone n takes n hours to arrive at the drainage basin outlet and stems from rain that has fallen from $t - (n - 1)\Delta t$ to $t - (n - 2)\Delta t$. By reasoning in this way, we introduce the **travel-time concept** in our model. Further, the quickflow contributions from time zones 1, 2, 3, ..., n can be added to give the total quickflow at the drainage basin outlet (the **superposition principle**). Following all this, we may write for the total quickflow proportion of the discharge at the outlet of a drainage basin with n time zones ($n = 5$ in Figure 5.35):

$$Q_{q,t+\Delta t} = c_1 A I_{t \to t+\Delta t} + c_2 A_2 I_{t-\Delta t \to t} + \dots + c_n A_n I_{t-(n-1)\Delta t \to t-(n-2)\Delta t} \qquad (5.69)$$

Equation 5.69 enables us to investigate the effects that changing the value of the runoff coefficient c has on the hydrograph at the drainage basin outlet for different rainfall input sequences and differently shaped drainage basins. For instance, increasing the value of c simulates the effect of deforestation or urbanization.

Figure 5.36 shows the effect that drainage basin shape has on the quickflow hydrograph form at the drainage basin outlet: Figure 5.36 is simply the outcome of calculating Equation 5.69 in a spreadsheet for differently shaped drainage basins that receive 8 and 10 length units of rain during two consecutive hours, and have a total area of 150 square length units, five time zones with $c = 0.5$, and areas A_1 to A_5 as defined in Table 5.2.

Figure 5.36 shows the hydrograph peak to be highest for drainage basin A, where a large part of the drainage basin is located near the drainage basin outlet, and thus where a large portion of water is quick to arrive at the outlet; and for drainage basin B, where a large part of the drainage basin is located far away from the drainage basin outlet, and thus where a large portion of water is late to arrive at the outlet. Drainage basin C, with two separated larger areas, shows two separate lower hydrograph peaks. The elongated shape of drainage basin D causes a relatively low, flat, and prolonged peak. Finally, drainage basin E, with a circularly shaped boundary, also causes a relatively high quickflow peak, as the basin shape will cause a large portion of the water to arrive at approximately the same time at the drainage basin outlet.

Note further that for all hydrographs the **time base**, defined as the time interval from the start to the end of quickflow, logically has to equal 7 hours, as this equals the sum of the duration of precipitation (2 hours) and the time of concentration (5 hours, as the number of time zones $n = 5$).

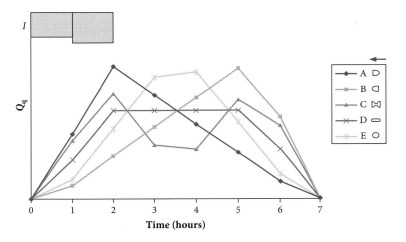

Figure 5.36. The significance of drainage basin shape for the quickflow hydrograph form (I = rainfall intensity; Q_q = quickflow)

Table 5.2 Areas (in square length units) of time zones 1–5 of drainage basins A–E in Figure 5.36; time zone 1 is nearest to the drainage basin outlet (as in Figure 5.35)

	1	2	3	4	5	Total
A	50	40	30	20	10	150
B	10	20	30	40	50	150
C	45	25	10	25	45	150
D	30	30	30	30	30	150
E	15	35	50	35	15	150

If you want to apply the time–area model to a real basin, you must at least have some idea of the length of each time zone as measured along the stream channel. For this, Equation B5.8.1, suggested by Kirpich (1940), may be helpful, as explained in Box 5.8.

The time–area model is a very simple model that assumes constant travel times. In reality, when discharge increases, water levels rise and flow velocities increase, shortening the travel time. Because of this, the time–area model is best applied in situations where travel times for changing discharges do not differ a great deal, and thus in small urban areas, or for hard surfaces such as parking lots and airports.

The linear reservoir model

Figure 5.37 shows a cross-section of a river recharged by groundwater. Assuming that the effects of evaporation on groundwater storage are negligible, then, if there was no more precipitation, the groundwater storage above the level of the river bed would

---

BOX 5.8 Estimation of the length of a time zone as measured along the stream channel for use in a time-area model

Kirpich (1940) suggested the following empirical equation to determine the time of concentration of a drainage basin (Chow *et al.* 1988; Mata-Lima 2006):

t_c = time of concentration of a drainage basin (minutes)

L = length of the main channel in the drainage basin (m)

$$t_c = 0.0195 \left(\frac{L}{\sqrt{S}} \right)^{0.77} \qquad \text{(B5.8.1)}$$

S = average drainage basin slope (m m^{-1})

The time of concentration t_c has been defined as the time (in minutes) required for surface water (or other water contributing to the quickflow of a stream) to flow from the farthest point in the watershed to the outlet. The average drainage basin slope S equals $\Delta H_{max} / L$, where ΔH_{max} is the difference in altitude between the highest and lowest point in the drainage basin, and L is as defined above; for instance, if $\Delta H_{max} = 9.5$ m and $L = 3070$ m, then t_c for the drainage basin equals 87 minutes = 1 hour and 27 minutes (Dastane 1978).

Equation B5.8.1 can be rewritten as follows:

L = length of the main channel of a subcatchment (m)

t_c = time of concentration of a subcatchment (minutes)

$$L = 166.24 \; t_c^{1.299} \sqrt{S} \qquad \text{(B5.8.2)}$$

S = average subcatchment slope (m m^{-1})

Note that the meaning of the variables has been changed in comparison with Equation B5.8.1. This is to use Equation B5.8.1 for determining L for different subcatchments (drainage basins located within the larger drainage basin). For instance, if you want to know the length of a time zone as measured along the stream channel in hours, simply insert values of 60 minutes, 120 minutes, and so on for t_c in Equation B5.8.2 and include the relevant values for the average drainage basin slopes S (m m^{-1}) in the equation; this will give you L for a subcatchment covering a time zone of 1 hour, L for a subcatchment covering a time zone of 2 hours, and so on.

Equation B5.8.1 has been developed for rural drainage basins in Tennessee (southern United States) with a well-defined channel and steep slopes (Chow *et al.* 1988) and is recommended for use in small rural drainage basins of up to 45 hectares (0.45 km^2) (Mata-Lima 2006). When used for overland flow on concrete or asphalt surfaces, Chow *et al.* (1988) recommend multiplying the result for t_c in Equation B5.8.1 by 0.4, and for concrete channels by 0.2; they further state that no adjustments need to be made for overland flow on bare soil or flow in roadside ditches. One way of translating their recommendations to Equation B5.8.2 is that results for L are to be multiplied by 2.5 for overland flow on concrete or asphalt surfaces, and by 5 for concrete channels.

As Equations B5.8.1 and B5.8.2 are empirical equations, the constants 0.0195 and 166.24 are not dimensionless, but contain hidden units: therefore, make sure that you insert variable values with the right units as given above.

The results of your calculations may be used to draw isochrones on a map of the drainage basin.

discharge completely into the river from the onset of the precipitationless, dry period (at $t = t$) to infinity ($t = \infty$). In mathematical notation:

S_t = storage above the level of the river bed (m^3)

$$S_t = \int_t^\infty Q_t \, dt \qquad (5.70)$$

As the groundwater storage depletes and the water table falls, the discharge of the river will gradually decline. Equation 5.62 gives the discharge Q_t at time t after the start of the recession (m^3 s^{-1}). Combining Equations 5.62 and 5.70 gives

Figure 5.37. A cross-section of a river recharged by groundwater, showing the groundwater storage above the level of the river bed

$$S_t = \int_t^\infty Q_0 e^{-\alpha t} dt = Q_0 \int_t^\infty e^{-\alpha t} dt = Q_0 \left[\frac{e^{-\alpha t}}{-\alpha} \right]_t^\infty = Q_0 \left(0 - \frac{e^{-\alpha t}}{-\alpha} \right) = \frac{Q_0 e^{-\alpha t}}{\alpha} = \frac{Q_t}{\alpha} \quad (5.71)$$

In short notation,

$$S = \frac{Q}{\alpha} \quad (5.72)$$

Equation 5.72 shows that groundwater flow (**effluent seepage**; section 3.12) into a river is described by a linear relation between the groundwater storage S (m³ or mm) above the level of the river bed and the river discharge Q (m³ s⁻¹ or mm s⁻¹), with the recession constant α (s⁻¹) as the connecting factor.

Equation 5.72 can be written in the general form that we encountered before as Equations 5.12 and 5.40 (see the two earlier examples of an artificial lake, in section 5.1):

> For groundwater depletion or a baseflow reservoir: $n = 1$

$$S = \frac{Q^n}{\alpha} \quad (5.73)$$

> **Exercise 5.4.1** At the start of a dry period with no precipitation, the baseflow of a river recharged by groundwater equals 100 m³ s⁻¹; $e^{-\alpha \times 1 \text{ month}} = 0.9$; the dry period lasts $1\frac{1}{2}$ months = 45 days.
>
> **a.** Determine the groundwater storage (m³) above the level of the river bed at the beginning of the dry period.
>
> **b.** Determine the baseflow (m³ s⁻¹) of the river after 1 and $1\frac{1}{2}$ months.
>
> **c.** Determine the groundwater storage (m³) above the level of the river bed after $1\frac{1}{2}$ months.

A **linear reservoir** is defined as a reservoir for which the relation between storage and outflow or discharge from the reservoir is linear, and thus with $n = 1$ in Equation 5.73. Note that groundwater flow is an important component of baseflow and that in many instances the two terms are used interchangeably. Outflow from a groundwater storage reservoir or baseflow reservoir may thus be modelled by a linear reservoir model.

We can expand the linear reservoir model by combining Equation 5.72 (the flow equation) and the continuity equation or water balance equation for a reservoir: a change in reservoir storage per unit of time (dS / dt) equals input I to the reservoir per unit of time minus the discharge Q from the reservoir (I = the precipitation rate minus the evaporation rate (and storage loss rates); further, a change in storage is positive when the storage increases and negative when the storage decreases):

> Input (m³ or mm per unit of time) = precipitation rate minus evaporation rate

> Output (m³ or mm per unit of time) = discharge Q

> Change in storage (m³ or mm per unit of time)

$$\frac{dS}{dt} = I - Q \quad (5.74)$$

Equation 5.72 (the flow equation) for a linear reservoir can also be written as follows:

$$dS = \frac{dQ}{\alpha} \qquad (5.75)$$

Combining Equations 5.74 (continuity) and 5.75 gives

$$\frac{dQ}{\alpha} = (I - Q)\, dt \qquad (5.76)$$

We can change the differential Equation 5.76 into a **simulation equation**, an equation that uses discrete time steps Δt to model changes through time, by replacing

$$\boxed{I_{\Delta t} = \text{input during time step } \Delta t}$$

dt by Δt, dQ by $\Delta Q = Q_{t+\Delta t} - Q_t$, Q by $\dfrac{Q_t + Q_{t+\Delta t}}{2}$, and I by $I_{\Delta t}$

This gives

$$Q_{t+\Delta t} - Q_t = \left(I_{\Delta t} - \frac{Q_t + Q_{t+\Delta t}}{2} \right) \alpha \Delta t \qquad (5.77)$$

which can be rewritten as

$$Q_{t+\Delta t} = \frac{2 - \alpha \Delta t}{2 + \alpha \Delta t} Q_t + \frac{2\alpha \Delta t}{2 + \alpha \Delta t} I_{\Delta t} \qquad (5.78)$$

Note that

$$\frac{2 - \alpha \Delta t}{2 + \alpha \Delta t} + \frac{2\alpha \Delta t}{2 + \alpha \Delta t} = 1$$

Thus, if we take

$$\beta = \frac{2 - \alpha \Delta t}{2 + \alpha \Delta t}$$

then

$$Q_{t+\Delta t} = \beta Q_t + (1 - \beta)\, I_{\Delta t} \qquad (5.79)$$

Equation 5.79 (or 5.78) constitutes a **rainfall–runoff simulation model,** a model of the reaction of the discharge (runoff) in a drainage basin to a precipitation event using discrete time steps Δt. **In effect, Equation 5.79 tells us that using the concept of a linear reservoir, the discharge $Q_{t+\Delta t}$ is influenced by the previous discharge Q_t, a time step Δt earlier, by a weighting factor β , and by the precipitation minus evaporation during time step Δt by a weighting factor $(1 - \beta)$.**

In applying the model, care should be taken to ensure that the value of α, essentially a decay constant with time^{-1} as unit, and the selected time step Δt, with (the reciprocal of time^{-1}, thus) time itself as unit do relate reasonably; if not, the model may give some very haphazard results. (For the same reason, combining Equations 5.63 and 5.78 for a period Δt with zero input, and thus $I_{\Delta t} = 0$, yields an equation

$$e^{-\alpha \Delta t} = \frac{2 - \alpha \Delta t}{2 + \alpha \Delta t}$$

that only holds as an approximation for certain sets of α and Δt; see Exercise 5.4.2.)

With the exception of baseflow, conditions in drainage basins are usually far from linear. Therefore, the rainfall–runoff simulation model based on a linear reservoir that we just discussed is best applied to model baseflow. Also, as the model in its present form does not take account of the travel time of water through a drainage basin, the model should be used for small drainage basins. Figure 5.38 shows the effect that using different values of the recession constant α has on the baseflow hydrograph modelled with the rainfall–runoff simulation model; as expected, a low value of α causes the baseflow to react slowly, which is noticeable both in the rising limb and falling limb of the hydrograph.

Exercise 5.4.2 During hydrograph recession, $I_{\Delta t} = 0 \Rightarrow$ Equation 5.78 for $I_{\Delta t} = 0$:

$$Q_{t+\Delta t} = \frac{2 - \alpha \Delta t}{2 + \alpha \Delta t} Q_t$$

Also during recession (Equation 5.63), $Q_{t+\Delta t} = Q_t e^{-\alpha \Delta t}$.

Combining the above two equations yields:

$$Q_{t+\Delta t} = e^{-\alpha \Delta t} Q_t = \frac{2 - \alpha \Delta t}{2 + \alpha \Delta t} Q_t \Rightarrow e^{-\alpha \Delta t} = \frac{2 - \alpha \Delta t}{2 + \alpha \Delta t}$$

The relation

$$e^{-\alpha \Delta t} = \frac{2 - \alpha \Delta t}{2 + \alpha \Delta t}$$

holds approximately for reasonable values of α and Δt, but certainly does not hold for all combinations of α and Δt (just try it out!). The reason for this is that the implicit assumptions made going from Equation 5.76 to Equation 5.77, which are that it is permissible to average the discharge data at the start and end of a time interval Δt, will often not be met when random combinations of α and Δt are used. However, instead of linking the simulation Equations 5.63 and 5.78, we may try to relate Equation 5.62 (without a discrete time interval Δt) to the differential Equation 5.76.

Use the differential Equation 5.76 with zero input, and thus $I = 0$, to prove that the recession curve is described by $Q_t = Q_0 e^{-\alpha t}$ (Equation 5.62).

High groundwater levels or high (pore) water pressures at (potential) slip surfaces in the subsurface can cause hill slopes to become unstable and to slide down the mountainside. As the location of the water table may play a crucial role in the

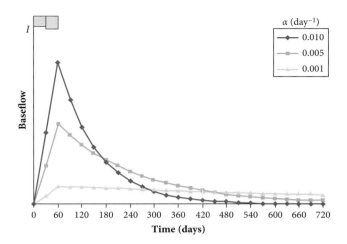

Figure 5.38. The effect on the baseflow hydrograph modelled with a linear reservoir model of using three different values of the recession constant α

initiation/reactivation of landslides, modelling the groundwater level – for instance, by using a linear reservoir model – is an important component of **landslide research**. As an example, Van Asch *et al.* (1996) successfully used a calibrated two-linear-reservoirs model in combination with precipitation data over a 30-year time period to show that deeper landslide movements in varved clay areas near Corps in the French Alps are mainly triggered by a maximum rise of water in vertical fissures in the clays and that complete saturation of the clays is not necessary; as a very practical spin-off, their calculations also showed that drainage of the shallow, more permeable colluvial cover overlying the varved clays is a very effective measure to stabilize these deeper landslides.

In their studies, Van Asch *et al.* (1996) used two linear reservoirs placed in series to represent the colluvial cover overlying the varved clays; thus the upper reservoir, representing the colluvial layer, is replenished (recharged) by precipitation, whilst water outflow from the upper reservoir is taken as water inflow to the lower reservoir representing the varved clays.

The exponential reservoir model

Figure 5.39 shows the core concept of another type of reservoir model. The rainfall rate minus the evaporation rate is taken as input I to the soil reservoir of a drainage basin. This soil reservoir or soil store loses water by **throughflow**, which is a water flow through the soil matrix in the direction of the stream, parallel to the sloping land surface; throughflow is caused by the soil being layered, as explained in sections 4.8 and 4.9. Above a soil layer that hinders vertical percolation, a perched water table may develop, the height of which is indicated by the water storage level h in Figure 5.39. If the quantity or volume (mm) of rain causes the storage level h (mm) to increase and then to exceed the water storage capacity h_c (mm) of the soil (Figure 5.39), water is forced to the surface: where such a perched water table cuts a sloping land surface, water will flow over the land surface as overland flow.

Varved clays consist of alternating thin layers of silt (or fine sand) and clay: they are formed in former glacial lakes by variations in sedimentation during the various seasons of the year. 'Colluvial cover' is a general term for sediment accumulations through the action of gravity at the foot of a slope.

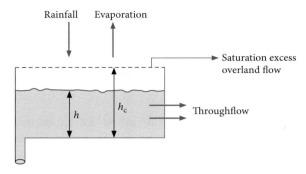

Figure 5.39. The conceptual soil moisture reservoir; h is the storage level in the soil (mm) and h_c is the storage capacity of the soil (mm); percolation to the groundwater is not modelled in STORFLO (after Kirkby *et al.* 1987)

Because the overland flow stems from the upper soil profile being saturated, overland flow produced in this way is called **saturation-excess overland flow**, or saturated overland flow for short. Thus, for saturation-excess overland flow to form, the quantity or volume of infiltrating water (mm) has to exceed the storage capacity (mm) of the soil, whereas for **infiltration-excess overland flow** or **Hortonian overland flow**, which we introduced in section 4.8, the rainfall intensity (mm hour^{-1}) has to exceed the soil's infiltration rate (mm hour^{-1}).

Kirkby *et al.* (1987) used the following flow equation to model the throughflow–outflow from the soil store in their **STORFLO model**:

$$\boxed{S_t = \text{soil moisture deficit (mm)}}$$

$$Q_t = Q_0 e^{\frac{S_t}{m}} \quad \boxed{m = \text{model parameter (mm)}} \tag{5.80}$$

S_t is the **soil moisture deficit**, the difference between the amount of water actually in the soil (mm) and the amount of water that the soil can hold, the **storage capacity** (mm): the soil moisture deficit has a negative value, but of course when the soil is saturated, $S_t = 0$ (and in Equation 5.80, $Q_t = Q_0$); m (mm) is a model parameter representing the soil characteristics.

Combining Equation 5.80 (the flow equation) with the continuity equation (or water balance equation), written as

$$\frac{dS_t}{dt} = I - Q_t \tag{5.81}$$

yields, after integrating and rewriting as a simulation equation (Kirkby 1975), the following simulation equation:

$$\boxed{I_{\Delta t} = \text{input during time step } \Delta t}$$

$$Q_{t+\Delta t} = \frac{I_{\Delta t}}{1 - e^{\frac{-I_{\Delta t}}{m}} + \frac{I_{\Delta t}}{Q_t} e^{\frac{-I_{\Delta t}}{m}}} \tag{5.82}$$

Equation 5.82 is the simulation equation of the exponential reservoir model STORFLO (Kirkby *et al.* 1987), which is a stripped-down version of the well-known **TOPMODEL** (e.g. Beven and Kirkby 1979; Beven and Moore 1994; Beven 1997; Beven 2001) with its characteristic built-in topographic wetness index $\ln(a / \tan\beta)$; a is the local upslope area draining through a certain point per unit contour length and $\tan\beta$ is the local slope at that point (Quinn *et al.* 1994, 1995; Sørensen *et al.* 2005).

STORFLO essentially models (slow) throughflow and saturated-excess overland flow from an exponential reservoir that represents the soil of a drainage basin. The model in the form presented here cannot cope with Hortonian overland flow or pipe-flow; also, the model is intended for use only on small drainage basins where travel times are small. For larger drainage basins, a so-called **routing routine** (Hornberger *et al.* 1998; Johnson 1999), a routine that simulates the movement of water through a channel, should be added.

Figure 5.40 stems from a printout of the author's younger 'modelling' days and provides an example of STORFLO model output. One can see how the discharge varies with m; m is in the denominator and not the numerator as α, and thus its effects are opposite to those for α (m, however, has another unit than α). Thus, if m is taken as small, the rising limb of a modelled hydrograph will show a steeper rise and the falling limb a steeper decline, indicating that depletion of the soil reservoir is fast; logically, if m is large, the modelled hydrographs show lesser rises and falls, and the depletion of the soil store is slow.

The **antecedent moisture conditions** in a drainage basin play an important role with regard to the height of the hydrograph peak and also the quickflow volumes. Figure 5.40 shows a more prominent and quicker reaction to rainfall as the drainage basin becomes wetter, to the right in Figure 5.40, which is especially evident for the more slowly reacting hydrograph with $m = 50$ mm: during the first rainfall

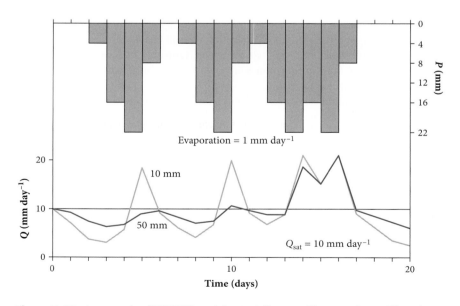

Figure 5.40. An example of STORFLO model output (for $m = 10$ mm and $m = 50$ mm)

event no saturation-excess overland flow is formed; during the second event, with the same amount and distribution of rainfall as the first one, some saturation-excess overland flow is formed (the portion above the horizontal 'Q_{sat} line'); whereas for the third event a lot of overland flow is produced. For the second part of the third event, under very wet conditions, the $m = 10$ mm and $m = 50$ mm hydrograph peaks even reach the same level, but with the throughflow decline of the more slowly reacting hydrograph with $m = 50$ mm (below the horizontal 'Q_{sat} line') from this peak being slower.

Figure 5.40 is clearly indicative of the transparent results that may already be established from using a very simple conceptual rainfall–runoff model.

Box 5.9 provides a short rationale on physically based models, whereas Box 5.10 introduces (non-physically based) statistical methods in hydrology that describe or estimate the magnitude of river runoff from recorded historical data.

5.5 Variable source area hydrology

The drainage basin hydrological system

Figure 5.41 shows the drainage basin hydrological system as we have largely come to understand it. Overland flow and rapid throughflow (pipeflow) deliver water to the stream rapidly, whereas groundwater flow and slow throughflow (matrix flow) deliver water to the stream slowly. Some remarks should, however, be added to this general, simplifying statement, especially with regard to the role of groundwater (in relief areas) and with regard to the spatial or topographic setting of the generation (production) of quickflow.

BOX 5.9 Physically based models

A **physically based model** uses known physical concepts or principles to model processes: a **conceptual model** is a physically based model in which one such physical concept or principle is highlighted and used. A model is called **lumped** when the modelled system is spatially averaged: a model is called **distributed** when it takes account of the spatial variation of the patterns of hydrological entities and processes.

Physically based models and conceptual models are of a **deterministic** nature; that is, the model outcomes are independent of chance and time. In other words, for the same initial and boundary conditions and input, a deterministic model will always produce the same output.

For prediction/forecasting purposes, model outcomes need to be compared with the real world and then the values of relevant model parameters need to be fine-tuned to establish a better correspondence between model and real world, a process called **model calibration**. After this, the calibrated model needs to be run under a new set of circumstances and again the model outcomes need to be compared with the real world, a process called **model validation/verification**. Only after this has been passed as satisfactory may a model be used as a predictive tool.

Other than this, models can be built and run as investigative tools for researchers to investigate the sensitivity of certain variables/parameters in a model and/or to sharpen their understanding of the workings of hydrological entities and processes.

BOX 5.10 Statistical methods

In an introductory text on physical hydrology, some non-physical methods may also be discussed, albeit briefly. The statistical methods discussed here provide tools for the description or estimation of the magnitude of river runoff from recorded historical data.

A flow-duration curve is a descriptive tool in which usually daily or weekly flows at a gauging station of a selected river are arranged according to their frequency of occurrence. Figure B5.10.1 shows the principle of making such a curve. As an example, Figure B5.10.1a shows a distribution of daily flows, where low flows (given in mm) occur more often than large flows. This is an example of a positively skewed statistical distribution: the magnitude of river flows generally shows this kind of distribution.

Figure B5.10.1b shows that by piling up the percentages of time for each consecutive runoff class, the same distribution of flows may also be presented as a cumulative distribution: along the vertical axis the cumulative percentage of time is given, whereas runoff classes, as before, are presented along the horizontal axis.

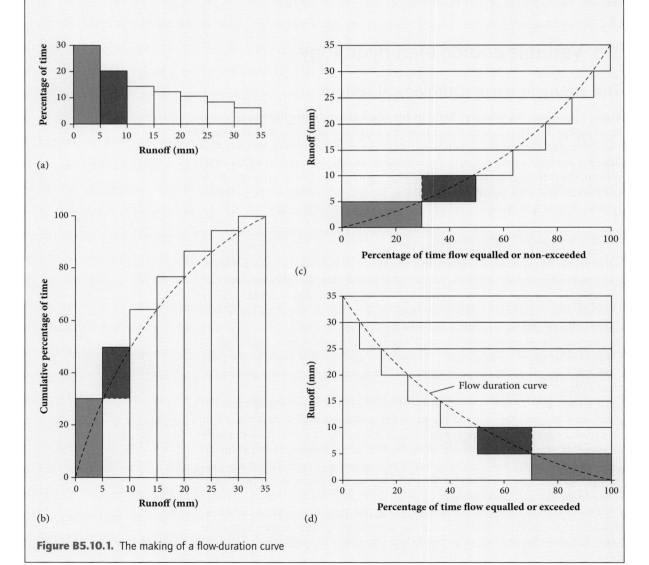

Figure B5.10.1. The making of a flow-duration curve

In Figure B.5.10.1c, the horizontal and vertical axes are switched. Importantly, the cumulative percentage of time along the horizontal axis can be interpreted as the percentage of time that flow is equalled or non-exceeded.

Finally, Figure B5.10.1d is a mirror image of Figure B.5.10.1c and the horizontal axis now gives the percentage of time that flow is equalled or exceeded; Figure 5.10.1d is an example of a flow-duration curve.

The vertical axis of a flow-duration curve may also be taken logarithmically. Usually, a flow-duration curve is transferred to a **dimensionless flow-duration curve**: for our example, this can be achieved by dividing the daily runoff values along the vertical axis by the river's average daily runoff for the period under study.

Figure B5.10.2 shows dimensionless flow-duration curves for three British rivers. The flow-duration curves of the River Tees and River Tamar have steep slopes, and their river flow can be interpreted as highly variable and consisting of a large quickflow component, whereas the gentle slope of the flow-duration curve of the Ver may be taken as indicative of a large baseflow component (Ward and Robinson 2000).

Yet another statistical method used in hydrology involves the construction of a **flood frequency curve**, which can be important for hydrological design purposes, such as the determination of the **design height** of levees or dykes that, for instance, are only to be flooded on average once in 1250 years. Once in 1250 years, $\frac{1}{1250} = 8 \times 10^{-4}$ year^{-1}, is a **probability of exceedence** P, while the reciprocal, 1250 years, is called a **return period** or recurrence interval T:

$$T = \frac{1}{P} \tag{B5.10.1}$$

In 1941, the German mathematician Emil J. Gumbel (1891–1966) developed a distribution that has been applied successfully in hydrology and is known as the **Gumbel distribution**. From this distribution, the probability of non-exceedence P' of a runoff peak can be determined as follows:

$$P' = e^{-e^{-y}} \tag{B5.10.2}$$

Equation B5.10.2 can be reworked to:

$$\ln P' = -e^{-y} \implies -\ln P' = e^{-y} \implies \ln\left(-\ln P'\right) = -y \implies$$

$$y = -\ln\left(-\ln P'\right) \tag{B5.10.3}$$

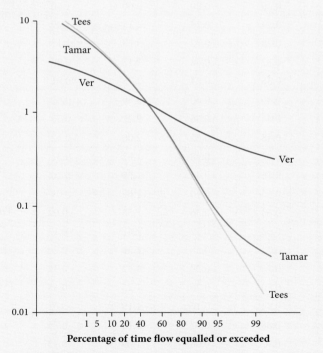

Figure B5.10.2. Dimensionless flow-duration curves for selected British rivers (after Ward and Robinson 2000)

y is called the reduced variate: as $P' = 1 - P$ and from Equation B5.10.1, it follows that

$$y = -\ln(-\ln P') = -\ln(-\ln(1-P)) = -\ln\left(-\ln\left(1-\frac{1}{T}\right)\right)$$

(B5.10.4)

The return period or recurrence interval T, or similarly, the probability of exceedence P, can be determined using the Weibull formula, named after the Swedish engineer and mathematician Ernst H.W. Weibull (1887–1979):

$$T = \frac{n+1}{m} \quad \text{or} \quad P = \frac{m}{n+1}$$

(B5.10.5)

As an example, Table B5.10 shows data from a 24-year annual maximum series for a hypothetical stream. In an annual maximum series, the single maximum runoff peaks Q_p for different years are listed. In Table B5.10, the number of data values equals the record length n of 24 years. Further, the listed runoff peaks are assigned an order number m by ranking them from largest to smallest. Using Equation B5.10.5, the probability of exceedence P and the return period T are calculated for each listed, single maximum runoff peak; then from Equation B5.10.4 the reduced variate y is calculated. The values of the reduced variate y for each single maximum runoff peak are shown in the last column of Table B5.10.

Table B5.10 The runoff peak Q_p, order number m, probability of exceedence P, probability of non-exceedence P', recurrence interval T, and reduced variate y from a 24-year annual maximum series for a hypothetical stream

Q_p ($m^3\,s^{-1}$)	m	P (fraction)	P' (fraction)	T (year)	y
300	1	0.04	0.96	25.00	3.20
270	2	0.08	0.92	12.50	2.48
252	3	0.12	0.88	8.33	2.06
234	4	0.16	0.84	6.25	1.75
228	5	0.20	0.80	5.00	1.50
216	6	0.24	0.76	4.17	1.29
210	7	0.28	0.72	3.57	1.11
204	8	0.32	0.68	3.13	0.95
201	9	0.36	0.64	2.78	0.81
192	10	0.40	0.60	2.50	0.67
186	11	0.44	0.56	2.27	0.55
183	12	0.48	0.52	2.08	0.42
174	13	0.52	0.48	1.92	0.31
168	14	0.56	0.44	1.79	0.20
162	15	0.60	0.40	1.67	0.09
159	16	0.64	0.36	1.56	-0.02
156	17	0.68	0.32	1.47	-0.13
150	18	0.72	0.28	1.39	-0.24
144	19	0.76	0.24	1.32	-0.36
141	20	0.80	0.20	1.25	-0.48
132	21	0.84	0.16	1.19	-0.61
126	22	0.88	0.12	1.14	-0.75
114	23	0.92	0.08	1.09	-0.93
102	24	0.96	0.04	1.04	-1.17

For statistical analysis, it is important that the selected runoff peaks in an annual maximum series are independent of one another; for instance, an annual runoff peak in January may be related to an annual runoff peak in the previous December. Therefore, preferably, a **hydrological year** (section 1.4) is used that starts and ends at the end of the dry season, when river runoff is lowest.

Yet another possibility, instead of using an annual maximum series, is to use a **partial duration series ('peaks over threshold' series).** This method uses runoff peaks above a selected threshold value, which ensures that there are more data values for analysis; however, there is then also a bigger chance of runoff peaks being statistically related, making the assumption of true independence less valid.

Figure B5.10.3 shows the relation between runoff peak Q_p and the reduced variate y for the data of Table B5.10. The relation is linear, which enables us to extrapolate from the recorded historical data. For instance, a runoff peak with a recurrence interval of 100 years can be calculated from Equation B5.10.4 to have a reduced variate y of 4.60, and then from the regression line of Figure B5.10.3 to have a Q_p value of 366 m³ s⁻¹.

Instead of calculating the reduced variate y, calculated values of P' (= 1 – P) or T (by using Equation B5.10.5) can also be plotted directly on so-called **Gumbel probability paper** (Figure B5.10.4), which is paper with unequal intervals along the x-axis, designed to take account of the Gumbel distribution given by Equation B5.10.2.

For the purpose of determining a flood frequency curve, one could also have tried other statistical distributions from extreme value theory (than the Gumbel distribution) or other plotting formulas (than the Weibull formula). The vertical axis with runoff peaks Q_p can be taken as linear as in our example, or may be scaled logarithmically if this produces a better linear fit of the data. Actually, whatever statistical method one selects, the main idea always is to have the data plot as a straight line on some sort of probability paper, as this enables us to extrapolate our findings to flows of a larger magnitude or a longer return period.

Returning to our example, the results could also have been extrapolated to show a runoff peak of 469 m³ s⁻¹ to have a return period of 1000 years; of course, the uncertainty of such an estimation from a data series of a mere 24 years is bound to be quite large.

A flood frequency curve offers us estimates of the probability of occurrence of events of a larger magnitude or a longer return period. The occurrence of an event with a recurrence interval of only 100 years implies that, on average, this event is estimated to occur once every hundred years; it by no means implies that we are safe from such an event happening again during the next 99 years. Also, when events of a very large magnitude happen, these may significantly alter the statistical relation. One should thus view flood frequency analysis simply as a means of providing a best estimate for hydrological design purposes, given the record of historical river runoff data available.

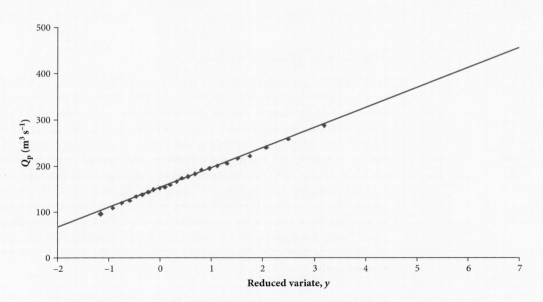

Figure B5.10.3. Runoff peak Q_p versus reduced variate y for the data of Table B5.10, using the Weibull formula

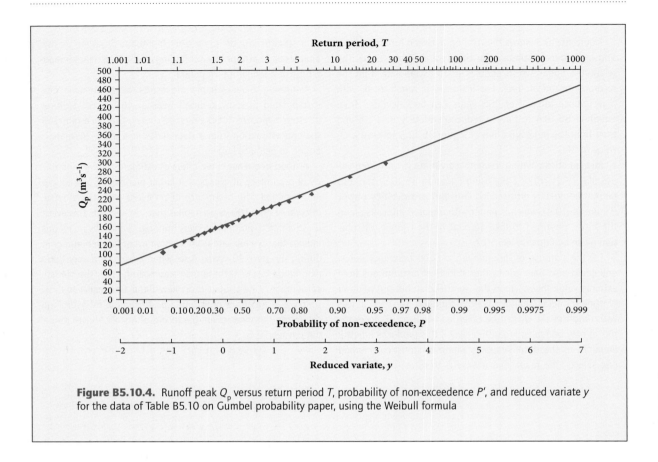

Figure B5.10.4. Runoff peak Q_p versus return period T, probability of non-exceedence P', and reduced variate y for the data of Table B5.10 on Gumbel probability paper, using the Weibull formula

The role of groundwater

With regard to groundwater flow, in sloping terrain in humid areas, rainfall events have been observed to raise the shallow water table bordering the stream above the water level of the stream, causing groundwater flow to contribute to the stream's quickflow (Hewlett 1961; Hewlett and Hibbert 1967). Also, during rainfall events in parts of The Netherlands with shallow water tables, groundwater may be transported rapidly through drains and ditches if these are deep enough to cut the rising water table (De Zeeuw 1966). The phenomenon of a rapid groundwater contribution via artificial drainpipes is even known to cause flooding problems in the higher, sandy parts of The Netherlands that lie above sea level and would therefore seem less prone to flooding; this is because the water table rises above the level of the artificial drain-pipes during flooding events. Of course, the Dutch examples just mentioned may equally well be labelled as pipeflow; however, in origin, the rapidly transported water is groundwater.

Infiltration-excess overland flow

For a long time, **infiltration-excess overland flow** (Hortonian overland flow) was thought to be responsible for the **flashy rainfall–runoff response** of drainage

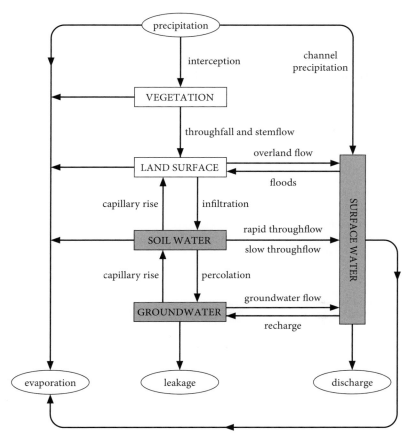

Figure 5.41. The drainage basin hydrological system
Ovals represent input or output processes; lower-case characters represent
hydrological processes; rectangles and upper-case characters show various kinds of
water storage; the blue background shows the major types of water storage within
a drainage basin (for average conditions)

basins, and this type of overland flow was thought to be the dominant process for
the generation of quickflow during precipitation events. While this is undoubtedly
true for (semi-)arid areas with soil crusts and surface sealing during rainfall, as well
as in urban areas, for dense soils or as a result of very high rainfall intensities, the
infiltration-excess overland flow mechanism cannot, however, explain the flashy
rainfall–runoff response of drainage basins in humid areas where the vegetation
maintains a soil structure that causes the **infiltration capacity**, the maximum rate
at which rain falling on the soil surface can infiltrate, to usually exceed the observed
rainfall intensities.

Throughflow and saturation-excess overland flow

Unconfined groundwater underlain by an unsaturated zone is called **perched groundwater** and its water table a **perched water table** (sections 3.4 and 4.8). Such a perched

water table may develop due to soil layering (section 4.8). A perched water table and throughflow may also develop as the result of the reducing hydraulic gradient as water moves downwards through a soil (section 4.8), even a relatively uniform one; this may cause the development of a saturated layer at some depth in the soil profile (Ward and Robinson 2000). Importantly for humid climates, when the perched water table reaches the land surface, any extra rain that falls will immediately become **saturation-excess overland flow**, which is transported over land and thus rapidly to the stream, contributing to the quickflow of the stream. Also, where an upper layer in the soil that supports throughflow becomes thinner, a perched water table may

Exercise 5.5 In the drainage basin of La Folie, near Entrechaux in the South of France, rain falls with a constant intensity on a smooth, gentle slope with vegetation and a rather dry soil profile. Figure E5.5 shows the infiltration rate f in mm hour^{-1} (vertical axis) against time in minutes (horizontal axis). The water table is located at considerable depth.

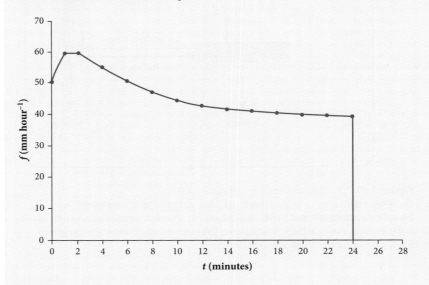

Figure E5.5

a. Determine the duration of the rainfall event.

b. Determine the rainfall intensity in mm hour^{-1}.

c. How long does it take for the interception storage to be at a maximum?

d. What percentage of the surface area is covered by vegetation?

e. Determine the time to ponding.

f. How long does it take for overland flow to occur?

g. What type of overland flow is this?

h. Determine the saturated hydraulic conductivity of the soil in mm hour^{-1}.

i. How long does it take for the overland flow to be at a maximum?

j. Determine the maximum overland flow rate in mm hour^{-1}.

develop and rise to the land surface. In conclusion, the **process of a rising water table** reaching the land surface is a very important mechanism for the generation of saturation-excess overland flow in humid climates. The saturated areas where quick-flow is generated, with saturation often occurring from below as just explained, are called **source areas**, usually with the adjective 'variable' added (as explained shortly) or **partial areas**.

Saturation-excess overland flow is sometimes referred to as **Dunnean overland flow** after Thomas Dunne, professor in geomorphology and hydrology; in, for example, Dunne and Black (1970) and Dunne et al. (1975).

Pipeflow (rapid throughflow)

Pipeflow, rapid throughflow bypassing the soil matrix – for instance, when water flows through burrows made by moles or voles, or when it flows through artificial drains (see also 'the role of groundwater' above) – may also be important as a process generating quickflow. This has been discussed at length for flow through natural pipes in section 4.9, with the undulating, forested Keuper marls in Luxembourg set as an example (Hendriks 1990, 1993). However, also in flat parts of The Netherlands, mole and vole burrows are observed as being important for the generation of quickflow: Rozemeijer and Van der Velde (2008) report that mole and vole burrows, preferably located in the relatively dry, well-drained, and least cultivated soils alongside ditches, play an important role in the rapid transport of water and pollutants such as phosphates and heavy metals towards these ditches from water ponded at the land surface; interestingly, the ponded areas in the arable land and meadows are also the result of a rising water table or sometimes even infiltration-excess overland flow.

Topographic convergence and variable source areas

Due to **topographic convergence**, the process of a rising water table just described can be amplified (Troch 2008). Topographic convergence zones can be hillslope hollows or topographic depressions (Anderson and Burt 1978); that is, slope concavities in plan, or slope concavities in section, as shown in Figure 5.42. For both types of concavity, throughflow can enter such a concavity more rapidly than it can leave down-slope, forcing water to the surface. Also the mechanisms of convergence caused by a thinner upper soil layer and as a result of the reducing hydraulic gradient as the vertical pathway of downward-moving water lengthens, both mentioned before, are shown in Figure 5.42.

The areas that become saturated from below expand up-slope and upstream if rainfall continues, as shown in Figure 5.43: this phenomenon has been observed *in situ* (Dunne et al. 1975) as well as through remote sensing techniques (Troch et al. 2000). These saturated areas, often wetted from below as explained earlier and contributing to quickflow, can expand and contract in size between storms and during the course of a single storm: they are called **variable source areas**, the adjective 'variable' now added because of its expanding and contracting nature. Often, but not necessarily so, variable source areas are located near the channel network that drains the area. The latter has to do with processes such as **colluviation**, the deposition of sediment through the action of gravity at the foot of a slope, causing the hillslope angle to diminish in the direction of the stream, and **alluviation**, the deposition of sediment by streams, causing areas near the stream

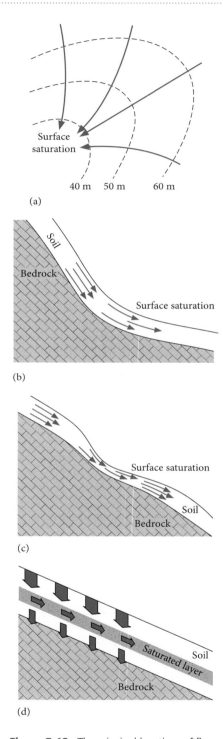

(a)

(b)

(c)

(d)

Figure 5.42. The principal locations of flow convergence in drainage basins (from Ward and Robinson 2000)

to be flat – that is, when viewed in cross-section: these 'flat' areas are called alluvial terraces and their slope is logically parallel to the (former) stream's gradient. Thus, processes such as colluviation and alluviation are responsible for having flattened and/or flattening the foot of hill slopes and thus for topographic convergence that amplifies the effect of a rising water table and may cause throughflow to exfiltrate.

Areas sensitive to pollutant transport

Variable source area hydrology is concerned with the detection and mapping of variable source areas; this is important as manure application, fertilizers, pesticides, and other human-applied substances can pollute streams, especially if applied to variable source areas when the ground is saturated (Cornell University 2005). Even in flat parts of The Netherlands with little topographic convergence, areas may become saturated as the result of a rising water table (or even infiltration-excess overland flow) and this has important implications for the transport of phosphates and heavy metals from arable land and meadows towards ditches and streams (Rozemeijer and Van der Velde 2008), as mentioned earlier. It is important to detect and map these **areas that are sensitive to pollutant transport**, as the rapid transport of pollutants towards the stream channel may have significant consequences for surface water quality.

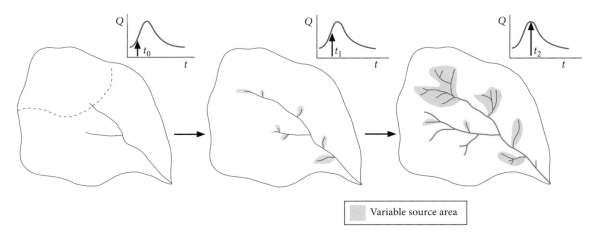

Figure 5.43. During rainfall events, the area that is saturated from below expands: we call this area the variable source area (from Troch 2008)

Summary

- Steady flow is flow for which the surface water velocities do not vary in time, or, in the case of turbulent flow, for which the statistical parameters (mean value and standard deviation of the water velocity) do not vary in time.

- Uniform flow is flow for which the surface water velocities are constant in the direction of flow, or, for an open channel with a constant cross-section, for which the water height H remains constant throughout all cross-sections.

- To describe the mechanical energy or total hydraulic head of surface water h_{total}, the velocity head $h_v = v^2/(2g)$ needs to be incorporated:

$$h_{total} = z + \frac{p}{\rho g} + \frac{v^2}{2g} \quad \text{(Bernoulli's law)}$$

- The propagation velocity of a surface wave or surface wave celerity v_w (m s^{-1}) equals $\sqrt{gH}$; v = water velocity (m s^{-1}). Throw a stone into a natural stream! When part of the ripples moves upstream from the point of contact of the stone with the water, the water flow is subcritical ($v < v_w$); when the ripples are totally swept downstream, the water flow is supercritical ($v > v_w$); when the upstream part of the ripple remains stationary at its point of initiation, the water flow is critical ($v = v_w$).

- The Froude number (Fr) is the ratio of the water velocity v over the surface wave propagation velocity v_w.

- The specific energy h_e is the energy per unit weight (J N^{-1} = m) of flowing water relative to the stream bottom. The specific discharge q_w (m^2 s^{-1}) is the discharge Q (m^3 s^{-1}) per unit channel width w (m) of a stream. With water height H presented on the vertical axis, the upper part of a specific energy diagram (for a selected specific discharge) holds for subcritical flow ($v < v_w$ or Fr < 1), and the lower part for supercritical flow ($v > v_w$ or Fr > 1). At the minimum specific energy, the flow is critical ($v = v_w$ or Fr = 1).

- Water flow over a natural step in a river (at the dip) is critical, as it changes from subcritical upstream of the dip to supercritical in the rapid.

- When supercritical flow reaches slower-moving, downstream water, the water is swept upwards and falls back on itself in an upstream direction; this is a hydraulic jump. At the hydraulic jump, kinetic energy is explosively converted into turbulence and potential energy.

- We can make good use of critical water flow to relate the discharge Q to the water height H slightly upstream

of a fixed construction such as a weir or flume. A general Q-H relation for a weir or flume is $Q = CH^n$, with n being or approximating $\frac{3}{2}$ or $\frac{5}{2}$.

- The exponent for H in Q-H relations for man-made fixed constructions such as weirs or flumes usually is a multiple of $\frac{1}{2}$ ($\frac{3}{2}$ or $\frac{5}{2}$). This has to do with the velocity head $h_v = v^2/(2g)$ playing an important role in surface water flow. This observation clearly distinguishes surface water flow equations from groundwater and soil water flow equations, where the exponent of ∇h in Darcy's law and the Darcy–Buckingham, Laplace, and Richards equations is always a full integer, 1 or 2, and never $\frac{1}{2}$ or a multiple of $\frac{1}{2}$.

- The bed load of a stream consists of particles (sand, stones) that are moved along the stream bed by rolling, pushing, and/or saltation (advancement by leaps); the suspended load of a stream consists of particles (clay, silt) that are carried in suspension (floating in the water column) above the stream bed. An advantage of a flume over a weir is that a flume largely cleans itself.

- The S-Q relation $S = Q^n/\alpha$ (S = reservoir storage) holds for an artificial lake with a lower opening in the dam for $n = 2$, for groundwater depletion or a baseflow reservoir for $n = 1$, and for an artificial lake with overflow for $n = \frac{2}{3}$.

- The Q-H relation is also called (in reversed terms) the stage–discharge relation. The stage or water level H of a stream can be measured by a pressure transducer or a shaft encoder and data logger – for instance, in a stilling well installation: the latter has the advantage that oscillations due to the turbulence of the water flow are subdued. The discharge Q can be determined with a number of techniques, for instance with an electromagnetic current meter using the velocity–area method or from spot gaugings using Acoustic Doppler Current Profilers (ADCPs).

- Salt dilution gauging is a well-suited technique for estimating the discharge in turbulent mountain streams: the discharge is determined from the degree of dilution by the flowing water of an added solution of sodium chloride (NaCl). Two methods of salt dilution gauging are constant rate injection and slug injection.

- Using constant rate injection, a salt solution is added to the stream with a constant rate or discharge. The salt load (mg s^{-1}) of a river equals the product of the river discharge (litre s^{-1}) and the river's salt concentration (mg litre^{-1}). The constant rate injection method is based on a simple mass balance or chemical mixing model: adding together the salt load (mg s^{-1}) of the river upstream of the injection point and the salt load (mg s^{-1}) from the injection point must yield the salt load (mg s^{-1}) at the downstream measuring location (if there is no downward or upward seepage along the measuring reach).

- Using slug injection, a slug or gulp of salt solution is emptied instantaneously into the stream, which causes a saltwater wave to pass downstream through the channel: by monitoring the electrical conductivity or EC (μS cm^{-1}) of the passing saltwater wave, the discharge can be determined.

- Measuring the EC at a large number of locations in the longitudinal profile of one or more streams and pencilling in the EC data thus obtained on a map or in a sketch of the longitudinal profile of a stream is called EC-routing. Such a routing can, for instance, provide useful information on the upward seepage of groundwater with a strongly differing chemical composition into a stream.

- A chemical mixing model can be used to extend discharge data to tributaries of a stream. For this, the discharge must be known at one location and one should measure the electrical conductivity or the concentration of a conservative ion – that is, a non-reactive ion – in the different tributaries. To apply the model successfully, the EC or concentration values of the tributaries should differ significantly.

- The stage–discharge relation or Q-H rating curve for a stream can often be presented as a power function: $Q = a(H - H_0)^b$, where H_0 is the maximum stage with zero discharge. The values of a and b can best be found after rewriting this power function in a logarithmic form.

- The Q-H rating curve is sensitive to hysteresis, the phenomenon of an equilibrium state being dependent on the history of the physical system. If enough data is available, it may be recommended to construe separate rating curves for rising and falling water levels; that is, the rising and falling limbs of a hydrograph.

- The discharge (or part of the discharge; m^3 s^{-1}) of a stream caused by processes delivering water rapidly to the stream is called quickflow: the quickflow volume is the discharge volume (m^3) that stems from these rapid processes.

- The discharge (or part of the discharge; m^3 s^{-1}) of a stream caused by processes delivering water slowly to the stream

is called baseflow: the baseflow volume is the discharge volume (m^3) that stems from these slow processes.

- Streams that hold water throughout the year are called perennial, whereas a channel that only holds water during and immediately after a rain or snowmelt event is called ephemeral: an ephemeral stream thus has no baseflow, only quickflow. In between these two extremes lie intermittent streams, streams that only carry water during the wet part of the year.

- A recession curve of a hydrograph for a perennial or intermittent stream can conceptually be interpreted as outflow from both a quickflow reservoir and a baseflow reservoir, with the larger outflow per unit of time from the quickflow reservoir. At the separation point, the quickflow reservoir has emptied, and from then on there is only outflow from the baseflow reservoir. As the outflow from the quickflow reservoir is dominant, the first part of the recession curve after the hydrograph peak can simply be interpreted as outflow from the quickflow reservoir alone.

- There are many methods for hydrograph separation into quickflow and baseflow: the constant-slope inflection-point method assumes an immediate response in baseflow to precipitation and connects the start of the rising limb with the inflection point (as separation point) on the recession curve: the position of the inflection point is found as the point in time at which the slope of the semi-logarithmic plot of ln Q_t versus t changes.

- The runoff coefficient of a stream can simply be determined by dividing the discharge volume in millimetres by the received precipitation volume (precipitation depth) in millimetres and then multiplying by 100 to convert from a fraction to a percentage. The term 'runoff' in 'runoff coefficient' should be taken as the discharge (volume) of the stream. The runoff coefficient on an annual basis is defined as the *total discharge volume* (mm or m^3) as a percentage of the precipitation volume (mm or m^3) on an annual basis, whereas the runoff coefficient on a storm basis is defined as only the *quickflow volume* (mm or m^3) as a percentage of the precipitation volume (mm or m^3) on a storm basis.

- A conceptual rainfall–runoff model is a model that uses a simple physical concept or principle to model the rainfall–runoff relation.

- The time of concentration is the time needed for an entire drainage basin to contribute to the quickflow.

- A time-area model incorporates the travel-time concept and the superposition principle to model the quickflow discharge Q_q from a drainage basin. The model is best applied in small urban areas or for hard surfaces such as parking lots and airports.

- For a river recharged by groundwater, the S–Q relation is linear ($n=1$): $S=Q/\alpha$ or $S_t=Q_t/\alpha$; α then equals the recession constant (time^{-1}) of the recession curve or falling limb of the river's hydrograph. Such a recession curve is mathematically described as $Q_t=Q_0 e^{-\alpha t}$, $t=0$ being the start of the recession.

- A linear reservoir is defined as a reservoir for which the relation between storage and outflow or discharge from the reservoir is linear; thus $S=Q^n/\alpha$ with $n=1$. A rainfall–runoff simulation model based on a linear reservoir may be used to model the baseflow in small drainage basins.

- An exponential reservoir model that incorporates the soil moisture deficit may be used to model throughflow in a small drainage basin: the soil moisture deficit (≤ 0) is the difference between the amount of water actually in the soil (mm) and the amount of water that the soil can hold, the storage capacity (mm).

- A flow-duration curve is a descriptive tool in which usually daily or weekly flows at a gauging station of a selected river are arranged according to their frequency of occurrence. Often, a flow-duration curve is transferred to a dimensionless flow-duration curve by dividing the runoff values by the river's average runoff for the period under study.

- A flood frequency curve is a graph showing the relationship between flood magnitude and return period or recurrence interval for a specific site; such a curve is important for hydrological design purposes; for instance, the determination of the design height of levees or dykes that are only to be flooded on average once in 1250 years; 1250 years is called a return period or recurrence interval.

- Overland flow may be produced as saturation-excess overland flow (Dunnean overland flow) or as infiltration-excess overland flow (Hortonian overland flow). For saturation-excess overland flow to occur, the quantity or volume of infiltrating water (mm) has to exceed the storage capacity (mm) of the soil. For infiltration-excess overland flow to occur, the rainfall intensity (mm hour^{-1}) has to exceed the soil's infiltration rate (mm hour^{-1}).

- The process of a rising water table reaching the land surface is a very important mechanism for the generation of saturation-excess overland flow in humid climates. Topographic convergence amplifies this process of a rising water table. Topographic convergence zones can be hillslope hollows or topographic depressions; that is, slope concavities in plan or slope concavities in section. Also, the mechanisms of convergence caused by a thinner upper soil layer and as a result of the reducing hydraulic gradient as the vertical pathway of downward-moving water lengthens, should be mentioned. Variable source areas are water-saturated areas that contribute to quickflow: these areas can expand and contract in size between storms and during the course of a single storm, and often, but not necessarily so, variable source areas are located near the channel network that drains the area.

Epilogue

The main focus in hydrology is on the terrestrial part of the **hydrological cycle**. **Hydrology** is concerned with the occurrence, movement, and composition of water below and at the earth's surface (Royal Netherlands Academy of Arts and Sciences 2005): it plays a central role within the study of the **Earth System** and touches upon many other areas, such as climatology, meteorology, fluid mechanics, sedimentology, hydrochemistry, land degradation, ecology, and so on.

In all respects, hydrology is a **no-regrets study**, whatever the background of a student or reader of this book may be, as water and water-related problems are all-important and will continue to be so.

The weather and, on another scale, **climate change** affect the **Water System** and vice versa, and because of this we have to continuously adapt. Floods and droughts, greenhouse gas emissions, plant and soil respiration, the carbon cycle, the availability of fresh water for agricultural purposes, the intrusion of salt water, water-related diseases, sustainable access to safe drinking water, adequate sanitation services, deterioration of ecosystems, declining groundwater quality, soil loss by erosion, pollution, bioremediation, preferential flow, radioactive waste disposal, natural gas transport in soil and groundwater systems, water management, and so on – these are, in random order, all topics related to hydrology.

This book on physical hydrology has introduced a number of physical concepts used in hydrology, as they are appreciated by the author. The book has covered a lot – and yet it has not, because there is quite a lot more! It is the author's hope that reading this book has provided a meaningful and pleasurable way of spending time for the reader, and a head-start to those wanting – in whatever specific way – to get more deeply entangled with hydrology.

Conceptual toolkit

C1 If you cannot do the maths

$a^0 = 1$ if $a \neq 0$

$\sqrt{a} = a^{\frac{1}{2}}$ if $a \geq 0$

$a^{-x} = \dfrac{1}{a^x}$ if $a \neq 0$

$\dfrac{1}{\sqrt{a}} = a^{-\frac{1}{2}}$ if $a > 0$

$a^x a^y = a^{x+y}$

$\dfrac{a^x}{a^y} = a^{x-y}$ if $a \neq 0$

$(a^x)^y = a^{xy}$

$(a+b)^2 = a^2 + 2ab + b^2$

$(a-b)^2 = a^2 - 2ab + b^2$

$a^2 - b^2 = (a+b)(a-b)$

$(ab)^x = a^x b^x$

$a(b+c) = ab + ac$

$ax^2 + bx + c = 0 \ (a \neq 0) \quad \Rightarrow x = \dfrac{-b \pm \sqrt{b^2 - 4ac}}{2a}$

$\log 10^a = a \quad \log 10 = \log 10^1 = 1 \quad \log 1 = \log 10^0 = 0$

$b = 10^a \quad \Rightarrow \quad \log b = \log 10^a = a$

$a = \log b \quad \Rightarrow \quad b = 10^a$

$\log ab = \log a + \log b$

$\log \dfrac{a}{b} = \log a - \log b$

$\log a^b = b \log a$

$\log \dfrac{1}{a} = \log a^{-1} = -\log a$

$e = 2.71828...$

$\ln e^a = a \quad \ln e = \ln e^1 = 1 \quad \ln 1 = \ln e^0 = 0$

$b = e^a \;\Rightarrow\; \ln b = \ln e^a = a$

$a = \ln b \;\Rightarrow\; b = e^a$

$\ln ab = \ln a + \ln b$

$\ln \dfrac{a}{b} = \ln a - \ln b$

$\ln a^b = b \ln a$

$\ln \dfrac{1}{a} = \ln a^{-1} = -\ln a$

$\ln a = (\ln 10) \times (\log a)$

$\pi = 3.14159...$

area of a circle $= \pi r^2$, with r = radius of circle

circumference of a circle $= 2\pi r$, with r = radius of circle

C2 Mathematical differentiation and integration

C2.1 Free fall

At $t = 0$ seconds, a water droplet starts to fall. We assume that the water droplet's fall is not disturbed by friction and that the droplet does not evaporate on its way down. All variables in the following exposé have absolute values (greater than or equal to zero) irrespective of the direction of the process. At $t = 0$ seconds, the velocity $v_0 = 0$ m s^{-1}. At $t = 1$ second, the acceleration due to gravity (≈ 9.8 m s^{-2}) causes v_1 to be 9.8 m s^{-1}, at $t = 2$ seconds, $v_2 = 9.8 + 9.8 = 2 \times 9.8 = 19.6$ m s^{-1}; and at $t = 3$ seconds, $v_3 = 19.6 + 9.8 = 3 \times 9.8 = 29.4$ m s^{-1}; and so on.

In equation form, this is the linear equation

g = acceleration due to gravity ≈ 9.8 m s^{-2}

v = velocity (m s^{-1}) $\quad\bullet\; v = gt \;\bullet\quad$ t = time (s)   (C2.1)

How far will the droplet fall in, let us say, 3 seconds?
The average velocity $\bar{v}$ over these three seconds is as follows:

v_0 = velocity (m s^{-1}) after 0 seconds $\qquad$ v_3 = velocity (m s^{-1}) after 3 seconds

$\bar{v}$ = average velocity (m s^{-1}) $\quad\bullet\; \bar{v} = \dfrac{v_0 + v_3}{2} = \dfrac{v_3}{2} = \dfrac{29.4}{2} = 14.7$ m s^{-1}

The vertical distance of fall z (m) equals the time of fall (seconds) times the average velocity (m s^{-1}):

$$z = t \; \bar{v} = t\frac{v_3}{2} = 3 \times \frac{29.4}{2} = 44.1 \text{ m}$$

In general,

t = time of fall (s) $\qquad$ v = velocity (m s^{-1}) after t seconds

z = vertical distance of fall (m) $\quad\bullet\; z = t\dfrac{v}{2}$ (C2.2)

Substituting Equation C2.1 in Equation C2.2 gives the quadratic equation

$$z = \frac{1}{2}gt^2 \tag{C2.3}$$

After a small increment in time, Δt, the distance z will have changed by a small increment Δz; after $t + \Delta t$ seconds, a distance $z + \Delta z$ is covered.

Substituting $z = z + \Delta z$ and $t = t + \Delta t$ in Equation C2.3 gives (some algebra rules are given in section C1)

$$z + \Delta z = \frac{1}{2}g(t + \Delta t)^2 = \frac{1}{2}g\left(t^2 + 2t\Delta t + (\Delta t)^2\right)$$

Combining this with Equation C2.3 gives

$$\Delta z = (z + \Delta z) - z = \frac{1}{2}g\left(t^2 + 2t\Delta t + (\Delta t)^2\right) - \frac{1}{2}g\left(t^2\right) = \frac{1}{2}g\left(2t\Delta t + (\Delta t)^2\right) = \frac{1}{2}g\Delta t(2t + \Delta t)$$

$\Delta z / \Delta t$ is the **(average) rate of change** of $z(t)$ over a period of time Δt. The above equation can be rewritten as follows:

$$\frac{\Delta z}{\Delta t} = \frac{1}{2}g(2t + \Delta t) = gt + \frac{1}{2}g\Delta t$$

When we take Δt smaller and still smaller again, we can define dz / dt as the limit of $\Delta z / \Delta t$ when Δt approaches a zero value. In mathematical notation,

$$\frac{dz}{dt} = \lim_{\Delta t \to 0} \frac{\Delta z}{\Delta t} \tag{C2.4}$$

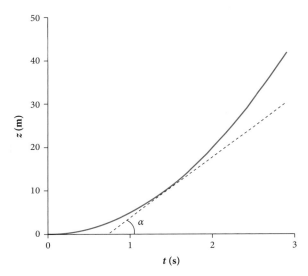

For our falling water droplet, this gives the following:

$$\frac{dz}{dt} = \lim_{\Delta t \to 0} \frac{\Delta z}{\Delta t} = \lim_{\Delta t \to 0}\left(gt + \frac{1}{2}g\Delta t\right) = gt \quad \Rightarrow \frac{dz}{dt} = gt$$

This is **(mathematical) differentiation**, and we have reworked Equation C2.3 for the vertical distance of fall z (m) to Equation C2.1 for the velocity v (m s^{-1}) of the water droplet after t seconds.

In mathematical notation,

$$v = \frac{dz}{dt} = gt \tag{C2.5}$$

dz / dt is the rate of change of $z(t)$ at a specific point in time t and equals the slope of the tangent line, a straight line touching – but not cutting – the curve (see the above figure).

Following earlier reasoning: after a small increment in time, Δt, the velocity v will have changed by a small increment Δv; after $t + \Delta t$ seconds, the velocity equals $v + \Delta v$.

Substituting $v = v + \Delta v$ and $t = t + \Delta t$ in Equation C2.1 gives

$$v + \Delta v = g(t + \Delta t) = gt + g\Delta t$$

Combining this with Equation C2.1 gives

$$\Delta v = (v + \Delta v) - v = gt + g\Delta t - gt = g\Delta t \quad \Rightarrow \frac{\Delta v}{\Delta t} = g$$

and

$$\frac{dv}{dt} = \lim_{\Delta t \to 0} \frac{\Delta v}{\Delta t} = \lim_{\Delta t \to 0} g = g \quad \Rightarrow \frac{dv}{dt} = g$$

Combining this with Equation C2.5 gives

$$\frac{dv}{dt} = \frac{d}{dt}\left(\frac{dz}{dt}\right) = \frac{d^2z}{dt^2} = g \approx 9.8 \text{ m s}^{-2} \text{ (constant)} \tag{C2.6}$$

In summary:

$$z = \frac{1}{2}gt^2$$

$$v = \frac{dz}{dt} = gt$$

$$g = \frac{dv}{dt} = \frac{d}{dt}\left(\frac{dz}{dt}\right) = \frac{d^2z}{dt^2}$$

Thus, when we differentiate the distance z (m) we obtain the velocity v (m s^{-1}), and when we differentiate the velocity v (m s^{-1}) we obtain the acceleration due to gravity g (m s^{-2}).

$\frac{dz}{dt}$ is called the **first derivative** of $z(t)$.

$\frac{dv}{dt}$ is the first derivative of $v(t)$.

$\frac{d}{dt}\left(\frac{dz}{dt}\right) = \frac{d^2z}{dt^2}$ is called the **second derivative** of $z(t)$.

C2.2 Common derivatives

The derivative of a variable is obtained by multiplying the variable by its exponent and then subtracting the value one from the exponent: the derivative of a constant (C) is zero.

In mathematical notation (n and C are constants):

$$\frac{d(x^n)}{dx} = nx^{n-1} \tag{C2.7}$$

$$\frac{dC}{dx} = 0 \tag{C2.8}$$

In analogy to section C2.1, but now also by directly applying Equations C2.7 and C2.8, derivatives of the following functions (of one variable) can be obtained (C and a are constants):

$$v = t^1 \quad \Rightarrow \frac{dv}{dt} = 1 \times t^0 = 1$$

$$v = C \quad \Rightarrow \frac{dv}{dt} = 0$$

$$v = t^1 + 3 \quad \Rightarrow \frac{dv}{dt} = 1 \times t^0 + 0 = 1$$

$$z = t^2 - 5 \quad \Rightarrow \frac{dz}{dt} = 2t^1 = 2t$$

$$z = at^2 + 7 \quad \Rightarrow \frac{dz}{dt} = 2at$$

$$z = \frac{1}{4}at^2 + 10 \quad \Rightarrow \frac{dz}{dt} = \frac{1}{2}at$$

$$y = x^3 - 3 \quad \Rightarrow \frac{dy}{dx} = 3x^2$$

$$y = x^4 + 2 \quad \Rightarrow \frac{dy}{dx} = 4x^3$$

$$h = \sqrt{x} + 3 = x^{\frac{1}{2}} + 3 \quad \Rightarrow \frac{dh}{dx} = \frac{1}{2}x^{-\frac{1}{2}} = \frac{1}{2\sqrt{x}}$$

$$h = x^3 + 4x^2 + 2x - 5 \quad \Rightarrow \frac{dh}{dx} = 3x^2 + 8x + 2 \quad \Rightarrow \frac{d^2h}{dx^2} = 6x + 8$$

A constant of multiplication (C) is placed outside the differential:

$$\frac{d(Cf(x))}{dx} = C\frac{d(f(x))}{dx} \tag{C2.9}$$

Another **common derivative** is as follows:

$$h = \ln r \quad \Rightarrow \frac{dh}{dr} = \frac{d(\ln r)}{dr} = \frac{1}{r} \tag{C2.10}$$

An example of using Equation C2.10 with R = constant (see the algebra rules in section C1):

$$\frac{d\left(\ln \frac{r}{R}\right)}{dr} = \frac{d(\ln r - \ln R)}{dr} = \frac{d(\ln r)}{dr} = \frac{1}{r}$$

Yet another **common derivative** is the following:

$$h = e^x \quad \Rightarrow \frac{dh}{dx} = \frac{d(e^x)}{dx} = e^x \tag{C2.11}$$

C2.3 The chain rule

An important tool for (mathematical) differentiation is the chain rule. The **chain rule** states that the derivative of $h(x)$ equals the derivative of $h(z)$ multiplied by the derivative of $z(x)$.

In mathematical notation,

$$\frac{dh}{dx} = \frac{dh}{dz}\frac{dz}{dx} \tag{C2.12}$$

Examples of applying the chain rule are as follows:

$$h = e^{\frac{x}{\lambda}} \Rightarrow z = \frac{x}{\lambda} \quad \frac{dh}{dz} = e^z \quad \frac{dz}{dx} = \frac{1}{\lambda} \Rightarrow \frac{dh}{dx} = e^z \frac{1}{\lambda} = \frac{1}{\lambda} e^{\frac{x}{\lambda}}$$

$$h = e^{\frac{-x}{\lambda}} \Rightarrow z = -\frac{x}{\lambda} \quad \frac{dh}{dz} = e^z \quad \frac{dz}{dx} = -\frac{1}{\lambda} \Rightarrow \frac{dh}{dx} = e^z \frac{-1}{\lambda} = -\frac{1}{\lambda} e^{\frac{-x}{\lambda}}$$

C_1 and C_2 are constants:

$$h^2 = C_1 x + C_2 \Rightarrow h = (C_1 x + C_2)^{\frac{1}{2}} = z^{\frac{1}{2}} \quad z = C_1 x + C_2 \quad \frac{dh}{dz} = \frac{1}{2} z^{-\frac{1}{2}} \quad \frac{dz}{dx} = C_1$$

$$\Rightarrow \frac{dh}{dx} = \frac{1}{2} z^{-\frac{1}{2}} C_1 = \frac{1}{2} C_1 (C_1 x + C_2)^{-\frac{1}{2}}$$

Alternatively:

$$y = h^2 = C_1 x + C_2 \Rightarrow \text{chain rule: } \frac{dh^2}{dx} = \frac{dh^2}{dh} \frac{dh}{dx} \quad \frac{dy}{dx} = \frac{dh^2}{dx} = C_1$$

$$\frac{dh^2}{dh} = 2h \Rightarrow \frac{dh^2}{dx} = 2h \frac{dh}{dx} \Rightarrow h \frac{dh}{dx} = \frac{1}{2} \frac{dh^2}{dx} = \frac{1}{2} C_1$$

$$h = (C_1 x + C_2)^{\frac{1}{2}} \Rightarrow \frac{dh}{dx} = \frac{h \frac{dh}{dx}}{h} = \frac{1}{2} C_1 (C_1 x + C_2)^{-\frac{1}{2}}$$

C2.4 The product and quotient rule

A product of two functions is differentiated by applying the **product rule**:

$$\frac{d(f(x)\,g(x))}{dx} = \frac{d(f(x))}{dx} g(x) + f(x) \frac{d(g(x))}{dx} \tag{C2.13}$$

A quotient of two functions is differentiated by applying the **quotient rule**:

$$\frac{d\left(\dfrac{f(x)}{g(x)}\right)}{dx} = \frac{\dfrac{d(f(x))}{dx} g(x) - f(x) \dfrac{d(g(x))}{dx}}{(g(x))^2} \tag{C2.14}$$

C2.5 A worked-out example of differentiation

An example of using the chain rule as well as both Equation C2.11 and C2.14 is the derivation of Equation 2.3 from Equation B2.2.3 in Chapter 2.

Equation B2.2.3 in Chapter 2 is as follows:

e_s = saturation vapour pressure (kPa)

T = air temperature (°C)

$$e_s = 0.6108\, e^{\frac{17.27T}{237.3+T}}$$

e = base of natural logarithm = 2.71828… (–)

$\Delta = de_s / dT$ is the gradient of the saturation pressure curve (kPa °C^{-1}).

$$\text{Chain rule: } \frac{de_s}{dT} = \frac{de_s}{dz}\frac{dz}{dT}$$

$$e_s = 0.6108\, e^z \quad z = \frac{17.27\, T}{237.3+T} = \frac{f(T)}{g(T)} \quad \frac{de_s}{dz} = 0.6108\, e^z = e_s$$

$$f(T) = 17.27\, T \quad g(T) = 237.3+T \quad \frac{d(f(T))}{dT} = 17.27 \quad \frac{d(g(T))}{dT} = 1$$

Inserting the quotient rule:

$$\frac{dz}{dT} = \frac{d\left(\dfrac{f(T)}{g(T)}\right)}{dT} = \frac{\dfrac{d(f(T))}{dT}g(T) - f(T)\dfrac{d(g(T))}{dT}}{(g(T))^2} = \frac{17.27\,(237.3+T) - (17.27\, T \times 1)}{(237.3+T)^2}$$

$$\frac{de_s}{dT} = \frac{de_s}{dz}\frac{dz}{dT} = e_s \times \frac{4098 + 17.27\, T - 17.27\, T}{(237.3+T)^2} = \frac{4098\, e_s}{(237.3+T)^2}$$

and we have obtained Equation 2.3 of Chapter 2:

Δ = gradient of the saturation pressure curve (kPa °C^{-1})

$$\Delta = \frac{4098\, e_s}{(237.3+T)^2}$$

C2.6 Mathematical notation for derivatives

$\dfrac{dh}{dx}$ = the **first derivative** of $h(x)$.

$\dfrac{dh^2}{dx}$ = the **first derivative** of $h^2(x)$.

$\dfrac{d^2h}{dx^2} = \dfrac{d}{dx}\left(\dfrac{dh}{dx}\right)$ = the **second derivative** of $h(x)$.

$\dfrac{d^2h^2}{dx^2} = \dfrac{d}{dx}\left(\dfrac{dh^2}{dx}\right)$ = the **second derivative** of $h^2(x)$.

C2.7 The minimum or maximum of a function

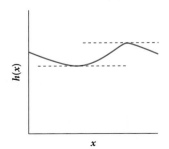

dh/dx is the **rate of change** of $h(x)$ at a specific distance x and equals the slope of a **tangent line** touching $h(x)$.

If $dh/dx = 0$ at a certain distance x, the tangent line at that certain distance has a flat slope; this indicates either a **(local) minimum** or **maximum** value for $h(x)$ at that distance x as shown in the above figure.

C2.8 Mathematical integration

If (mathematical) differentiation is 'walking down the stairs', then **(mathematical) integration** is 'walking up the stairs'.

Since the derivative of a constant (C) is zero (see Equation C2.8), the integral of 0 is a constant (C) with an unknown value. The integral of a variable is obtained by adding the value one to the exponent and by dividing the variable by this new exponent value, followed by adding a constant (C) (the integral of 0) to the total.

In mathematical notation (a, n, and C are constants):

$$\int ax^n dx = \int ax^n + 0\, dx = x^{n+1} \times \frac{a}{n+1} + C = \frac{a}{n+1} x^{n+1} + C \qquad \text{(C2.15)}$$

$$\int 0\, dx = \int 0 \times x^0 dx = x^1 \times \frac{0}{1} + C = 0 + C = C \qquad \text{(C2.16)}$$

Equation C2.15 is the opposite action of Equation C2.7; Equation C2.16 is the opposite action of Equation C2.8.

C2.9 Examples of integration

Some examples (compare these with the common derivatives in C2.2; C, C_1, C_2, g, and a are constants) are as follows:

$$\int 1\, dt = \int 1 \times t^0 dt = t^1 \times \frac{1}{1} + C = t + C$$

$$\int 2\, dt = \int 2 \times t^0 dt = t^1 \times \frac{2}{1} + C = 2t + C$$

$$\int C_1\, dt = \int C_1 \times t^0 dt = t^1 \times \frac{C_1}{1} + C_2 = C_1 t + C_2$$

$$\int g\, dt = gt + C$$

$$\int 2t\, dt = \int 2 \times t^1 dt = t^2 \times \frac{2}{2} + C = t^2 + C$$

$$\int 2at\, dt = \int 2a \times t^1 dt = t^2 \times \frac{2a}{2} + C = at^2 + C$$

$$\int gt\, dt = \frac{1}{2} gt^2 + C$$

$$\int \frac{1}{2}at\, dt = \int \frac{1}{2}a \times t^1 dt = t^2 \times \frac{\frac{1}{2}a}{2} + C = \frac{1}{4}at^2 + C$$

$$\int x^2 dx = x^3 \times \frac{1}{3} + C = \frac{1}{3}x^3 + C$$

$$\int 3x^2 dx = x^3 \times \frac{3}{3} + C = x^3 + C$$

$$\int x^3 dx = x^4 \times \frac{1}{4} + C = \frac{1}{4}x^4 + C$$

$$\int \sqrt{x}\, dx = \int x^{\frac{1}{2}} dx = x^{\frac{3}{2}} \times \frac{1}{\frac{3}{2}} + C = \frac{2}{3}x^{\frac{3}{2}} + C = \frac{2}{3}x\sqrt{x} + C$$

$$\int \frac{1}{2\sqrt{x}} dx = \int \frac{1}{2}x^{-\frac{1}{2}} dx = x^{\frac{1}{2}} \times \frac{1}{\frac{1}{2}} + C = \sqrt{x} + C$$

$$\int \frac{1}{\sqrt{x}} dx = \int x^{-\frac{1}{2}} dx = x^{\frac{1}{2}} \times \frac{1}{\frac{1}{2}} + C = 2\sqrt{x} + C$$

$$\int 6x + 8\, dx = 3x^2 + 8x + C_1$$

$$\int 3x^2 + 8x + C_1\, dx = x^3 + 4x^2 + C_1 x + C_2$$

C2.10 The definite integral

Let us return to the falling water droplet.

By differentiation, we obtain v (m s^{-1}) from z (m) as follows:

$$v = \frac{dz}{dt} = \frac{d(\frac{1}{2}gt^2)}{dt} = gt \qquad \text{(C2.17)}$$

By integration, we obtain z (m) from v (m s^{-1}) as follows:

$$z = \int v\, dt = \int gt\, dt = \frac{1}{2}gt^2 + C \qquad \text{(C2.18)}$$

From these equations, we can calculate that ($g \approx 9.8$ m s^{-2}):

At $t = 0$ seconds, $v_0 = 0$ m s^{-1} and $z_0 = 0 + C$ m.

At $t = 1$ seconds, $v_1 = 9.8$ m s^{-1} and $z_1 = 4.9 + C$ m.

At $t = 2$ seconds, $v_2 = 19.6$ m s^{-1} and $z_2 = 19.6 + C$ m.

At $t = 3$ seconds, $v_3 = 29.4$ m s^{-1} and $z_3 = 44.1 + C$ m.

Between $t = 0$ and $t = 1$ seconds, the water droplet has fallen $4.9 + C - (0 + C) = 4.9$ m.

Between $t = 1$ and $t = 2$ seconds, the water droplet has fallen $19.6 + C - (4.9 + C) = 14.7$ m.

Between $t = 2$ and $t = 3$ seconds, the water droplet has fallen $44.1 + C - (19.6 + C) = 24.5$ m.

When we regard time increments $\Delta t = (t_2 - t_1)$, we lose C, and between $t = t_1$ and $t = t_2$ seconds the water droplet has fallen $z(t_2) - z(t_1) = \Delta z$ m. These values of Δz are the values of a **definite integral**:

$$\Delta z = \int_{t_1}^{t_2} v \, dt = \int_{t_1}^{t_2} gt \, dt = \frac{1}{2} gt_2^2 - \frac{1}{2} gt_1^2 \qquad (C2.19)$$

For instance, between $t_1 = 1$ and $t_2 = 2$ seconds, we can write the following:

$$\Delta z = \int_1^2 v \, dt = \int_1^2 gt \, dt = (\frac{1}{2} g \times 2^2) - (\frac{1}{2} g \times 1^2) = 19.6 - 4.9 = 14.7 \text{ m}$$

The value of this definite integral is the same as the area under the (linear) curve $v = gt$ between $t = 1$ and 2 seconds, as is evident from the above figure.

In general, for all smooth curved functions $f(t)$:

$$\text{area under the curve } f(t) \text{ between } t_1 \text{ and } t_2 = \int_{t_1}^{t_2} f(t) \, dt \qquad (C2.20)$$

The area under a smooth curve can be reshaped to rectangle areas for small time increments of Δt. In the above figure, this **rectangle method** has been applied to $v(t)$ for time increments Δt of 1 second.

C2.11   Further examples of integration

Since integration is the opposite action to differentiation (see Equations C2.10 and C2.11):

$$\int \frac{1}{r} \, dr = \ln r + C \qquad (C2.21)$$

$$\int e^x dx = e^x + C \qquad (C2.22)$$

The opposite of applying the **chain rule** (section C2.3) is as follows:

$$\int e^{\frac{x}{\lambda}}\, dx = \lambda\, e^{\frac{x}{\lambda}} + C$$

$$\int e^{\frac{-x}{\lambda}}\, dx = -\lambda\, e^{\frac{-x}{\lambda}} + C$$

Since mathematical differentiation and integration are opposing actions, the outcome of any such action can be checked afterwards by applying the opposite action, since this should deliver the original function.

C3 Quick reference to some differentiation rules

$f = f(x)$	$f' = \dfrac{df}{dx}$
C	0
x	1
x^2	$2x$
x^3	$3x^2$
$-\dfrac{1}{x} = -x^{-1}$	$x^{-2} = \dfrac{1}{x^2}$
x^n	nx^{n-1}
$\dfrac{x^{n+1}}{n+1}$	x^n

constant rule	$C f$	$C f'$
sum rule	$f \pm g$	$f' \pm g'$
product rule	$f g$	$f' g + f g'$
quotient rule	$\dfrac{f}{g}$	$\dfrac{f' g - f g'}{g^2}$
chain rule	$y(z(x))$	$\dfrac{dy}{dx} = \dfrac{dy}{dz}\dfrac{dz}{dx}$
	e^x	e^x
	$^e\log x = \ln x$	$\dfrac{1}{x}$

Mathematics toolboxes

M1 Steady groundwater flow in a confined aquifer between two parallel, fully penetrating canals with different water levels

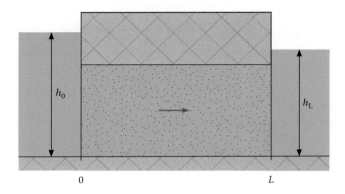

Darcy's law: $Q' = -KD\,(dh\,/\,dx)$.

Continuity: $Q' = $ constant $\Rightarrow - KD(dh\,/\,dx) = $ constant.

Because K and D are constants: $dh\,/\,dx = $ constant $= C_1 \Rightarrow h = C_1 x + C_2$.

Boundary conditions:

$$x = 0, \text{ then } \quad h = h_0$$

$$x = L, \text{ then } \quad h = h_L$$

This yields

$$h = \frac{h_L - h_0}{L} x + h_0$$

In summary, for steady groundwater flow in a confined aquifer between two parallel, fully penetrating canals with different water levels:

$$h = C_1 x + C_2$$

M2 Steady groundwater flow in an unconfined aquifer between two parallel, fully penetrating canals with different water levels

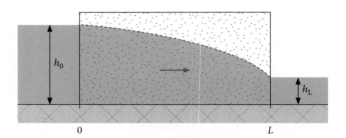

Darcy's law: $Q' = -Kh \, (dh / dx)$.

Continuity: $Q' = \text{constant} \Rightarrow -Kh(dh / dx) = \text{constant}$.

Because K is constant (but h is not):

$$h\frac{dh}{dx} = \text{constant} \Rightarrow \frac{d}{dx}\left(h\frac{dh}{dx} \right) = 0$$

Because

$$\frac{dh^2}{dx} = 2h\frac{dh}{dx}$$

(the chain rule

$$\frac{dy}{dx} = \frac{dy}{dh}\frac{dh}{dx}$$

with $y = h^2$ gives

$$\frac{dh^2}{dx} = \frac{dh^2}{dh}\frac{dh}{dx} = 2h\frac{dh}{dx}$$

– see the Conceptual Toolbox):

$$h\frac{dh}{dx} = \frac{1}{2}\frac{dh^2}{dx}$$

and thus

$$\frac{d}{dx}\left(h\frac{dh}{dx} \right) = \frac{d}{dx}\left(\frac{1}{2}\frac{dh^2}{dx} \right) = 0;\ \frac{1}{2}\frac{d^2h^2}{dx^2} = 0;\ \frac{d^2h^2}{dx^2} = 0;\ \frac{dh^2}{dx} = C_1;\ h^2 = C_1 x + C_2$$

Boundary conditions:

$$x = 0, \text{ then } \quad h = h_0$$

$$x = L, \text{ then } \quad h = h_L$$

This yields

$$h^2 = \frac{h_L^2 - h_0^2}{L}x + h_0^2$$

In summary, for steady groundwater flow in an unconfined aquifer between two parallel, fully penetrating canals with different water levels:

$$h^2 = C_1 x + C_2$$

M3 Steady groundwater flow in a leaky aquifer

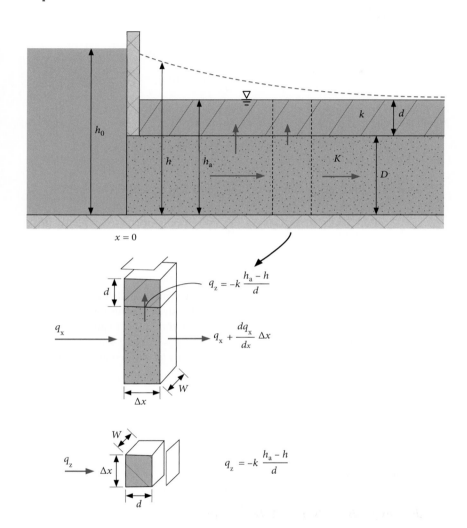

The water balance (continuity) for the volume of ground between the striped vertical lines is as follows:

$$Q'_{in, x} = Q'_{out, x} + Q'_{out, z}$$

The ingoing volume flux in the x direction, $Q'_{in, x} = q_x D$.

dq_x / dx is the rate in which q_x changes with x (due to water seepage in the polder). By multiplying this rate of change by the section Δx where the change takes place, we obtain the total change $[(dq_x / dx)\Delta x]$ over this section.

The outgoing volume flux in the x direction is as follows:

$$Q'_{out,\,x} = \left(q_x + \frac{dq_x}{dx} \Delta x \right) D$$

The outgoing volume flux in the vertical direction z is as follows:

$$Q'_{out,\,z} = q_z \Delta x = -k \frac{h_a - h}{d} \Delta x$$

Thus the water balance equation (continuity) is as follows:

$$q_x D = \left(q_x + \frac{dq_x}{dx} \Delta x \right) D - k \frac{h_a - h}{d} \Delta x \text{ with } q_x = -K \frac{dh}{dx} \text{ and } c = \frac{d}{k}$$

which gives

$$\frac{d\left(-K \frac{dh}{dx} \right)}{dx} = \frac{h_a - h}{Dc}; -K \frac{d^2 h}{dx^2} = \frac{h_a - h}{Dc}; \frac{d^2 h}{dx^2} = \frac{h - h_a}{KDc}$$

If we define λ, the leakage factor, as $\sqrt{KDc}$ (unit: length) and f as $h - h_a$, then

$$\frac{d^2 h}{dx^2} = \frac{f}{\lambda^2}$$

Because $h_a = $ constant,

$$\frac{df}{dx} = \frac{dh}{dx} \text{ and thus } \frac{d^2 h}{dx^2} = \frac{d^2 f}{dx^2} = \frac{f}{\lambda^2}$$

We want to solve $d^2 f / dx^2 = f / \lambda^2$ in order to determine the hydraulic head h of the leaky aquifer as a function of distance x from the dyke. Because there is seepage into the polder, we may expect the hydraulic head h in the aquifer to gradually diminish in the x direction. Possible solutions may thus be $f = e^{x/\lambda}$ or $f = e^{-x/\lambda}$.

If the solution is $f = e^{x/\lambda}$, then reasoning backwards by mathematical differentiation gives

$$\frac{df}{dx} = \frac{e^{\frac{x}{\lambda}}}{\lambda}$$

and, as a follow-up,

$$\frac{d^2 f}{dx^2} = \frac{e^{\frac{x}{\lambda}}}{\lambda^2} = \frac{f}{\lambda^2}$$

which is the equation to be solved.

If the solution is $f = e^{-x/\lambda}$, then by the same reasoning,

$$\frac{df}{dx} = \frac{e^{\frac{-x}{\lambda}}}{-\lambda}$$

and

$$\frac{d^2 f}{dx^2} = \frac{e^{\frac{-x}{\lambda}}}{\lambda^2} = \frac{f}{\lambda^2}$$

which is also the equation to be solved.

Since both answers are possible, the complete answer in mathematical terms is

$$f = h - h_a = C_1 e^{\frac{x}{\lambda}} + C_2 e^{\frac{-x}{\lambda}}$$

which we can rewrite as

$$h = h_a + C_1 e^{\frac{x}{\lambda}} + C_2 e^{\frac{-x}{\lambda}}$$

If the boundary conditions are:

$x = 0$, then $h = h_0 : h_0 = h_a + C_1 e^0 + C_2 e^0 = h_a + C_1 + C_2$; thus $C_1 + C_2 = h_0 - h_a$

$x = \infty$, then $h = h_a : h_\infty = h_a + C_1 e^\infty + C_2 e^{-\infty}$; thus $C_1 e^\infty + C_2 e^{-\infty} = h_\infty - h_a = h_a - h_a = 0$

then $C_1 e^\infty + C_2 e^{-\infty} = (C_1 \times \infty) + (C_2 \times 0) = 0 \Rightarrow C_1 = 0$ and thus $C_2 = h_0 - h_a$.

Substituting $C_1 = 0$ and $C_2 = h_0 - h_a$ in

$$h = h_a + C_1 e^{\frac{x}{\lambda}} + C_2 e^{\frac{-x}{\lambda}}$$

gives

$$h = h_a + \left(h_0 - h_a\right) e^{\frac{-x}{\lambda}}$$

In summary, for steady flow in a leaky aquifer:

$$h = h_a + C_1 e^{\frac{x}{\lambda}} + C_2 e^{\frac{-x}{\lambda}}$$

M4 Steady groundwater flow in a recharged, unconfined aquifer bordered by two parallel, fully penetrating canals with equal water levels

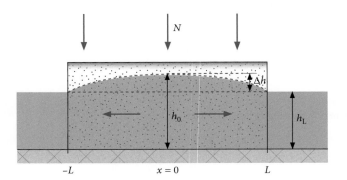

Darcy's law: $Q' = - Kh\,(dh\,/\,dx)$.
Continuity: $Q' = Nx$.
Combination of these two equations gives

$$Nx = - Kh\frac{dh}{dx}$$

Reworking this:

$$Nx\ dx = - Kh\ dh; \int Nx\ dx = \int -Kh\ dh; \frac{1}{2}Nx^2 = -\frac{1}{2}Kh^2 + C$$

$$h^2 = -\frac{N}{K}x^2 + C$$

Boundary conditions:

$$x = 0,\ \text{then}\quad h = h_0; h_0^2 = -\frac{N}{K}0^2 + C = C; C = h_0^2;\quad \text{thus } h^2 = -\frac{N}{K}x^2 + h_0^2$$

$$x = L,\quad \text{then } h = h_L; h_L^2 = -\frac{N}{K}L^2 + h_0^2;\quad \text{thus } L^2 = \frac{K}{N}\left(h_0^2 - h_L^2\right)$$

Δh is the convexity or differential head, which is the difference between the water table at the water divide and the water level in the canals: $\Delta h = h_0 - h_L$.
The mean depth of the aquifer $\overline{D}$ may be approximated by

$$\overline{D} = \frac{h_0 + h_L}{2}; h_0 + h_L = 2\overline{D}$$

$$L^2 = \frac{K}{N}\left(h_0^2 - h_L^2\right) = \frac{K}{N}\left(h_0 + h_L\right)\left(h_0 - h_L\right) = \frac{K}{N}2\overline{D}\Delta h = \frac{2K\overline{D}\Delta h}{N}; N = \frac{\Delta h}{\left(\dfrac{L^2}{2K\overline{D}}\right)}$$

In summary, for steady groundwater flow in a recharged, unconfined aquifer bordered by parallel, fully penetrating canals with equal water levels:

$$h^2 = -\frac{N}{K}x^2 + C$$

M5 Steady groundwater flow in a recharged, unconfined aquifer bordered by two parallel, fully penetrating streams with different water levels

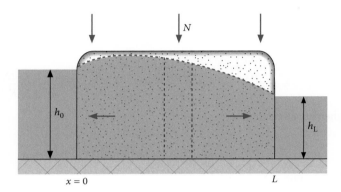

Darcy's law:

$$Q'_x = -K\left(h\frac{dh}{dx}\right)_x \; ; Q'_{x+\Delta x} = -K\left(h\frac{dh}{dx}\right)_{x+\Delta x}$$

The rate in which $h(dh/dx)$ changes with x can be expressed as

$$\frac{d\left(h\dfrac{dh}{dx}\right)}{dx}$$

By multiplying this rate of change by the section Δx where the change takes place, we obtain the total change

$$\frac{d\left(h\dfrac{dh}{dx}\right)}{dx}\Delta x$$

over this section (The same method was used with dq_x/dx in the derivation under M3). Thus,

$$Q'_{x+\Delta x} - Q'_x = -K\frac{d\left(h\dfrac{dh}{dx}\right)}{dx}\Delta x$$

Continuity: $Q'_{x+\Delta x} - Q'_x = N\Delta x$.
Combination of these two equations gives

$$-K\frac{d\left(h\dfrac{dh}{dx}\right)}{dx} = N$$

Because

$$\frac{dh^2}{dx} = 2h\frac{dh}{dx}$$

(the chain rule

$$\frac{dy}{dx} = \frac{dy}{dh}\frac{dh}{dx}$$

with $y = h^2$ gives

$$\frac{dh^2}{dx} = \frac{dh^2}{dh}\frac{dh}{dx} = 2h\frac{dh}{dx}$$

– see C2.3):

$$h\frac{dh}{dx} = \frac{1}{2}\frac{dh^2}{dx}$$

and thus

$$-K\frac{d\left(h\dfrac{dh}{dx}\right)}{dx} = -K\frac{d\left(\dfrac{1}{2}\dfrac{dh^2}{dx}\right)}{dx} = -\frac{1}{2}K\frac{d^2h^2}{dx^2} = N; \frac{d^2h^2}{dx^2} = \frac{-2N}{K}$$

By mathematical integration,

$$\frac{dh^2}{dx} = -2\frac{N}{K}x + C_1 \text{ and } h^2 = -\frac{N}{K}x^2 + C_1x + C_2$$

Boundary conditions:

$$x = 0, \quad \text{then } h = h_0 \text{ and } h_0^2 = C_2$$

$$x = L, \quad \text{then } h = h_L \text{ and } h_L^2 = -\frac{N}{K}L^2 + C_1L + h_0^2; h_L^2 - h_0^2 + \frac{N}{K}L^2 = C_1L$$

and thus

$$C_1 = \frac{h_L^2 - h_0^2}{L} + \frac{N}{K}L$$

$$h^2 = -\frac{N}{K}x^2 + C_1x + C_2 \text{ with } C_1 = \frac{h_L^2 - h_0^2}{L} + \frac{N}{K}L \text{ and } C_2 = h_0^2$$

yields

$$h^2 = -\frac{N}{K}x^2 + \frac{(h_L^2 - h_0^2)}{L}x + \frac{NL}{K}x + h_0^2$$

In summary, for steady groundwater flow in a recharged, unconfined aquifer bordered by two parallel, fully penetrating streams with different water levels:

$$h^2 = -\frac{N}{K}x^2 + C_1x + C_2$$

For $N = 0$, this equation reduces to:

$$h^2 = C_1x + C_2$$

M6 Radial-symmetric, steady groundwater flow in a confined aquifer to a fully penetrating well in the centre of a circular island

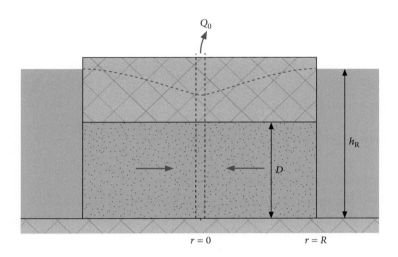

In the steady groundwater flow cases M1 to M5, the hydraulic head h decreases with increasing positive x in the direction of flow, linking $Q' > 0$ with $dh / dx < 0$. Because of this, a minus sign is present in Darcy's law. However, in this radial-symmetric steady groundwater flow case, the hydraulic head h decreases with decreasing r in the direction of flow ($dh / dr > 0$). As water is pumped up, thus flowing upwards (in a positive z direction) in a pumping well, it may be preferable to assign a positive sign to this volume flux or discharge Q_0 at the centre of the pumping well ($Q_0 > 0$). Because then both Q_0 and dh / dr are positive in the direction of flow, the minus sign must be dropped in this radial-symmetric version of Darcy's law. Also, in Darcy's law describing the volume flux density q_r, the minus sign should then be dropped. Thus:

$$q_r = K \frac{dh}{dr}$$

The volume flux or discharge Q_0 from a fully penetrating well in a confined aquifer in the centre of a circular island is as follows:

$$Q_0 = q_r \, 2\pi r D$$

Combination of these two equations gives

$$Q_0 = K \frac{dh}{dr} 2\pi r D \; ; \frac{dh}{dr} = \frac{Q_0}{2\pi KD} \frac{1}{r} \; ; \int \frac{dh}{dr} \, dr = \int \frac{Q_0}{2\pi KD} \frac{1}{r} \, dr$$

$$h = \frac{Q_0}{2\pi KD} \ln r + C$$

The boundary condition $r = R$ gives $h = h_R$ and

$$h_R = \frac{Q_0}{2\pi KD} \ln R + C$$

Hence

$$C = h_R - \frac{Q_0}{2\pi KD} \ln R$$

This yields

$$h = \frac{Q_0}{2\pi KD} \ln r + h_R - \frac{Q_0}{2\pi KD} \ln R; \; h = h_R + \frac{Q_0}{2\pi KD} \ln \frac{r}{R};$$

$$h - h_R = \frac{Q_0}{2\pi KD} \ln \frac{r}{R} \; \text{for } r_w \le r \le R$$

where r_w is the radius of the pumped well.

Because $r \le R$: $\frac{r}{R} \le 1$; $\ln \frac{r}{R} \le 0$; $2\pi KD > 0$

In a pumping well: $h < h_R \Rightarrow h - h_R < 0$ and thus $Q_0 > 0$

In a recharge well: $h > h_R \Rightarrow h - h_R > 0$ and thus $Q_0 < 0$

In summary, for radial-symmetric, steady groundwater flow in a confined aquifer to a fully penetrating well in the centre of a circular island:

$$h = h_R + \frac{Q_0}{2\pi KD} \ln \frac{r}{R} \; \text{for } r_w \le r \le R$$

Another possible solution of

$$\frac{dh}{dr} = \frac{Q_0}{2\pi KD} \frac{1}{r}$$

as derived earlier is (h_1 = hydraulic head at radial distance r_1 from the well; h_2 = hydraulic head at radial distance r_2 from the well):

$$\int_{h_1}^{h_2} \frac{dh}{dr} dr = \int_{r_1}^{r_2} \frac{Q_0}{2\pi KD} \frac{1}{r} dr \Rightarrow [h]_{h_1}^{h_2} = \frac{Q_0}{2\pi KD} [\ln r]_{r_1}^{r_2} \Rightarrow h_2 - h_1 = \frac{Q_0}{2\pi KD} (\ln r_2 - \ln r_1)$$

$$\Rightarrow h_2 - h_1 = \frac{Q_0}{2\pi KD} \ln \frac{r_2}{r_1} \; \text{or } h_2 - h_1 = \frac{Q_0}{2\pi T} \ln \frac{r_2}{r_1}$$

M7 Radial-symmetric, steady groundwater flow in an unconfined aquifer to a fully penetrating well in the centre of a circular island

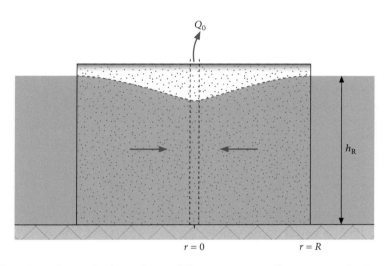

$r = 0$ $r = R$

The volume flux or discharge from a fully penetrating well in an unconfined aquifer in the centre of a circular island is as follows:

$$Q_0 = q_r \, 2\pi \, rh$$

Darcy's law: $q_r = K \, (dh/dr)$.
Combination of these two equations gives

$$Q_0 = K \frac{dh}{dr} 2\pi rh \; ; \frac{dh}{dr} = \frac{Q_0}{2\pi Khr} \; ; \frac{Q_0}{r} \, dr = 2\pi \, Kh \, dh$$

where h is not constant:

$$\int \frac{Q_0}{r} \, dr = \int 2\pi \, Kh \, dh \; ; \; Q_0 \ln r = \pi Kh^2 + C \; ; \; h^2 = \frac{Q_0}{\pi K} \, \ln r + C$$

The boundary condition $r = R$ gives $h = h_R$ and

$$h_R^2 = \frac{Q_0}{\pi K} \ln R + C \; ; \; C = h_R^2 - \frac{Q_0}{\pi K} \ln R$$

This yields

$$h^2 = \frac{Q_0}{\pi K} \ln r + h_R^2 - \frac{Q_0}{\pi K} \ln R \; ; \; h^2 = h_R^2 + \frac{Q_0}{\pi K} \ln \frac{r}{R}$$

$$h^2 - h_R^2 = \frac{Q_0}{\pi K} \, \ln \frac{r}{R} \quad \text{for} \quad r_w \leq r \leq R$$

Because $r \leq R$: $\dfrac{r}{R} \leq 1$; $\ln \dfrac{r}{R} \leq 0$; $\pi K > 0$.

In a pumping well: $h^2 < h_R^2$; $\Rightarrow h^2 - h_R^2 < 0$ and thus $Q_0 > 0$.

In a recharge well: $h^2 > h_R^2$; $\Rightarrow h^2 - h_R^2 > 0$ and thus $Q_0 < 0$.

In summary, for radial-symmetric, steady groundwater flow in an unconfined aquifer to a fully penetrating well in the centre of a circular island:

$$h^2 = h_R^2 + \frac{Q_0}{\pi K}\ln\frac{r}{R} \ \text{ for } \ r_w \leq r \leq R$$

M8 Derivation of the Richards equation

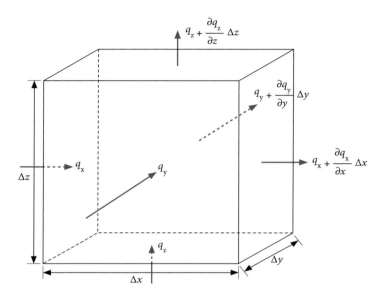

(dq_x / dx is the rate in which q_x changes with x; by multiplying this rate of change by the section Δx where the change takes place, we obtain the total change $[(dq_x / dx)\Delta x]$ over this section: the same reasoning is applied to dq_y / dy and dq_z / dz in the y and z directions.)

Inflow:

$$Q_{in} = q_x \Delta y \Delta z + q_y \Delta x \Delta z + q_z \Delta x \Delta y$$

Outflow:

$$Q_{out} = \left(q_x + \frac{\partial q_x}{\partial x} \Delta x \right) \Delta y \Delta z + \left(q_y + \frac{\partial q_y}{\partial y} \Delta y \right) \Delta x \Delta z + \left(q_z + \frac{\partial q_z}{\partial z} \Delta z \right) \Delta x \Delta y$$

Net inflow:

$$Q_{in} - Q_{out} = -\left(\frac{\partial q_x}{\partial x} + \frac{\partial q_y}{\partial y} + \frac{\partial q_z}{\partial z} \right) \Delta x \Delta y \Delta z$$

Change of storage:

$$\frac{\partial \theta}{\partial t} \Delta x \, \Delta y \, \Delta z$$

Continuity: net inflow = change of storage:

$$-\left(\frac{\partial q_x}{\partial x} + \frac{\partial q_y}{\partial y} + \frac{\partial q_z}{\partial z} \right) \Delta x \Delta y \Delta z = \frac{\partial \theta}{\partial t} \Delta x \Delta y \Delta z$$

Dividing both sides by $\Delta x \Delta y \Delta z$ yields the **continuity equation** (Equation 4.14):

$$\frac{\partial q_x}{\partial x} + \frac{\partial q_y}{\partial y} + \frac{\partial q_z}{\partial z} = -\frac{\partial \theta}{\partial t}$$

The **Darcy–Buckingham equations** for the x, y, and z directions can be written as follows:

$$q_x = -K(\psi)\frac{\partial h}{\partial x}, \quad q_y = -K(\psi)\frac{\partial h}{\partial y} \quad \text{and} \quad q_z = -K(\psi)\frac{\partial h}{\partial z}$$

Inserting these equations in the continuity equation yields the **Richards equation**:

$$\frac{\partial}{\partial x}\left(K(\psi)\frac{\partial h}{\partial x}\right) + \frac{\partial}{\partial y}\left(K(\psi)\frac{\partial h}{\partial y}\right) + \frac{\partial}{\partial z}\left(K(\psi)\frac{\partial h}{\partial z}\right) = \frac{\partial \theta}{\partial t} \qquad \text{(Equation 4.15)}$$

M9 Other forms of the Richards equation

The Richards equation (4.15) can be written, presented here for the z direction (1-D), as follows:

$$\frac{\partial \theta}{\partial t} = \frac{\partial}{\partial z}\left(K(\psi)\frac{\partial h}{\partial z}\right) = \frac{\partial}{\partial z}\left(K(\psi)\frac{\partial(\psi+z)}{\partial z}\right) = \frac{\partial}{\partial z}\left(K(\psi)\left(\frac{\partial \psi}{\partial z}+1\right)\right)$$

$$\Rightarrow \frac{\partial \theta}{\partial t} = \frac{\partial}{\partial z}\left(K(\psi)\frac{\partial \psi}{\partial z}\right) + \frac{\partial K(\psi)}{\partial z}$$

– this is the **mixed form** of the Richards equation.

The specific water capacity $C(\psi) = \partial \theta / \partial \psi$ can be determined from the soil moisture characteristic (no hysteresis): using the chain rule the left-hand side of the mixed form of the Richards equation can be rewritten as follows:

$$\frac{\partial \theta}{\partial t} = \frac{\partial \theta}{\partial \psi}\frac{\partial \psi}{\partial t} = C(\psi)\frac{\partial \psi}{\partial t} \Rightarrow C(\psi)\frac{\partial \psi}{\partial t} = \frac{\partial}{\partial z}\left(K(\psi)\frac{\partial \psi}{\partial z}\right) + \frac{\partial K(\psi)}{\partial z}$$

– this is the **pressure head based form**, **ψ-based form**, or **capacitance form** of the Richards equation.

The hydraulic diffusivity (or soil water diffusivity) is defined as

$$D(\theta) = K(\psi)\frac{\partial \psi}{\partial \theta} \quad \text{(no hysteresis)}$$

Using the chain rule the first part of the right-hand side of the above mixed form of the Richards equation can be rewritten as follows:

$$\frac{\partial}{\partial z}\left(K(\psi)\frac{\partial \psi}{\partial z}\right) = \frac{\partial}{\partial z}\left(K(\psi)\frac{\partial \psi}{\partial \theta}\frac{\partial \theta}{\partial z}\right) = \frac{\partial}{\partial z}\left(D(\theta)\frac{\partial \theta}{\partial z}\right)$$

Inserting this in the above mixed form of the Richard equation yields (of course $K(\psi) = K(\theta)$):

$$\frac{\partial \theta}{\partial t} = \frac{\partial}{\partial z}\left(D(\theta)\frac{\partial \theta}{\partial z}\right) + \frac{\partial K(\theta)}{\partial z}$$

– this is the **moisture content based form**, **θ-based form**, or **diffusivity form** of the Richards equation.

The (1-D) Richards equation is a highly non-linear partial differential equation; no exact solutions to it are known for general boundary and initial conditions. Because flow domains usually are heterogeneous and do not have a simple geometry, the Richards equation can best be solved using numerical methods, whereby the flow domain is divided into grid cells or elements (see section 3.15, 'Rationale').

In numerical modelling, applying a specific form of the Richards equation as presented above has distinct advantages and disadvantages. Numerical methods solving the **ψ-based form** of the Richards equation have the advantage of being applicable to near-saturated and unsaturated conditions, but may have slowly convergent solutions and may experience important mass balance errors. On the other hand, numerical methods solving the **θ-based form** generally have rapidly convergent solutions and an improved mass balance, but are strictly limited to unsaturated conditions and homogeneous media (because θ is discontinuous at layer interfaces; see Figure 4.34a).

The **mixed form** of the Richards equation is most robust with respect to the mass balance, but this alone does not always guarantee acceptable solutions.

Celia *et al.* (1990) proposed a **mass-conservative solution** of the Richards equation based on the mixed form, and this numerical solution is to be preferred for hydrology applications at and near the land surface (Troch 2008).

Nowadays, **temporal and spatial adaptive approaches** to the Richards equation are also applied that use alternative forms of the Richards equation – where, for instance, the θ-based form may be used when soil moisture content is low and the ψ-based form or mixed form when soil moisture content is high (Jendele 2002). Similarly, when domains are variably saturated, an 'extended' formulation of the Richards equation including, respectively, the part of the domain that is saturated or unsaturated, may be used as part of the solution of the problem.

A well-known model for modelling water flow in the unsaturated zone is HYDRUS (Šimůnek and Van Genuchten 2008).

M10 Open-channel flow

Open-channel flow is channel flow with a free water surface, such as flow in a river or in a partially full pipe (Chow 1988). Primary hydraulic equations for steady, uniform, open-channel flow are:

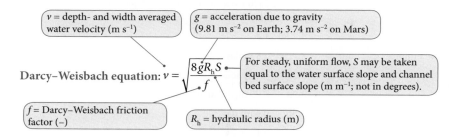

v = depth- and width averaged water velocity (m s^{-1})

g = acceleration due to gravity (9.81 m s^{-2} on Earth; 3.74 m s^{-2} on Mars)

Darcy–Weisbach equation: $v = \sqrt{\dfrac{8 g R_h S}{f}}$

For steady, uniform flow, S may be taken equal to the water surface slope and channel bed surface slope (m m^{-1}; not in degrees).

f = Darcy–Weisbach friction factor (–)

R_h = hydraulic radius (m)

The Darcy–Weisbach equation is named after French engineer Henri P.G. Darcy (1803–1858), who we already know from Darcy's law, and German mathematician and engineer Julius L. Weisbach (1806–1871), who obtained the same result.

A similar equation, obtained earlier, is the following:

C = Chézy roughness coefficient (m$^{0.5}$ s^{-1})

Chézy equation: $v = C\sqrt{R_h S}$

The Chézy equation is named after French hydraulic engineer Antoine de Chézy (1718–1798).

Another well-known hydraulic equation for steady, uniform, open-channel flow is as follows:

Manning equation: $v = \dfrac{R_h^{\frac{2}{3}} S^{\frac{1}{2}}}{n}$

n = Manning roughness coefficient

The Manning equation is named after Irish engineer Robert Manning (1816–1897). The Manning equation is valid for fully turbulent flows ($n^6 \sqrt{R_h S} \geq 1.1 \times 10^{-13}$) (Henderson 1966, in Chow 1988), in which the Darcy–Weisbach friction factor f is independent of the Reynolds number Re (Chow 1988).

The Chézy (C) and Manning (n) roughness coefficients are related to the Darcy–Weisbach friction factor f in the following way:

$$C = \sqrt{\frac{8g}{f}} \text{ and } n = R_h^{\frac{1}{6}} \sqrt{\frac{f}{8g}}$$

The Darcy–Weisbach friction factor f is a complicated outcome of the interactions between flow, flow turbulence, the added resistance of transported sediment, and – most importantly – roughness elements such as grains on the channel bed and bed forms (Kleinhans 2005).

Answers to the exercises

Chapter 1

1.4.1a yearly averages $\Rightarrow \overline{\Delta S} = 0 \Rightarrow \overline{P} = \overline{Q} + \overline{G} + \overline{E}$ ($\overline{G}$ = average groundwater flow to the sea)

1.4.1b 500 mm year^{-1}; 37.5×10^8 m^3 year^{-1}

1.4.2 $P = Q + \dfrac{\Delta S}{\Delta t}$; 30 mm hour^{-1} in 40 minutes $= \dfrac{2}{3}$ hour

$\Rightarrow P = \dfrac{2}{3} \times 30 = 20$ mm in 40 minutes

$Q = \dfrac{150 \text{ m}^3}{10^4 \text{ m}^2} = 15 \times 10^{-3}$ m $= 15$ mm in 40 minutes

5 mm; 5×10^{-3} m $\times 10^4$ m$^2 = 50$ m^3

Chapter 3

3.7.1a water flow is perpendicular to both screens

3.7.1b 0.24 m^3 day^{-1} (distance of flow = 50 cm)

3.7.1c 1 hour and 12 minutes

3.7.2 2100

3.7.3 0.275 m day^{-1}; 0.881 m day^{-1}

B3.2 558 mm year^{-1}

3.7.4 6.25×10^{-2} m day^{-1}; 1.79×10^{-2} m day^{-1}

3.7.5a

3.7.5b 0 cm if the bottom of the sample is taken as the reference level

3.7.5c 7 cm if the bottom of the sample is taken as the reference level

3.7.5d 1.4

3.7.5e 4.8 m day^{-1}

3.7.5f sandy soil of the Luxembourg sandstone

3.9a 9.75 m; 9.5 m; 9.25 m

3.9b 0.4 m^2 day^{-1}

3.9c 0.15 m day^{-1}

3.9d 667 days

3.10.2a 9.75 m; 9.5 m; 9.25 m

3.10.2b 0.7 m^2 day^{-1}

3.10.2c 0.1 m day^{-1} in the upper layer; 0.05 m day^{-1} in the lower layer ($q_u > q_l$)

3.10.2d 0.25 m day^{-1} in the upper layer; 0.5 m day^{-1} in the lower layer ($v_{e,u} < v_{e,l}$)

3.10.2e 200 days (by the fastest route)

3.10.3 horizontal, 16.5 m day^{-1}; vertical, 3.03 m day^{-1}

3.10.4a 10^{-3} m day^{-1}; 8000 days

3.10.4b 2×10^{-3} m day^{-1}; 2000 days

3.10.4c the effective velocity increases with a factor of 2; the residence time decreases with a factor of 4

3.10.5a 1500 days

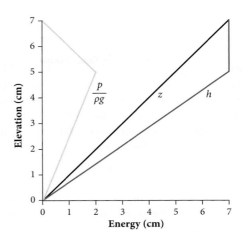

3.10.5b continuity $\Rightarrow Q_1 = Q_2$; the area perpendicular to the water flow is constant

$\Rightarrow q_1 = q_2; -k_1 i_1 = -k_2 i_2$

$-5 \times 10^{-3} \dfrac{h_{2.5} - 18.1}{2.5 - 0} = -10^{-2} \dfrac{17.5 - h_{2.5}}{12.5 - 2.5} \Rightarrow h_{2.5} = 17.9$ m

(0.4 m above the free water surface of the lake)

3.10.5c v_e for layer 1 = 4×10^{-3} m day^{-1}; v_e for layer 2 = 2×10^{-3} m day^{-1}; residence time = 5625 days (15.4 years)

3.10.6 $c' = c_1' + c_2' + c_3' \Rightarrow \dfrac{L}{K} = \dfrac{L_1}{K_1} + \dfrac{L_2}{K_2} + \dfrac{L_3}{K_3}$

$\Rightarrow \dfrac{500}{K} = \dfrac{200}{5} + \dfrac{200}{10} + \dfrac{100}{5}$

$\Rightarrow$ substitute hydraulic conductivity $K = 6.25$ m day^{-1};

$i = \dfrac{-0.5}{500} = -1 \times 10^{-3}$

$Q' = Q_1' = Q_2' = Q_3' \Rightarrow -KDi = -K_1 Di_1 = -K_2 Di_2 = -K_3 Di_3$

$D =$ constant $\Rightarrow -Ki = -K_1 i_1 = -K_2 i_2 = -K_3 i_3$

3.10.6a 0.15 m^2 day^{-1}

3.10.6b 32 000 days (87.7 years)

3.10.6c -1.25×10^{-3}; -0.625×10^{-3}; -1.25×10^{-3}

3.10.6d i and K are inversely proportional:

$i_1 : i_2 : i_3 = \dfrac{1}{K_1} : \dfrac{1}{K_2} : \dfrac{1}{K_3} = 2 : 1 : 2$

3.11.1 $h_A = 3 - 0.66 = 2.34$ m above mean sea level; $h_B = 3 - 0.54 = 2.46$ m above mean sea level; $h_C = 3 - 0.86 = 2.14$ m above mean sea level

Where is $h_A = 2.34$ m above mean sea level located on the line BC? Alternatively:

Gradients are vectors

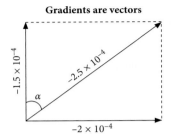

3.11.1a -2.5×10^{-4}

3.11.1b $\alpha = 53°$

3.11.2a $\dfrac{2}{5} \times 600$ mm $+ \dfrac{3}{5} \times 420$ mm $= 492$ mm

(weighted average), or alternatively:

$\dfrac{(600 \times 10^{-3}\,\text{m} \times 2 \times 10^6\,\text{m}^2) + (420 \times 10^{-3}\,\text{m} \times 3 \times 10^6\,\text{m}^2)}{5 \times 10^6\,\text{m}^2} =$

0.492 m = 492 mm

3.11.2b $P +$ seepage = pump $+ E_a + \Delta S$; $\Delta S = 0 \Rightarrow$ seepage = 142 mm year^{-1} = 0.4 mm day^{-1}

3.11.2c $P +$ seepage = pump $+ E_a + \Delta S$; ΔS open water =

$\dfrac{2}{5} \times 200$ mm = 80 mm; storage coefficient = 0.4 $\Rightarrow$ 4 mm of added water gives a 10 mm rise of the water table and therefore a 200 mm rise of the water table is caused by an addition of

80 mm of water $\Rightarrow \Delta S_{\text{land}} = \dfrac{3}{5} \times 80$ mm = 48 mm $\Rightarrow \Delta S_{\text{total}}$

= 128 mm $\Rightarrow$ seepage = 270 mm year^{-1} = 0.7 mm day^{-1}

3.12 the groundwater contribution to the river is from both sides (r and l): $|Q_r| = |Q_l| = KAi = KDLi$

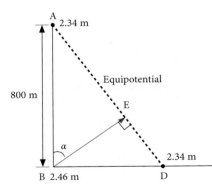

$i_{BA} = \dfrac{2.34 - 2.46}{800} = -1.5 \times 10^{-4};$

$i_{BC} = \dfrac{2.14 - 2.46}{1600} = -2 \times 10^{-4}$

$\Rightarrow Q_B - Q_A = |Q_r| + |Q_l| = 2KAi = 2KDLi \Rightarrow KD = 4320$ m^2 day^{-1}

3.15.1.1 $h^2 = C_1 x + C_2$

$x = 0 \Rightarrow h_0 = 6 \Rightarrow 6^2 = C_1 \times 0 + C_2 \Rightarrow C_2 = 36$

$x = 25 \Rightarrow h_{25} = 3 \Rightarrow 3^2 = C_1 \times 25 + C_2 =$
$C_1 \times 25 + 36 \Rightarrow C_1 = -1.08$

$h^2 = -1.08x + 36$

3.15.1.1a 5.53 m; 5.02 m; 4.45 m; 3.79 m

3.15.1.1c $Q' = -Kh\dfrac{dh}{dx}; h^2 = C_1 x + C_2$

$\Rightarrow h = (C_1 x + C_2)^{\frac{1}{2}}$

$\Rightarrow \dfrac{dh}{dx} = \dfrac{1}{2}(C_1 x + C_2)^{-\frac{1}{2}} \times C_1$

$\Rightarrow h\dfrac{dh}{dx} = (C_1 x + C_2)^{\frac{1}{2}} \times \dfrac{1}{2}(C_1 x + C_2)^{-\frac{1}{2}} \times C_1 = \dfrac{1}{2}C_1$

$Q' = -\dfrac{1}{2}C_1 K = 0.07 \text{ m}^2 \text{ day}^{-1}$; alternatively: $h^2 = C_1 x + C_2$

$\Rightarrow \dfrac{dh^2}{dx} = C_1$; $\dfrac{dy}{dx} = \dfrac{dy}{dh}\dfrac{dh}{dx}$ with $y = h^2$ gives

$\dfrac{dh^2}{dx} = \dfrac{dh^2}{dh}\dfrac{dh}{dx} = 2h\dfrac{dh}{dx} = C_1 \Rightarrow h\dfrac{dh}{dx} = \dfrac{1}{2}C_1$;

$Q' = -\dfrac{1}{2}C_1 K = 0.07 \text{ m}^2 \text{ day}^{-1}$

3.15.1.2 $h^2 = C_1 x + C_2$

$x = 10 \Rightarrow h_{10} = 6.75 \Rightarrow C_2 = -10 \times C_1 + 6.75^2$

$x = 75 \Rightarrow h_{75} = 6.25 \Rightarrow C_1 = -0.1 \Rightarrow C_2 = 46.5625$

$h^2 = -0.1x + 46.5625$

check: $x = 10 \Rightarrow h^2 = 45.5625 \Rightarrow h = 6.75$

check: $x = 75 \Rightarrow h^2 = -7.5 + 46.5625 = 39.0625 \Rightarrow h = 6.25$

$x = 0 \Rightarrow h^2 = 46.5625 \Rightarrow h = 6.82$ m

$x = 100 \Rightarrow h^2 = 36.5625 \Rightarrow h = 6.05$ m

3.15.1.3a for the left compartment: $h^2 = C_1 x + C_2$;
$x = 0 \Rightarrow h_0 = 6 \Rightarrow C_2 = 36$

$x = 40 \Rightarrow h_{40} = h_{40} \Rightarrow C_1 = \dfrac{h_{40}^2 - 36}{40} \Rightarrow h^2 = \dfrac{h_{40}^2 - 36}{40}x + 36$

for the right compartment: $h^2 = C_3 x + C_4$;

$x = 40 \Rightarrow h_{40} = h_{40} \Rightarrow C_4 = h_{40}^2 - 40C_3$

$x = 200 \Rightarrow h_{200} = 3 \Rightarrow C_3 = \dfrac{9 - h_{40}^2}{40}$

$\Rightarrow h^2 = \dfrac{9 - h_{40}^2}{160}x + h_{40}^2 - 40\left(\dfrac{9 - h_{40}^2}{160}\right)$

continuity: $Q'_{\text{left}} = Q'_{\text{right}} \Rightarrow -K_1\left(h\dfrac{dh}{dx}\right)_{\text{left}} = -K_2\left(h\dfrac{dh}{dx}\right)_{\text{right}}$

$\Rightarrow h_{40} = 4.09$ m

3.15.1.3c 0.24 m^2 day^{-1}

3.15.1.3d yes; $h_{40} = \sqrt{\dfrac{144K_1 + 9K_2}{4K_1 + K_2}}$: h_{40} is a function of

both K_1 and K_2; a lower value for K leads to a steeper hydraulic gradient i

3.15.1.4a for the left compartment: $h = C_1 x + C_2$;
$x = 0 \Rightarrow h_0 = 10 \Rightarrow C_2 = 10$

$x = ? \Rightarrow h_? = 8 \Rightarrow C_1 = \dfrac{-2}{?} \Rightarrow h = \dfrac{-2}{?}x + 10$

for the right compartment: $h^2 = C_3 x + C_4$;

$x = ? \Rightarrow h_? = 8 \Rightarrow C_4 = 64 - ?C_3$

$x = 100 \Rightarrow h_{100} = 6 \Rightarrow C_3 = \dfrac{-28}{100 - ?}$

$\Rightarrow h^2 = \dfrac{-28}{100 - ?}x + \left(64 - ?\left(\dfrac{-28}{100 - ?}\right)\right)$

continuity: $Q'_{\text{left}} = Q'_{\text{right}} \Rightarrow -KD\left(\dfrac{dh}{dx}\right)_{\text{left}} = -K\left(h\dfrac{dh}{dx}\right)_{\text{right}}$

$\Rightarrow ? = 53.33$ m (irrespective of the value of the hydraulic conductivity K)

3.15.1.4b 3.33 m day^{-1}

3.15.2.1 $h = h_a + C_1 e^{\frac{x}{\lambda}} + C_2 e^{\frac{-x}{\lambda}}$; $c = \dfrac{d}{k}$ = 1000 days and

$\lambda = \sqrt{KDc}$ = 800 m

$x = 0 \Rightarrow h_0 = 44 \Rightarrow C_2 = 2 - C_1$

$x = \infty \Rightarrow h_\infty = h_a = 42 \Rightarrow C_1 = 0 \Rightarrow C_2 = 2$; $h = 42 + 2 e^{\frac{-x}{800}}$

3.15.2.1a 43.98 m; 43.76 m; 43.46 m; 43.07 m; 42.57 m; 42.31 m

3.15.2.1c

$q_z = -k\dfrac{h_a - h}{d} = \dfrac{h - h_a}{c} = \dfrac{42 + 2 e^{\frac{-x}{800}} - 42}{1000} = 0.002 e^{\frac{-x}{800}}$

$\Rightarrow$1.98 mm day^{-1}; 1.76 mm day^{-1}; 1.46 mm day^{-1}; 1.07 mm day^{-1}; 0.57 mm day^{-1}; 0.31 mm day^{-1}

3.15.2.1d $Q'_z = Q'_{x=0} = -KD\left(\dfrac{dh}{dx}\right)_{x=0}$

$= -KD \times \left(\dfrac{-1}{\lambda} \times 2 e^{\frac{-x}{\lambda}}\right)_{x=0} = 1.6 \text{ m}^2 \text{ day}^{-1}$

$Q'_z = \int_0^\infty q_z \, dx = \int_0^\infty 0.002 e^{\frac{-x}{\lambda}} dx = \left[-\lambda \times 0.002 e^{\frac{-x}{\lambda}}\right]_0^\infty = 0.002\lambda$

$= 1.6 \text{ m}^2 \text{ day}^{-1}$

$Q_z = 1.6 \text{ m}^2 \text{ day}^{-1} \times 1000 \text{ m} = 1600 \text{ m}^3 \text{ day}^{-1}$

3.15.2.1e

$$Q_z' = Q_{x=0}' - Q_{x=100}' = -KD\left(\frac{dh}{dx}\right)_{x=0} - \left(-KD\left(\frac{dh}{dx}\right)_{x=100}\right)$$

$$= 1.6 - 1.412 = 0.188 \text{ m}^2 \text{ day}^{-1}$$

$$Q_z' = \int_0^{100} q_z \, dx = \int_0^{100} 0.002 \, e^{\frac{-x}{\lambda}} \, dx = \left[-\lambda \times 0.002 \, e^{\frac{-x}{\lambda}}\right]_0^{100}$$

$$= -\lambda \times 0.002 \, e^{\frac{-100}{\lambda}} + 0.002\lambda = -1.412 + 1.6 = 0.188 \text{ m}^2 \text{ day}^{-1}$$

$$Q_z = 0.188 \text{ m}^2 \text{ day}^{-1} \times 1000 \text{ m} = 188 \text{ m}^3 \text{ day}^{-1}$$

3.15.2.2a $h = h_a + C_1 e^{\frac{x}{\lambda}} + C_2 e^{\frac{-x}{\lambda}}$; $c = \dfrac{d}{k} = 1500$ days and

$$\lambda = \sqrt{KDc} = 600 \text{ m}$$

$$x = 0 \Rightarrow h_0 = 16 \Rightarrow C_2 = 3 - C_1; \ x = 1200 \Rightarrow h_{1200} = 15$$

$$\Rightarrow C_1 = \frac{2 - 3e^{-2}}{e^2 - e^{-2}}$$

$$C_1 = 0.220 \Rightarrow C_2 = 2.780 \Rightarrow h = 13 + 0.220 \, e^{\frac{x}{600}} + 2.780 \, e^{\frac{-x}{600}}$$

3.15.2.2b $Q_z' = |Q_{x=0}'| + |Q_{x=1200}'| =$

$$\left|-KD\left(\frac{dh}{dx}\right)\right|_{x=0} + \left|-KD\left(\frac{dh}{dx}\right)\right|_{x=1200} \quad \text{or}$$

$$Q_z' = \int_0^{1200} q_z \, dx = \int_0^{1200} -k\frac{h_a - h}{d} \, dx = 1.52 \text{ m}^2 \text{ day}^{-1}$$

3.15.2.2c $Q_x' = 0$ or $\dfrac{dh}{dx} = 0$

$$\Rightarrow \frac{0.220}{600} \, e^{\frac{x}{600}} + \frac{2.780}{-600} \, e^{\frac{-x}{600}} = 0$$

$$\Rightarrow \frac{0.220}{600} \, e^{\frac{x}{600}} = \frac{2.780}{600} \, e^{\frac{-x}{600}} \Rightarrow 0.220 \, e^{\frac{x}{600}} = 2.780 \, e^{\frac{-x}{600}}$$

$$\ln\left(0.220 \, e^{\frac{x}{600}}\right) = \ln\left(2.780 \, e^{\frac{-x}{600}}\right)$$

$$\Rightarrow \ln 0.220 + \ln\left(e^{\frac{x}{600}}\right) = \ln 2.780 + \ln\left(e^{\frac{-x}{600}}\right)$$

$$\Rightarrow \ln 0.220 + \frac{x}{600} = \ln 2.780 + \frac{-x}{600}$$

$$\frac{2x}{600} = \ln 2.780 - \ln 0.220 \Rightarrow x = 761 \text{ m}$$

3.15.2.3a $h = h_a + C_1 e^{\frac{x}{\lambda}} + C_2 e^{\frac{-x}{\lambda}}$; $c = \dfrac{d}{k} = 500$ days and

$$\lambda = \sqrt{KDc} = 500 \text{ m}$$

$$x = 0 \Rightarrow h_0 = 26 \Rightarrow C_2 = -1 - C_1$$

$$x = 500 \Rightarrow h_{500} = 26 \Rightarrow C_1 = \frac{-1 + e^{-1}}{e - e^{-1}}$$

$$C_1 = -0.269 \Rightarrow C_2 = -0.731 \Rightarrow h = 27 - 0.269 \, e^{\frac{x}{500}} - 0.731 \, e^{\frac{-x}{500}}$$

3.15.2.3b 26.04 m; 26.07 m; 26.10 m; 26.11 m; 26.11 m

3.15.2.3c note that water through the leaky confining layer flows downwards; therefore

$$q_z = -k\frac{h - h_a}{z - z_a} = -k\frac{h - h_a}{-d} = \frac{h - h_a}{c} =$$

$$\frac{27 - 0.269 \, e^{\frac{x}{500}} - 0.731 \, e^{\frac{-x}{500}} - 27}{500} =$$

$$\frac{-0.269 \, e^{\frac{x}{500}} - 0.731 \, e^{\frac{-x}{500}}}{500} \Rightarrow q_z \text{ at } x = 250 \text{ m equals } -1.8 \times 10^{-3}$$

m day^{-1}

3.15.2.3d −0.92 m^2 day^{-1}

3.15.2.3e pump + replenishment by precipitation excess = leakage; replenishment by precipitation excess = 10^{-3} m day^{-1} $\times$ 500 m = 0.5 m^2 day^{-1}; the leakage term in the water balance must be taken as a positive number; pump = 0.42 m^2 day^{-1}

3.15.2.3f for $n_e = 0.3$: −6 × 10^{-3} m day^{-1}

3.15.2.4 $h = 14 - 0.376 \, e^{\frac{x}{600}} - 1.625 \, e^{\frac{-x}{600}}$

3.15.2.4a 12.0 m; 11.0 m

3.15.2.4b −1.52 m^2 day^{-1}

3.15.2.4c 439 m

3.15.2.5a $0 \le x \le 10 : h = C_1 x + C_2$; $10 \le x \le \infty$:

$$h = h_a + C_1 e^{\frac{x}{\lambda}} + C_2 e^{\frac{-x}{\lambda}}$$

$$0 \le x \le 10 : h = \frac{h_{10} - 30}{10} x + 30 \ ; \ 10 \le x \le \infty :$$

$$h = 26 + \frac{h_{10} - 26}{e^{\frac{-10}{200}}} e^{\frac{-x}{200}}$$

continuity: $Q'_{0 \le x \le 10} = Q'_{10 \le x \le \infty}$

$$\Rightarrow \left(\frac{dh}{dx}\right)_{0 \le x \le 10} = \left(\frac{dh}{dx}\right)_{10 \le x \le \infty} \quad \text{at } x = 10$$

$$\Rightarrow h_{10} = 29.81 \text{ m} \Rightarrow 0 \le x \le 10 : h = -0.019x + 30$$

$$10 \le x \le \infty : h = 26 + 4.005 \, e^{\frac{-x}{200}}$$

3.15.2.5b 29.90 m; 29.72 m; 29.53 m

3.15.2.5c 9.52 m^2 day^{-1}

3.15.2.5d 0 mm day^{-1}; 46.4 mm day^{-1}; 44.2 mm day^{-1}

3.15.3.1 $h^2 = -\dfrac{N}{K}x^2 + C; x = 0 \Rightarrow h = h_0 \Rightarrow C = h_0^2$

$\Rightarrow h^2 = -\dfrac{N}{K}x^2 + h_0^2$

$x = L \Rightarrow h = h_L \Rightarrow L^2 = \dfrac{K}{N}\left(h_0^2 - h_L^2\right)$

$\Delta h = h_0 - h_L$ and $\overline{D} = \dfrac{h_0 + h_L}{2} \Rightarrow 2L = 2\sqrt{\dfrac{2K\overline{D}\Delta h}{N}}$

$\Rightarrow$ distance between both canals $= 2L = 60$ m

3.15.3.2 $h^2 = -\dfrac{N}{K}x^2 + C_1 x + C_2$

$h^2 = -\dfrac{N}{K}x^2 + \left(\dfrac{h_L^2 - h_0^2}{L} + \dfrac{NL}{K}\right)x + h_0^2$; using

one of two methods (answer 3.15.1.1b or M5):

$h\dfrac{dh}{dx} = -\dfrac{N}{K}x + \dfrac{h_L^2 - h_0^2}{2L} + \dfrac{NL}{2K}$

$\Rightarrow Q' = \dfrac{-K(h_L^2 - h_0^2)}{2L} + N\left(x - \dfrac{L}{2}\right)$; at the water divide:

$Q' = 0$ or $h\dfrac{dh}{dx} = 0$ with x = distance d'

$\Rightarrow$ distance $d' = \dfrac{L}{2} + \dfrac{K(h_L^2 - h_0^2)}{2NL}$;

at the water divide $h = h_{max}$ and $x = d'$:

$h_{max} = \sqrt{h_0^2 + \dfrac{(h_L^2 - h_0^2)d'}{L} + \dfrac{N}{K}(L - d')d'}$

3.15.4.1a interpolate between the given hydraulic head values; also extrapolate: hydraulic head of 8 m at line x = 450 m; hydraulic head of 7 m at line x = 1450 m

3.15.4.1b 1 cm day^{-1}

3.15.4.1c 2.5 cm day^{-1}

3.15.4.1d y = 625 m

3.15.4.1e $Q = qA = qWD$, with W taken as the width of the waste dump: Q = 50 m^3 day^{-1}

3.15.4.2a 1:1 (continuity for steady flow)

3.15.4.2b $Q_0 = q_r 2\pi r D = $ constant: if r is halved, q_r is doubled $\Rightarrow$ 1:2

3.15.4.3a 1000 m

3.15.4.3b $h - h_R = \dfrac{Q_0}{2\pi KD}\ln\dfrac{r}{R} = -0.1$ m $\Rightarrow$ 606 m

3.15.4.4a 875 m^3 day^{-1}

3.15.4.4b 930 m^3 day^{-1}

3.15.4.4c $2\overline{D} = 50$ and $h + h_R = 47$; thus the answer for b is a factor 50/47 too high

3.15.4.5a on the right-hand side of Figure 3.51: the hydraulic

gradient i along the line W equals $\frac{1}{1000}$; continuity: $|Q_0| = |Q_W|$; $Q_W = qA = qWD = -KiWD \Rightarrow W = 628$ m

3.15.4.5b $h_{tot} = h_r + h_x = h_R + \dfrac{Q_0}{2\pi T}\ln\dfrac{r}{R} + ir + C$

$\Rightarrow \dfrac{dh_{tot}}{dr} = \dfrac{Q_0}{2\pi T}\left(\dfrac{1}{r}\right) + i$

$\dfrac{dh_{tot}}{dr} = 0 \Rightarrow (x,y) = (-100\text{ m}, 0\text{ m})$

3.15.5.1a 9.58 m (Q_3 has no effect, as r > R)

3.15.5.1b 9.91 m (Q_3 has no effect, as r > R)

3.15.5.2a 1100 m^2 day^{-1} (section 3.10)

3.15.5.2b lowering of the hydraulic head $h_R - h = 0.25$ m (r = 500 m)

3.15.5.2c it will be less due to the constant hydraulic head of the canal

3.15.5.2d 0.03 m (r of the image recharge well $= \sqrt{(400 + 2 \times 50)^2 + 300^2} = 583.10$ m); yes

3.15.5.2e 2 : 9 (same as the transmissivities $T = KD$ of both layers)

3.15.5.2f 1 : 3 (because the effective porosities are the same for both layers: same as the volume flux densities q of both layers $\Rightarrow$ same as the hydraulic conductivities K of both layers)

3.15.5.3a R = 1105 m $\Rightarrow$ drawdown $h - h_R = -1.02$ m

3.15.5.3b r of the image pumping well = 400 m $\Rightarrow h - h_R = 2 \times -1.02 = -2.03$ m

Chapter 4

4.1.1a 0 atm

4.1.1b 1 atm

4.1.1c 1 atm $\approx$ 1000 hPa = $10^3\ 10^2$ N m^{-2} = 10^5 N m^{-2}; divide pressure p in 10^5 N m^{-2} by ρg in 10^3 kg m^{-3} 10 m s^{-2}; this gives 10^5 N m^{-2} 10^{-3} kg^{-1} m^3 10^{-1} m^{-1} s^2 = 10 N kg^{-1} s^2 = 10 <kg m s^{-2}> kg^{-1} s^2 = 10 m; thus 1 atm $\approx$ 10 m

4.1.1d 2 atm

4.1.1e 3 atm

4.1.1f 1 atm $\approx$ 10 m $\Rightarrow$ 1000 hPa $\approx$ 10 m $\Rightarrow$ 1 hPa $\approx$ 1 cm

4.1.2 $pF = \log(-\psi) = 2 \Rightarrow -\psi = 10^2$ cm $\Rightarrow \psi = -10^2$ cm = -1 m; using the bottom (b) of the clay layer as the reference level ($z_b = 0$ m): $h_b = z_b + \psi = 0 - 1 = -1$ m; top (t) of the clay layer ($z_t = 2$ m): $h_t = z + \dfrac{p}{\rho g} = 2 + 1 = 3$ m

$\Delta h = h_b - h_t = -1 - 3 = -4$ m; $\Delta z = z_b - z_t = 0 - 2 = -2$ m;

$i = \dfrac{\Delta h}{\Delta z} = \dfrac{-4\text{ m}}{-2\text{ m}} = 2$

$K = 2 \times 10^{-3}$ m day^{-1} $\Rightarrow q = -Ki = -4 \times 10^{-3}$ m day^{-1}

4.2a   determine the values of ψ ($= \psi_C$) from the ψ_M tensiometer readings: left tensiometer, $z = -60$ cm, $\psi = -10$ cm $\Rightarrow h = z + \psi = -70$ cm; right tensiometer, $z = -80$ cm, $\psi = +10$ cm $\Rightarrow h = z + \psi = -70$ cm

connect the z s for the left and right tensiometers with a straight line

connect the ψ s for the left and right tensiometers with a straight line

connect the h s for the left and right tensiometers with a straight line

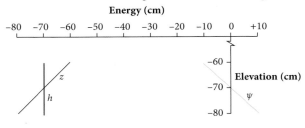

Energy (cm)

h is constant with depth $\Rightarrow$ hydrostatic equilibrium (no water flow)

4.2b $\psi = 0$ cm $\Rightarrow$ depth of the water table $= -70$ cm

4.4.1   3 (for the upper 20 cm) + 2 (for the lower 20 cm) = 5 cm

4.4.2b   A, sand; B, clay

4.4.2c   soil A, 4.0 cm; soil B, 6.4 cm; the sandy soil (A) has less available soil water for plants than the clay soil (B)

4.8.1 $S_f = 60$ mm $\Rightarrow \psi_f = -60$ mm

at the wetting front: $z = L$ $(L < 0)$ and $\psi = \psi_f$ $(\psi_f < 0)$
$\Rightarrow h_f = L + \psi_f$

at the surface: the hydraulic head $h_0 = 20$ mm

example for $L = -20$ mm: $h_f = L + \psi_f = -20 + (-60) = -80$ mm

$\Delta h = h_f - h_0 = -80 - 20 = -100$ mm;

$\Delta z = L - 0 = L = -20$ mm;

$i = \dfrac{\Delta h}{\Delta z} = 5$ $q = -Ki = -20 \times 5 = -100$ mm hour^{-1}

$\Rightarrow f = 100$ mm hour^{-1}

$\theta_s = n_e = 45\%$ and $\theta_i = 20\% \Rightarrow \theta_s - \theta_i = 25\% \Rightarrow 25\%$ of the soil can still be filled with water $\Rightarrow$ the cumulative infiltration (mm) $= 25\%$ of $|L|$ (mm); for $L = -20$ mm the cumulative infiltration (mm) = 5 mm; the infiltration rates f (mm hour^{-1}) when the wetting front is 20, 40, 80, 160, 320, 640, and 1280 mm below the soil surface are 100, 60, 40, 30, 25, 23, and 21 mm hour^{-1}; the corresponding cumulative infiltrations F (mm) are 5, 10, 20, 40, 80, 160, and 320 mm

4.8.1b   see the answers for 4.8.1a; see the main text

4.8.2   from $f_t - f_c = (f_0 - f_c)\,e^{-\alpha t}$, derive

$$\frac{f_{t+\Delta t} - f_c}{f_t - f_c} = e^{-\alpha \Delta t}$$

$\Delta t = 30$ minutes $\Rightarrow$

$$e^{-\alpha \Delta t} = \frac{f_{t+\Delta t} - f_c}{f_t - f_c} = \frac{25 - 5 \text{ mm hour}^{-1}}{45 - 5 \text{ mm hour}^{-1}} = \frac{20}{40} = 0.5$$

$$\frac{f_{t+2\Delta t} - f_c}{f_t - f_c} = e^{-\alpha 2\Delta t} = \left(e^{-\alpha \Delta t}\right)^2 = (0.5)^2 = 0.25$$

$$\frac{f_{t+\frac{1}{2}\Delta t} - f_c}{f_t - f_c} = e^{-\alpha \frac{1}{2}\Delta t} = \left(e^{-\alpha \Delta t}\right)^{\frac{1}{2}} = (0.5)^{\frac{1}{2}} = \sqrt{0.5} = 0.707$$

$$\frac{f_{t+\frac{1}{6}\Delta t} - f_c}{f_t - f_c} = e^{-\alpha \frac{1}{6}\Delta t} = \left(e^{-\alpha \Delta t}\right)^{\frac{1}{6}} = (0.5)^{\frac{1}{6}} = \sqrt[6]{0.5} = 0.891$$

$f_t - f_c$ after 90 minutes $= 0.5 \times (f_t - f_c)$ after 60 minutes $= 0.5 \times 20$ mm hour^{-1} = 10 mm hour^{-1}; f_t after 90 minutes $= 10 + 5 = 15$ mm hour^{-1}

$f_t - f_c$ after 120 minutes $= 0.25 \times (f_t - f_c)$ after 60 minutes $= 0.25 \times 20$ mm hour^{-1} = 5 mm hour^{-1}; f_t after 120 minutes $= 5 + 5 = 10$ mm hour^{-1}

$f_t - f_c$ after 75 minutes $= \sqrt{0.5} \times (f_t - f_c)$ after 60 minutes $= 0.707 \times 20$ mm hour^{-1} = 14.1 mm hour^{-1}; f_t after 75 minutes $= 14.1 + 5 = 19.1$ mm hour^{-1}

$f_t - f_c$ after 65 minutes $= \sqrt[6]{0.5} \times (f_t - f_c)$ after 60 minutes $= 0.891 \times 20$ mm hour^{-1} = 17.8 mm hour^{-1}; f_t after 65 minutes $= 17.8 + 5 = 22.8$ mm hour^{-1}

4.8.3a $K = 0.4$ mm min^{-1}; at $t = 1$ minute: $f = 2.5$ mm min^{-1} $\Rightarrow S = 4.2$ mm min$^{-0.5}$

4.8.3b $f(1) = \dfrac{1}{2}S + K = 2.5 \Rightarrow K = 2.5 - \dfrac{1}{2}S$

$f(60) = 0.4 = \dfrac{1}{2}S\,(60)^{-0.5} + K = \dfrac{1}{2}S\,(60)^{-0.5} + 2.5 - \dfrac{1}{2}S$

$S = \dfrac{2.5 - 0.4}{\frac{1}{2} - \frac{1}{2}(60)^{-0.5}} = 4.82$ mm min$^{-0.5} \Rightarrow K = 2.5 - \dfrac{1}{2}\,(4.82)$

$= 0.09$ mm min^{-1}

4.8.4a   slope between $t = 60$ and 30 minutes

$\Rightarrow K = \dfrac{26 - 17}{60 - 30} = 0.3$ mm min^{-1}

at $t = 1$ minute: $F = 2.0$ mm $\Rightarrow S = 1.7$ mm min$^{-0.5}$

4.8.4b $F(1) = 2 = S + K \Rightarrow K = 2 - S$

$F(60) = 26 = S\sqrt{60} + 60K =$

$S\sqrt{60} + 60 \times (2 - S) = S\sqrt{60} + 60 \times 2 - 60S$

$\Rightarrow S = \dfrac{120 - 26}{60 - \sqrt{60}} = 1.8$ mm min$^{-0.5}$

$\Rightarrow K = 2 - 1.8 = 0.2$ mm min^{-1}

4.8.5a   4.6 mm min$^{-0.5}$

4.8.5b $L = \dfrac{F}{\theta_s - \theta_i} = \dfrac{S\sqrt{t}}{\theta_s - \theta_i} = \left(\dfrac{S}{\theta_s - \theta_i}\right) \times \sqrt{t}$; the length of

the wetting front thus scales with the square root of time:

$\dfrac{L_2}{L_1} = \dfrac{\sqrt{t_2}}{\sqrt{t_1}} = \sqrt{\dfrac{t_2}{t_1}} \Rightarrow$

$L(t_2) = \sqrt{\dfrac{t_2}{t_1}} \times L(t_1) \Rightarrow L(6) = \sqrt{\dfrac{6}{2}} \times 20$

$= 34.6 \text{ cm}; \; L(24) = \sqrt{\dfrac{24}{2}} \times 20 = 69.3 \text{ cm}$

4.8.6a

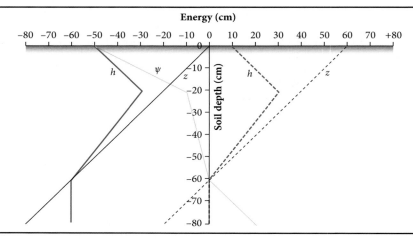

4.8.6b   −60 cm

4.8.6c   −20 cm

4.8.6d   0 cm day^{-1}

4.8.6e   1

4.8.6f   0.19 cm day^{-1}

4.8.6g   0.19 cm day^{-1}

4.8.6h $K(\psi)$ is very much smaller for evaporation than for percolation, while the hydraulic gradient for evaporation is only

slightly larger (by a factor of $\dfrac{4}{3}$)

4.8.7a   continuity: $q_u = q_1 \Rightarrow -K_u i_u = -K_1 i_1$; $\dfrac{K_u}{K_1} = \dfrac{3}{8} = \dfrac{i_1}{i_u}$

$h_{60} = z_{60} + \psi_{60} = 60 + 10 = 70$; $h_0 = z_0 + \psi_0 = 0 + 0 = 0$

$i_u = \dfrac{h_{40} - h_{60}}{z_{40} - z_{60}} = \dfrac{h_{40} - 70}{40 - 60} = \dfrac{h_{40} - 70}{-20}$

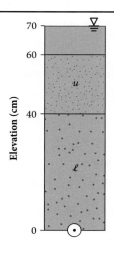
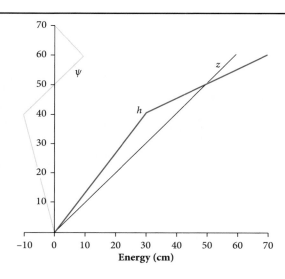

$$i_1 = \frac{h_0 - h_{40}}{z_0 - z_{40}} = \frac{0 - h_{40}}{0 - 40} = \frac{h_{40}}{40}$$

$$\frac{i_1}{i_u} = \frac{3}{8} = \frac{\left(\dfrac{h_{40}}{40}\right)}{\left(\dfrac{h_{40} - 70}{-20}\right)} \Rightarrow h_{40} = 30 \text{ cm} \Rightarrow i_u = \frac{30 - 70}{-20} = 2$$

and $i_1 = \dfrac{30}{40} = \dfrac{3}{4}$

4.8.7b i the pressure head in the drainage pipe is zero
ii for the flow process to be maintained, there must be
continuity of water flow (no stagnation as in Figure 4.34)
iii the matric suction must be less than the air-entry suction
for both layers; if not, the hydraulic conductivity will not be
equal to the (constant) saturated hydraulic conductivity, but
a function of the matric potential (in Figure 4.34, the hydraulic
conductivity is a function of the matric potential!)

4.8.7c -0.9 mm min^{-1}

4.9a   m: $\dfrac{3}{0.5} = 6$ mm hour$^{-1} \Rightarrow 6$ mm; p: $\dfrac{3}{0.4} = 7.5$ mm
hour$^{-1} \Rightarrow 7.5$ mm

4.9b   m: $\dfrac{3}{0.5} = 6$ mm hour$^{-1} \Rightarrow 6$ mm; excess water flows into
preferential flow domain $\Rightarrow$ p: $\dfrac{6 + 3 \times \left(\dfrac{0.8}{0.2}\right)}{0.4} = 45$ mm

4.9c P = precipitation depth (mm); 100 mm hour^{-1}
$= P + \left((P-3) \times 4\right) \Rightarrow P = 22.4$ mm; infiltration depth
$= 250$ mm

4.9d $n_e = (0.8 \times 0.5) + (0.2 \times 0.4) = 0.48$

$K = (0.8 \times 3) + (0.2 \times 100) = 22.4$ mm hour^{-1}

4.9e $\dfrac{3}{0.48} = 6.25$ mm

4.9f $\dfrac{6}{0.48} = 12.5$ mm

4.9g P = precipitation depth = 22.4 mm; infiltration depth
$= \dfrac{22.4}{0.48} = 46.7$ mm

4.9h   taking the soil as homogeneous leads to a strong
underestimation of the maximum depth of infiltration under
high rainfall intensities

Chapter 5

5.1.1   18 m^3 s^{-1}

5.1.2   Bernoulli's law: $\dfrac{v_1^2}{2g} + \dfrac{p_1}{\rho g} + z_1 = \dfrac{v_2^2}{2g} + \dfrac{p_2}{\rho g} + z_2$ and

$z_1 = z_2 \Rightarrow$

$$\frac{v_1^2}{2g} + \frac{p_1}{\rho g} = \frac{v_2^2}{2g} + \frac{p_2}{\rho g} \Rightarrow p_1 - p_2 = \frac{\rho}{2}\left(v_2^2 - v_1^2\right)$$

continuity: $Q = v_1 A_1 = v_2 A_2 \Rightarrow v_1 = \dfrac{Q}{A_1}$ and $v_2 = \dfrac{Q}{A_2}$

combine Bernoulli's law and continuity:

$$p_1 - p_2 = \frac{\rho}{2}\left(\frac{Q^2}{A_2^2} - \frac{Q^2}{A_1^2}\right) \Rightarrow p_1 - p_2 = \frac{\rho}{2}\left(\frac{A_1^2 Q^2}{A_1^2 A_2^2} - \frac{A_2^2 Q^2}{A_1^2 A_2^2}\right)$$

$$\Rightarrow \frac{2 A_1^2 A_2^2}{\rho}\left(p_1 - p_2\right) = Q^2\left(A_1^2 - A_2^2\right)$$

$$\Rightarrow Q^2 = \frac{2 A_1^2 A_2^2 \left(p_1 - p_2\right)}{\left(A_1^2 - A_2^2\right)\rho} \Rightarrow Q = A_1 A_2 \sqrt{\frac{2\left(p_1 - p_2\right)}{\left(A_1^2 - A_2^2\right)\rho}}$$

$$\Rightarrow Q = A_1 A_2 \sqrt{\frac{2g\left(\dfrac{p_1}{\rho g} - \dfrac{p_2}{\rho g}\right)}{\left(A_1^2 - A_2^2\right)}}$$

5.1.3   Equation 5.29 holds: $H_2 = \dfrac{2}{3}H_1$; inserting this equation
in Equation 5.42:

$A = H_2^2 \tan\left(\dfrac{\theta}{2}\right)$ yields $A = \dfrac{4}{9}H_1^2 \tan\left(\dfrac{\theta}{2}\right)$; combining this

equation with $Q = v_2 A$ and

$v_2 = \sqrt{\dfrac{2}{3}gH_1}$ (Equation 5.31) yields

$$Q = \sqrt{\frac{2}{3}gH_1}\,\frac{4}{9}H_1^2 \tan\left(\frac{\theta}{2}\right) = \left(\frac{32}{243}g\right)^{\frac{1}{2}} \tan\left(\frac{\theta}{2}\right)H_1^{\frac{5}{2}},$$

which can be simplified to a general Q–H relation for a V-notch
weir for different angles θ of the V-notch:

$$Q = C \tan\left(\frac{\theta}{2}\right)H_1^{\frac{5}{2}}$$

5.2.1   15.5 litre s^{-1}

5.2.2   18 litre s^{-1}

5.2.3   47.8 litre s^{-1}

5.2.4   45.7 litre s^{-1}

5.2.5 $Q_1 = 10$ litre s^{-1}; $Q_2 = 7.5$ litre s^{-1}

5.2.6   4.3 m^3 s^{-1}

5.2.7a   2007: $H_0 = 20$ cm; 2008: $H_0 = 30$ cm

5.2.7b   the Q–H relation is given in Figure 5.29

5.2.7c   take $\log\left(H - H_0\right) = 0 \Rightarrow \log Q = \log a = -1.86$

$\Rightarrow a = 0.0138$; $\log\left(H - H_0\right) = 1.5 \Rightarrow \log Q = 1.9$

$\Rightarrow \log Q = 1.9 \quad \Rightarrow \quad \log Q = \log a + b \log(H - H_0)$ gives:

$1.9 = -1.86 + 1.5b \Rightarrow b = 2.5$

$\Rightarrow Q = 0.0138(H - H_0)^{2.5}$

$Q = 0.0138(H - H_0)^{2.5}$, with Q in litre s^{-1} and H in cm

5.2.7d 0.0138 has the units of $\dfrac{Q}{(H - H_0)^{2.5}} =$ litre s^{-1} cm$^{-2.5}$

$=$ dm^3 s^{-1} cm$^{-2.5}$ $= 10^{-3}$ m^3 s^{-1} cm$^{-2.5}$ (1 cm $= 10^{-2}$ m $\Rightarrow$
1cm$^{0.5}$ $= 10^{-1}$ m$^{0.5}$ $\Rightarrow$ 1cm$^{2.5}$ $= 10^{-5}$ m$^{2.5}$ $\Rightarrow$ 1 cm$^{-2.5}$ $= 10^5$ m$^{-2.5}$);

0.0138 litre s^{-1} cm$^{-2.5}$ $= 0.0138 \times 10^{-3}$ m^3 s^{-1} cm$^{-2.5}$ $= 0.0138 \times$
10^{-3} m^3 s^{-1} $\times 10^5$ m$^{-2.5}$ $= 1.38$ m$^{0.5}$ s^{-1}; thus 0.0138 litre s^{-1} cm$^{-2.5}$
$= 1.38$ m$^{0.5}$ s^{-1}

when $H - H_0$ is given in m, $(H - H_0)^{2.5}$ is in m$^{2.5}$; with 1.38
having units m$^{0.5}$ s^{-1}, Q in the equation $Q = 1.38(H - H_0)^{2.5}$
must be in m^3 s^{-1}; thus $Q = 1.38(H - H_0)^{2.5}$, with Q in m^3 s^{-1}
and H in m

5.3.1a 5 mm

5.3.1b 10^4 m^3

5.3.1c 5 litre s^{-1}

5.3.1d 90 m^3

5.3.1e 0.045 mm

5.3.1f 0.9%

5.3.1g 1% of the drainage basin area is occupied by the stream
channel $\Rightarrow$ 1% of 5 mm rainfall depth $=$ 0.05 mm channel
precipitation; calculated quickflow volume (5.3.1e) $=$ 0.045 mm
$\Rightarrow$ channel precipitation (mm) and quickflow volume (mm) are
approximately equal $\Rightarrow$ channel precipitation is the only process
that contributes to the quickflow of this rainfall event

5.3.2a 4%

5.3.2b 0.5 mm

5.3.2c 5×10^3 m^3

5.4.1a 2.46×10^9 m^3

5.4.1b 90 m^3 s^{-1}; 85.4 m^3 s^{-1}

5.4.1c 2.1×10^9 m^3 ($= 85.4\%$ of 2.46×10^9 m^3)

5.4.2 during recession, $I = 0$; Equation 5.76 becomes
$\dfrac{dQ}{\alpha} = -Q \, dt$

$\dfrac{dQ}{\alpha} = -Q \, dt \Rightarrow \dfrac{1}{Q} dQ = -\alpha dt \Rightarrow \displaystyle\int_{Q_0}^{Q_t} \dfrac{1}{Q} dQ = \int_0^t -\alpha dt$

$\Rightarrow [\ln Q]_{Q_0}^{Q_t} = [-\alpha t]_0^t$

$\Rightarrow \ln Q_t - \ln Q_0 = -\alpha t \Rightarrow \ln \dfrac{Q_t}{Q_0} = -\alpha t \Rightarrow \dfrac{Q_t}{Q_0} = e^{-\alpha t}$

$\Rightarrow Q_t = Q_0 e^{-\alpha t}$

5.5a 24 minutes

5.5b 60 mm hour^{-1}

5.5c 1 minute

5.5d at the beginning, 50 mm hour^{-1} of rain infiltrates
$\Rightarrow 60 - 50 = 10$ mm hour^{-1} of rain is intercepted by vegetation
$\Rightarrow \dfrac{10}{60} \times 100\% = 17\%$ of the surface area is covered by
vegetation

5.5e 2 minutes

5.5f 2 minutes

5.5g Hortonian overland flow $=$ infiltration excess overland
flow

5.5h 40 mm hour^{-1}

5.5i 22 minutes

5.5j $60 - 40 = 20$ mm hour^{-1}

References

4th World Water Forum (2006). Middle East and North Africa regional document. http://www.worldwatercouncil.org/fileadmin/ wwc/World_Water_Forum/WWF4/Regional__process/MIDDLE _EAST_AND_NORTH_AFRICA.pdf

Allen, R.G., Pereira, L.S., Raes, D., and Smith, M. (1998). Crop evapotranspiration. Guidelines for computing crop water requirements. FAO Irrigation and drainage paper 56. Food and Agriculture Organization of the United Nations (FAO), Rome. http://www.fao.org/docrep/X0490E/x0490e00.htm#Contents

Anderson, M.G. and Burt, T.P. (1978). The role of topography in controlling throughflow generation. *Earth Surface Processes*, 3, 331–44.

Anderson, M.P. and Woessner, W.W. (1992). *Applied Groundwater Modeling: Simulation of Flow and Advective Transport*. Academic Press.

Appelo, C.A.J. (2008). Geochemical experimentation and modeling are tools for understanding the origin of arsenic in groundwater in Bangladesh and elsewhere. In: *Arsenic in Groundwater – A World Problem*. Proceedings of a Seminar, Utrecht, 29 November 2006. Netherlands National Committee of the IAH (NNC-IAH), pp. 33–50.

Bear, J. (1969). *Dynamics of Fluids in Porous Media*. Elsevier, Amsterdam.

Beven, K.J. (1997). (ed.) *Distributed Hydrological Modelling: Applications of the TOPMODEL Concept*. Wiley, 348 pp.

Beven, K.J. (2001). *Rainfall–Runoff Modelling – the Primer*. Wiley. http://www.es.lancs.ac.uk/hfdg/publications/hfdg_publications_ book.htm

Beven, K. (2004). Robert E. Horton's perceptual model of infiltration processes. *Hydrological Processes*, 18, 3447–60.

Beven, K. and Germann, P. (1982). Macropores and water flow in soils. *Water Resources Research*, 18, 1311–25.

Beven, K.J. and Kirkby, M.J. (1979). A physically-based variable contributing area model of basin hydrology. *Hydrological Sciences Bulletin*, 24, 43–69.

Beven, K.J. and Moore, I.D. (eds) (1994). *Terrain Analysis and Distributed Modelling in Hydrology*. Advances in Hydrological Processes. Wiley, 249 pp.

Bierkens, M.F.P. and Van den Hurk, B. (2008). Feedback mechanisms: precipitation and soil moisture. Chapter 9 in Bierkens, M.F.P., Dolman, A.J., and Troch, P. (eds), *Climate and the Hydrological Cycle*. International Association of Hydrological Sciences (IAHS) Special Publication 8, 175–93.

Bogaard, T.A. and Hendriks, M.R. (2001). Hydrological pilot study of the Ijen caldera and Asembagus irrigation area, East Java, Indonesia. ICG Internal Report 01/01, 13 pp.

Bonell, M., Hendriks, M.R., Imeson, A.C., and Hazelhoff, L. (1984). The generation of storm runoff in a forested clayey drainage basin in Luxembourg. *Journal of Hydrology*, 71, 53–77.

Bos, M.G. (1989). *Discharge Measurement Structures*. Working Group on Small Hydraulic Structures, International Institute for Land Reclamation and Improvement (ILRI), Publication 20 (3rd rev. edn), Wageningen, The Netherlands.

Bouma, J. (1977). Soil survey and the study of water in the unsaturated zone. Soil Survey Paper 13. Netherlands Soil Survey Institute, Wageningen, 106 pp.

Broecker, W.S. (1997). Thermohaline circulation, the Achilles heel of our climate system: Will man-made CO_2 upset the current balance? *Science*, 278, 1582–8.

Broecker, W.S. (2006). Was the Younger Dryas triggered by a flood? *Science*, 312, 1146–8.

Bruggeman, G.A. (1999). *Analytical Solutions of Geohydrological Problems*. Developments in Water Science 46. Elsevier, Amsterdam.

Cammeraat, L.H. and Kooijman, A.M. (2009). Biological control of pedological and hydro-geomorphological processes in a deciduous forest ecosystem. *Biologia*, 64/3, 428–432.

Celia, M.A., Bouloutas, E.T., and Zarba, R.L. (1990). A general mass-conservative numerical solution for the unsaturated flow equation. *Water Resources Research*, 26, 1483–96.

Chen, D., Pyrak-Nolte, L.J., Griffin, J., and Giordano, N.J. (2007). Measurement of interfacial area per volume for drainage and imbibition. *Water Resources Research*, 43, W12504, doi:10.1029/2007WR006021, 1–6.

CHO-TNO (1986). *Verklarende Hydrologische Woordenlijst [Explanatory Hydrological Glossary]*. Gespreksgroep Hydrologische Terminologie (red. J.C. Hooghart). Commissie voor Hydrologisch Onderzoek TNO (CHO-TNO), 's-Gravenhage, 130 pp. SISO 568 UDC 556 (038).

Chow, V.T., Maidment, D.R., and Mays, L.W. (1988). *Applied Hydrology*. McGraw-Hill.

Cohen, J.E. and Small, C. (1998). Hypsographic demography: the distribution of human population by altitude. *Proceedings of the National Academy of Sciences, USA*, 95, 14 009–14.

Cornell University (2002). Preferential flow, extension and educational module. Department of Biological and Environmental Engineering, Soil and Water Laboratory. http://soilandwater. bee.cornell.edu/research/pfweb/index.htm

Cornell University (2005). Variable source area hydrology. Department of Biological and Environmental Engineering, Soil and Water Laboratory. http://soilandwater.bee.cornell.edu/ research/VSA/index.html

Dastane, N.G. (1978). Effective rainfall in irrigated agriculture. FAO Irrigation and drainage paper M-56. Food and Agriculture Organization of the United Nations (FAO), Rome. http:// www.fao.org/DOCREP/X5560E/x5560e00.htm

De Jong, S.M., Van der Kwast, J., Addink, E.A., and Su, B. (2008). Remote sensing for hydrological studies. Chapter 15 in Bierkens, M.F.P., Dolman, A.J., and Troch, P. (eds), *Climate and the Hydrological Cycle*. International Association of Hydrological Sciences (IAHS) Special Publication 8, pp. 297–320.

De Vries, J.J. (1980). *Inleiding tot de Hydrologie van Nederland* [*Introduction to the Hydrology of The Netherlands*]. Rodopi, Amsterdam.

De Vries, J.J. (1982). *Anderhalve Eeuw Hydrologisch Onderzoek in Nederland* [*One and a Half Centuries of Hydrological Research in The Netherlands*]. Rodopi, Amsterdam.

De Vries, J.J. (1994). From speculation to science: the founding of groundwater hydrology in the Netherlands. In: Touret, J.L.R. and Visser, R.P.W. (eds), *Dutch pioneers of the Earth sciences*. History of Science and Scholarship in the Netherlands, vol. 5, pp. 139–64. Royal Netherlands Academy of Arts and Sciences. http://www.knaw.nl/publicaties/pdf/20021148.pdf

De Vries, J.J. and Cortel, E.A. (1990). Introduction to hydrogeology. Lecture notes, Institute of Earth Sciences, VU University Amsterdam, The Netherlands.

De Zeeuw, J.W. (1966). *Hydrograph Analysis of Areas with Prevailing Groundwater Discharge*. Veenman en Zonen, Wageningen (with English summary).

De Zeeuw, J.W. (1973). Hydrograph analysis for areas with mainly groundwater runoff. In: *Drainage Principle and Applications*, vol. II, Chapter 16, Theories of field drainage and watershed runoff, pp. 321–58. International Institute for Land Reclamation and Improvement (ILRI), Publication 16, Wageningen, The Netherlands.

Doerr, S.H., Shakesby, R.A. and Walsh, R.P.D. (2000). Soil water repellency: its causes, characteristics and hydro-geomorphological significance. *Earth-Science Reviews*, 51, 33–65.

Doorenbos, J. and Pruitt, W.O. (1977). Crop water requirements. FAO Irrigation and drainage paper 24. Food and Agriculture Organization of the United Nations (FAO), Rome.

Drijfhout, S. (2007). Stopt de Golfstroom in de 21ᵉ eeuw? [Will the Gulf Stream come to a halt in the 21st century?] Presentation at the Koninklijk Nederlands Meteorologisch Instituut (KNMI), 6 September 2007.

Dunne, T. and Black, R.D. (1970). Partial area contributions to storm runoff in a small New England watershed. *Water Resources Research*, 6, 1296–311.

Dunne, T., Moore, T.R., and Taylor, C.H. (1975). Recognition and prediction of runoff-producing zones in humid regions. *Hydrological Sciences Bulletin*, 20, 305–27.

Dupuit, J. (1863). *Études théoriques et practiques sur le mouvement des eaux dans les canaux découvert et à travers les terrains perméables* [*Theoretical and practical studies on the movement of water through subsoil pathways in permeable terrain*], 2nd edn. Dunod, Paris.

Ellison, C.R.W., Chapman, M.R., and Hall, I.R. (2006). Surface and deep ocean interactions during the cold climate event 8200 years ago. *Science*, 312, 1929–32.

Emblanch, C., Zuppi, G. M., Mudry, J., Blavoux, B., and Batiot, C. (2003). Carbon 13 of TDIC to quantify the role of the unsaturated zone: the example of the Vaucluse karst systems (southeastern France). *Journal of Hydrology*, 279, 262–74.

Engelen, G.B. and Kloosterman, F.H. (1996). *Hydrological Systems Analysis: Methods and Applications*. Water Science and Technology Library 20. Dordrecht, Kluwer.

Evans, R.G. (1999). Frost protection in orchards and vineyards. Northern Plains Agricultural Research Laboratory, USDA – Agricultural Research Service, Sidney, MT. http://www.sidney. ars.usda.gov/Site_Publisher_Site/pdfs/personnel/Frost%20 Protection%20in%20Orchards%20and%20Vineyards.pdf

Fetter, C.W. (2001). *Applied Hydrogeology*, 4th edn. Prentice Hall.

Fitts, C.R. (2002). *Groundwater Science*. Academic Press, Elsevier Science.

Forchheimer, P. (1886). Über die Ergebigkeit von Brunnen-Anlagen und Sickerslitzen [Springs and seepage]. *Zeitschrift des Architecten und Ingenieurs Vereins zu Hannover*, 32, 539–64.

Ford, D. and Williams, P. (2007). *Karst Hydrogeology and Geomorphology*. Wiley.

Gray, W.G. and Hassanizadeh, S.M. (1991a). Paradoxes and realities in unsaturated flow theory. *Water Resources Research*, 27, 1847–54.

Gray, W.G. and Hassanizadeh, S.M. (1991b). Unsaturated flow theory including interfacial phenomena. *Water Resources Research*, 27, 1855–63.

Green, W.H. and Ampt, G.A. (1911). Studies on soil physics: I. Flow of air and water through soils. *Journal of Agricultural Science*, 4, 1–24.

Gregory, K.J. and Walling, D.E. (1973). *Drainage Basin Form and Process: a geomorphological approach*. Edward Arnold, London, 458 pp.

Haitjema, H.M. (1995). *Analytical Element Modeling of Groundwater Flow*. San Diego, CA, Academic Press, 394 pp.

Haitjema, H.M. and Mitchell-Bruker, S. (2005). Are water tables a subdued replica of the topography? *Ground Water*, 43(6), 781–6.

Hassanizadeh, S.M. and Gray, W.G. (1990). Mechanics and thermodynamics of multiphase flow in porous media including interphase boundaries. *Advances in Water Resources*, 13(4), 169–86.

Hassanizadeh, S.M., Celia, M.A., and Dahle, H.K. (2002). Dynamic effect in the capillary pressure–saturation relationship and its impacts on unsaturated flow. *Vadose Zone Journal*, 1, 38–57.

Held, R.J. and Celia, M.A. (2001). Modelling support of functional relationships between capillary pressure, saturation, interfacial area and common lines. *Advances in Water Resources*, 24, 325–43.

Henderson, F.M. (1966). *Open Channel Flow*. Macmillan, New York.

Hendriks, M.R. (1990). Regionalisation of hydrological data: effects of lithology and land use on storm runoff in east Luxembourg. PhD thesis, VU University Amsterdam, The Netherlands. Also available as Netherlands Geographical Studies 114, Royal Dutch Geographical Society (KNAG), Utrecht.

Hendriks, M.R. (1993). Effects of lithology and land use on storm runoff in east Luxembourg. *Hydrological Processes*, 7, 213–26.

Hewlett, J.D. (1961). Soil moisture as a source of base flow from steep mountain watersheds. Southeastern Forest Experiment Station, Paper 132. US Forest Service, Ashville, 11 pp.

Hewlett, J.D. and Hibbert, A.R. (1967). Factors affecting the response of small watersheds to precipitation in humid areas. In: Sopper, W.E. and Lull, H.W. (eds), *Forest Hydrology*. Pergamon Press, Oxford, pp. 275–90.

Hils, M. (1988). Einfluss des langfristiger Klimaschwankungen auf die Abflüsse des Rheins unter besonderer Berücksichtigung der Lufttemperatur [The influence of long-term climatic fluctuations on discharges of the Rhine with special consideration of the air temperature]. Diplomarbeit, Bundesamt für Gewasserkunde, Koblenz, 137 pp.

Hornberger, G.M., Raffensperger, J.P., Wiberg, P.L., and Eshleman, K.N. (1998). *Elements of Physical Hydrology*. Johns Hopkins University Press.

Horton R.E. (1939). Analysis of runoff-plot experiments with varying infiltration capacity. *Transactions, American Geophysical Union*, 20, 693–711.

Hubbert, M.K. (1940). The theory of groundwater motion. *Journal of Geology*, 48, 785–944.

IPCC (2001). *Climate Change 2001: the Scientific Basis*. Intergovernmental Panel on Climate Change (IPCC)/Cambridge University Press, Cambridge. http://www.grida.no/climate/ipcc_tar/wg1

IPCC (2007). *Climate Change 2007: The Physical Science*. Summary for Policymakers. Contribution of Working Group I to the Fourth Assessment Report of the Intergovernmental Panel on Climate Change (IPCC). http://ipcc-wg1.ucar.edu/index.html

Jendele, L. (2002). An improved numerical solution of multiphase flow analysis in soil. *Advances in Engineering Software*, 33, 659–68.

Johnson, D.L. (1999). Channel routing. Hydrometeorology Course 00-3, 9–24 May 2000. http://www.comet.ucar.edu/class/hydromet/07_Jan19_1999/html/routing/index.htm

Kirkby, M.J. (1975). Hydrograph modelling strategies. In: Peel, R.F., Chisholm, M.D., and Haggett, P. (eds), *Processes in Physical and Human Geography*. Academic Press, London, pp. 69–90.

Kirkby, M.J., Naden, P.S., Burt, T.P., and Butcher, D.P. (1987). *Computer Simulation in Physical Geography*. Wiley, 227 pp.

Kirpich, Z.P. (1940). Time of concentration of small agricultural watersheds. *Civil Engineering*, 10(6), 362.

Kleinhans, M.G. (2005). Flow discharge and sediment transport models for estimating a minimum timescale of hydrological activity and channel and delta formation on Mars. *Journal of Geophysical Research*, 110, E12003, doi:1029/2005JE002521.

KNMI (2002). *Klimaatatlas van Nederland: de Normaalperiode 1971–2000 [Climate Atlas of The Netherlands: the Standard Period 1971–2000]*. Koninklijk Nederlands Meteorologisch Instituut (KNMI; the Royal Dutch Meteorological Institute), Elmar. http://www.knmi.nl/klimatologie/normalen1971-2000/index.html

Kosman, H. (1988). *Drinken uit de Plas 1888–1988. Honderd jaar Amsterdamse Plassenwaterleiding [One Hundred Years of Lake Drinking-water for Amsterdam]*. Gemeentewaterleidingen Amsterdam.

Kruseman, G.P. and De Ridder, N.A. (1994) (with assistance from Verweij, J.M.). *Analysis and Evaluation of Pumping Test Data*. International Institute for Land Reclamation and Improvement (ILRI), Publication 47 (2nd edn). Wageningen, The Netherlands.

Kwadijk, J.C.J. (1991). Sensitivity of the River Rhine discharge to environmental change, a first tentative assessment. *Earth Surface Processes and Landforms*, 16, 627–37.

Löhr, A.J. (2005). Ecological effects and environmental impact of the extremely acid Banyupahit–Banyuputih river and the Kawah Ijen crater lake in Indonesia. PhD thesis, VU University Amsterdam, The Netherlands.

Löhr, A.J., Bogaard, T.A., Heikens, A., Hendriks, M.R., Sumarti, S., Van Bergen, M.J., Van Gestel, C.A.M., Van Straalen, N.M.,

Vroon, P.Z., and Widianarko, B. (2004). Natural pollution caused by the extremely acid crater lake Kawah Ijen, East Java, Indonesia. *Environmental Science and Pollution Research*, 7 pp. http://dx.doi.org/10.1065/espr2004.09.118

Lonely Planet (2006). *San Francisco City Guide*. Lonely Planet Publications.

Maidment, D.R. (1993). Hydrology. In Maidment, D.R. (ed.), *Handbook of Hydrology*. McGraw-Hill, New York.

Mata-Lima, H. (2006). Hydrologic design that incorporates environmental, quality, and social aspects. *Environmental Quality Management*, 15(3), 51–60, doi:10.1002/tqem.20092.

McCarthy, E.L., (1934). Mariotte's bottle. *Science*, 80, 100.

McDonald, M.G. and Harbaugh, A.W. (1988). A modular three-dimensional finite difference ground-water flow model. US Geological Survey Techniques of Water Resources Investigations 6, Chapter, 586 pp.

Monteith, J.L. (1973). *Principles of Environmental Physics*. New York: American Elsevier, 241 pp.

Nienhuis, Ph. and Hemker, K. (2009). MicroFEM tutorial. http://www.microfem.com/

Penman, H.L. (1948). Natural evaporation from open water, bare soil and grass. *Proceedings of the Royal Society, A* (London), 193, 120–45.

Philip, J. (1957). The theory of infiltration: 1. The infiltration equation and its solution. *Soil Science*, 83, 345–57.

Philip, J. (1964). The gain, transfer and loss of soil water. *Water Resources Use and Management*, Melbourne University Press, pp. 257–75.

Piper, A.M. (1944). A graphic procedure in the geochemical interpretation of water analyses. *Transactions of the American Geophysical Union*, 25, 914–23.

Quinn, P.F., Beven, K.J., and Lamb, R. (1995). The ln(a/tan beta) index: how to calculate it and how to use it within the TOPMODEL framework. *Hydrological Processes*, 9, 161–82.

Quinn, P., Beven, K., Chevallier, P., and Planchon, O. (1994). The prediction of hillslope flow paths for distributed hydrological modelling using digital terrain models. In: Beven, K.J. and Moore, I.D. (eds), *Terrain Analysis and Distributed Modelling in Hydrology*. Wiley.

Rawls, W.J., Brakensiek, D.L., and Miller, N. (1983). Green–Ampt infiltration parameters from soils data. *Journal of the Hydraulics Division, American Society of Civil Engineers*, 109(1), 62–70.

Robinson, M. and Beven, K.J. (1983). The effect of mole drainage on the hydrological response of a swelling clay soil. *Journal of Hydrology*, 64, 205–23.

Rolston, D.E. (2007). Historical development of soil-water physics and solute transport in porous media. *Water Science and Technology: Water Supply*, 7(1), 59–66. http://www.iwaponline.com/ws/00701/0059/007010059.pdf

Royal Netherlands Academy of Arts and Sciences (2005). *Turning the Water Wheel Inside Out: Foresight Study on Hydrological Science in The Netherlands*. Verkenningen, deel 7, Amsterdam. http://www.knaw.nl/cfdata/publicaties/detail.cfm?boeken__ordernr=20041090

Rozemeijer, J. and Van der Velde, Y. (2008). Oppervlakkige afstroming ook van belang in het vlakke Nederland [Water flow at the surface is also important in the flat Netherlands]. H_2O, 19, 92–4.

Rubin, J. (1966). Theory of rainfall uptake by soils initially drier than their field capacity and its applications. *Water Resources Research*, 2, 739–94.

Schmidt, F.H. (1976). *Inleiding tot de Meteorologie* [*Introduction to Meteorology*]. Aula-boeken 112. Het Spectrum.

Schoeller, H. (1955). Géochimie des eaux souterraines [Geochemistry of subsoil waters]. *Revue de l'Institut Français du Pétrole*, 10, 230–44.

Schultz, D.M. and Friedman, R.M. (2007). Tor Harold Percival Bergeron. In: Koertge, N. (ed.), *New Dictionary of Scientific Biography*. Charles Scribner's Sons. http://www.cimms.ou.edu/~schultz/papers/TorBergeron.pdf

Schuurmans, J.M., Bierkens, M.F.P., Pebesma, E.J., and Uijlenhoet, R. (2007). Automatic prediction of high-resolution daily rainfall fields for multiple extents: the potential of operational radar. *Journal of Hydrometeorology*, 8, 1204–24.

Shuttleworth, W.J. (1993). Evaporation. In: Maidment, D.R. (ed.), *Handbook of Hydrology*. McGraw-Hill, New York.

Schwartz, F.W. and Zhang, H. (2003). *Fundamentals of Ground Water*. Wiley.

Šimůnek, J. and Van Genuchten, M.Th. (2008). Modeling non-equilibrium flow and transport processes using HYDRUS. *Vadose Zone Journal*, 7, 782–97. http://vzj.scijournals.org/cgi/reprint/7/2/782

Sørensen, R., Zinko, U., and Seibert, J. (2005). On the calculation of the topographic wetness index: evaluation of different methods based on field observations. *Hydrology and Earth System Sciences, Discussions*, 2, 1807–34.

Stiff, H.A. (1951). The interpretation of chemical water analysis by means of patterns. *Journal of Petroleum Technology*, 3, 15–17.

Strack, O.D.L. (1989). *Groundwater Mechanics*. Prentice Hall.

Strack, O.D.L. (1999). Principles of the analytical element method. *Journal of Hydrology*, 226(3–4), 128–38.

Torkzaban, S., Hassanizadeh, S.M., Schijven, J.F., and Van den Berg, H.H.J.L. (2006). Role of air–water interfaces on retention of viruses under unsaturated conditions. *Water Resources Research*, 42, W12S14, doi:10.1029/2006WR004904

Tóth, J. (1963). A theoretical analysis of groundwater flow in small drainage basins. *Journal of Geophysical Research*, 68(16), 4795–812.

Travis, D.J., Carleton, A.M., and Lauritsen, R.G. (2002). Contrails reduce daily temperature range. *Nature*, 418, 601.

Troch, P.A. (2008). Land surface hydrology. Chapter 5 in: Bierkens, M.F.P., Dolman, A.J., and Troch, P. (eds), *Climate and the Hydrological Cycle*. IAHS (International Association of Hydrological Sciences) Special Publication 8, pp. 99–115.

Troch, P.A., Verhoest, N., Gineste, P., Paniconi, P., and Merot, P. (2000). Variable source areas, soil moisture and active microwave observations at Zwalmbeek and Coët-Dan. Chapter 8 in: Grayson, R. and Blöschl, G. (eds), *Catchment Hydrology: Observations and Modelling*. Cambridge University Press, pp. 187–208.

Vachaud, G., Vauclin, M., Khanji, D., and Wakil, M. (1973). Effects of air pressure on water flow in an unsaturated stratified vertical column of soil. *Water Resources Research*, 9, 160–73.

Van Asch, Th.W.J., Hendriks, M.R., Hessel, R., and Rappange, F.E. (1996). Hydrological triggering conditions of landslides in varved clays in the French Alps. *Engineering Geology*, 42, 239–51.

Van den Akker, C. (2007). On the spreading mechanism of shallow groundwater in the Hinterland of the Dutch Dune hill area. In: *Slope Transport Processes and Hydrology: a Tribute to Jan Nieuwenhuis. Engineering Geology*, 91(1), 72–7.

Van der Kwast, J. and De Jong, S.M. (2004). Modelling evapotranspiration using the Surface Energy Balance System (SEBS) and Landsat TM data (Rabat region, Morocco). In: *EARSeL Workshop on Remote Sensing for Developing Countries*, Cairo, pp. 1–11.

Van der Perk, M. (2006). *Soil and Water Contamination: from Molecular to Catchment Scale*. Taylor & Francis, London, 389 pp.

Van Rijn, L.C. (1994). *Principles of Fluid Flow and Surface Waves in Rivers, Estuaries, Seas and Oceans*, 2nd edn. Aqua Publications, Oldemarkt, 335 pp. SISO 533.3 UDC 532 NUGI 831.

Van Schaik, N.L.M., Hendriks, R.F.A., and Van Dam, J.C. (2007). Determination of matrix and macropore flow characteristics (using tracer infiltration profiles and inverse modeling in SWAP). *Geophysical Research Abstracts*, 9, 10385. SRef-ID: 1607-7962/gra/EGU2007-A-10385, European Geosciences Union.

Van Til, M. and Mourik, J. (1999). *Hieroglyfen van het Zand: Vegetatie en Landschap van de Amsterdamse Waterleidingduinen* [*Vegetation and Landscape of the Amsterdam Drinking-water Dunes*]. Gemeentewaterleidingen Amsterdam.

Von Hoyer, M. (1971). *Hydrogeologische und hydrochemische Untersuchungen im Luxemburger Sandstein* [*Hydrogeological and Hydrochemical Investigations in the Luxembourg Sandstone*]. Publications Service Géologique de Luxembourg, vol. XXI.

Wang, Z., Feyen, J., Van Genughten, M.Th., and Nielsen, D.R. (1998). Air entrapment effects on infiltration rate and flow instability. *Water Resources Research*, 34, 213–22.

Ward, R.C. and Robinson, M. (2000). *Principles of Hydrology*, 4th edn. McGraw-Hill, 450 pp.

Weaver, A.J. and Hillaire-Marcel, C. (2004). Global warming and the next ice age. *Science*, 304, 400–2.

Wellings, S.R. and Bell, J.P. (1982). Physical controls of water movement in the unsaturated zone. *Quarterly Journal of Engineering Geology*, 15(3), 235–41.

Wuebbles, D.J. (2007). Evaluating the impacts of aviation on climate change. Department of Atmospheric Sciences, University of Illinois at Urbana-Champaign. World University Network (WUN) video conference Horizons in Earth Systems, third annual running: Climate Change Science: Towards an Earth System Context. 28 March 2007. http://www.wun.ac.uk/horizons/earthsystems/index.html

Index

soil pipes 191
soil water 1, 141
 adsorption 141
soil water flow 9
soil water potential *see* total potential
soil water repellancy 194
solid soil matrix 59
sonar 228
sorptivity 178–82
source areas 271, 273
specific discharge 67, 213
 critical 215
 see also volume flux density
specific energy 212
 minimum 214
specific energy diagrams 213, 215
specific yield 137
splash 184
splash erosion 184
sprinkling 18
S–Q relations
 artificial lake with lower opening in dam 207
 artificial lake with overflow 221
 groundwater depletion or baseflow reservoir 257
squares method 98
staff gauge 225
stage 225
 measurement 225–7
stage–discharge (Q–H) relations 225, 239–43
 artificial lake with lower opening in dam 207
 artificial lake with overflow 220
 critical flow at dip 217
 flume 224
 rectangular weir 222
 V-notch weir 223
stagnation
 in percolation process 189
 wetting front 192
stagnation point 125–6
stand-alone wells 120–2
standard atmospheric pressure 16, 52
stationary groundwater flow *see* steady groundwater flow
statistical methods 31, 264–8
steady flow, surface water 201
steady groundwater flow 49, 52
 one-dimensional
  in confined aquifer 80–3, 136, 291
  in leaky aquifer and leaky confining layer 110–14, 136, 293–5

 in recharged unconfined aquifer 117–19, 136, 296–8
 in unconfined aquifer 107–10, 136, 292
 radial-symmetric
  in confined aquifer 120–1, 136, 299–300
  in unconfined aquifer 121–2, 136, 301–2
Steinmergelkeuper marls 192
stemflow 8
Stiff diagrams 102–3
stilling well installation 226–7
storage capacity 170, 261
storage coefficient 95
STORFLO model 261, 262
'storm' 250
Strack, Otto D.L. 138
streamline 52, 77
streamline refraction 79
stream tube 97
stress 84
structural porosity 190
subcatchments 256
subcritical flow 208, 211, 215
sublimation 15
subsidence 84
substitute hydraulic conductivity 87, 89
subsurface flow 9
suction 144
 and capillary pore diameter 149–51
 and volumetric moisture content 152
 at wetting front 173–4
sum rule 290
sun's shortwave radiation incident at top of earth's atmosphere 36–7
supercooled water 21
supercritical flow 208, 211, 215
superposition (principle) 127–9, 253
surface resistance 42
surface roughness 40, 41
surface storage 9, 200
surface tension of water 150
surface water 1, 200
surface water flow 9
surface wave celerity 208–10
suspended load 224
symbols, in cross-sections 56

T

Tamar, River 265
tangent line 281, 286

Tees, River 265
Tennessee, rural drainage basins 256
tensiometer 146–8
TERRA ASTER satellite configuration 46
textural porosity 190
thalassocline water 102
Theissen polygons 30–1
The Netherlands *see* Netherlands, The
thermal expansion 3
thermohaline 5
Thiem equation 122
Thornthwaite method 45
throat 223–4
throughfall 8
throughflow 188, 260
 see also pipeflow
time–area model 252–5
time base 254
time of concentration 245, 252
 determination 255–6
tipping bucket raingauge 29
Tollbaach drainage basin 102
TOPMODEL 262
topographic convergence 271–3
topography 31
topography-controlled water table 99, 100
Torricelli, Evangelista 204
Torricelli's experiment 204–5
total discharge volume 250
total dissolved solids (*TDS*) 233
total head 202
total hydraulic resistance 89
total potential 143, 164
 determining 146–8
 evaporation effects on 169
total pressure 84
total transmissivity 86
total water potential *see* total potential
transient groundwater flow 52, 137
transmissivity 85, 96–7
 total 86
transpiration 8, 32
travel time 91, 253
travel-time concept 253
travertine 74
trichloroethylene 129
tsunami
 propagation velocity 208, 210
 wave height 210
turbulent flow 73–4, 200, 201